14 December,
1992

Dear Doctor Spence,

I hope that this book might
tempt you to venture South!
You will always be welcome.
Thank you for everything.

Yours,
Daniel Joyce.

THE WRITERS' LANDSCAPE
SETTLEMENT

THE WRITERS' LANDSCAPE

Settlement

SUZANNE FALKINER

CONCEPT BY LESLEY MCKAY

SIMON & SCHUSTER
AUSTRALIA

THE WRITERS' LANDSCAPE — SETTLEMENT

First published in Australasia in 1992 by
Simon & Schuster Australia
20 Barcoo Street, East Roseville NSW 2069

A Paramount Communications Company
Sydney New York London Toronto Tokyo Singapore

National Library of Australia
Cataloguing in Publication Data

Falkiner, Suzanne, 1952– .
  Settlement.

  Bibliography.
  Includes index.
  ISBN 0 7318 0145 8.

  1. Australian literature - History and criticism. 2. Landscape in literature.
  3. Australia in literature. I. Title. (Series: Writers' landscape).

A820.93294

Photographs supplied by Wildlight Photo Agency
1/165 Hastings Parade, Bondi Beach, NSW Australia 2026
Telephone (02) 30 1737, 30 2822. After hours (02) 32 1616
Facsimile (02) 30 2466 Telex S.Y. 2495

Designed by Steven Dunbar
Jacket photograph supplied by Wildlight Photo Agency
Author photograph by Philip Quirk
Imageset in Australia by Everysize Typeart Services, Sydney
Printed in Singapore by Kyodo Printing Co. (Singapore) Pte Ltd

CONTENTS

Acknowledgements

My thanks must go first of all to Lesley McKay for inviting me to write this book; to Kirsty Melville, without whose imaginative and flexible response to our ideas the execution would have been much more difficult; to Jacqueline Kent and Susan Morris-Yates for invaluable editing and to Steven Dunbar for the design. My appreciation as always to Rose Creswell, and to Julian Leatherdale, Julia Cain, the Mitchell Library and State Library of New South Wales, the Woollahra Library, and Phil Quirk and Lisa Selsby at Wildlight. Thanks also to those many Australian writers who in conversation and correspondence discussed their work, advised, and pointed me in the direction of new sources.

INTRODUCTION

'Do you go much into your country?' asked Voss, who had found some conviction to lean upon.

'Not really. Not often,' said Laura Trevelyan. 'We drive out sometimes, for picnics, you know. Or we ride out on horseback. We will spend a few days with friends, on a property. A week in the country makes a change, but I am always happy to return to this house.'

'A pity that you huddle,' said the German. 'Your country is of great subtlety.'

With rough persistence he accused her of the superficiality which she herself suspected. At times she could hear her own voice. She was also afraid of the country which, for lack of any other, she supposed was hers. But this fear, like certain dreams, was something to which she would never have admitted.

Patrick White, *Voss*, 1957[1]

The greatest and most frequently remarked-upon paradox in Australian life has always been that the predominant cultural image is of a nation of bush dwellers, while Australia is and always has been a highly urbanised society. While Skippy the bush kangaroo hops across Italian television screens daily and Australian film-makers stride the streets of New York wearing Akubra bush hats and Drizabone riding raincoats, Australians have determinedly gone on living in cities.

While this paradox has been exhaustively examined since critical analysis of Australian literature began, it is only relatively recently that the gap between the image and the reality has begun to close and Australian writers have come to recognise themselves in their literature as predominantly urban dwellers. As that literature has reached a more mature stage of development, however, the closing of that gap has had an important bearing on the parallel development of what has been termed Australian regional writing.

The capital cities of the states of Australia were founded in an order defined by the spread of European occupation, and consequently they mirror the political, economic and sociological development of the nation. The penal settlements of the east coast came first: Sydney in 1788, followed sixteen years later by Hobart on the island of Tasmania in 1804, and Brisbane (Moreton Bay) twenty years after that. Perth was founded to establish a strategic foothold on the barren and isolated western coast in 1829. The prosperous southern mainland cities of Melbourne and Adelaide, relative latecomers in 1835 and 1836 respectively, were founded as a response to economic expansion and the discovery of new pastoral lands, with Melbourne, stimulated by the gold rushes, temporarily overtaking Sydney in size in the 1850s. The settlement at Darwin (Palmerston), culmination of a series of unsuccessful attempts to maintain an outpost in the tropical north to frustrate the perceived territorial ambition of Britain's rivals, finally took permanent hold in 1864. The last of the cities to be founded was Canberra, founded in 1911 with the specific intention of establishing a national seat of government.

With the exception of Canberra, the only inland capital, these cities were established in the higher rainfall areas of the coast. The largest—Brisbane, Sydney, Melbourne and Adelaide — were all situated within the narrow strip of fertile coastal plain that borders the south and east of the continent, and that today contains more than two-thirds of Australia's population.

Because most of the Australian capital cities existed as urban settlements before the continent was arbitrarily divided into states, with a few temporary exceptions such as areas of gold discovery, the cities have always proven more effective magnets of population and immigration than geographical regions. With the growth of these urban centres came the development of regional characteristics, literary and otherwise. As the cities provided a home terrain to the writers who documented both them and their surrounding areas, these writers subsequently also assumed, willingly or otherwise, the mantle of regionalism. The degree of political and cultural difference that has built up around the capital cities and their hinterland states has varied. Western Australia, Tasmania and Queensland have at various times in their history perceived themselves as sufficiently politically, economically or socially distant from the Sydney/Melbourne population axes to make gestures towards secession, and for this reason literary regionalism has probably been strongest in these states. Melbourne and Sydney, the largest urban centres, have tended to define themselves against each other as cultural rivals on a north versus south basis, in much the same manner as New York and Los Angeles/San Francisco on the east and west coasts of the United States.

Many Australians have argued that these geographical factors, along with the socio-economic and political functions of the original settlements, have also influenced the psychological outlook of the inhabitants of the resulting urban regions. The differences may be subtle, but are very apparent to natives; to outsiders, particularly Europeans, the Australian cities have in contrast often seemed tediously homogeneous in culture and architecture. Both perceptions are probably equally true.

One point of similarity between all Australian cities, except perhaps Darwin, is that until recently they were relatively homogenous in their Anglo-Celtic populations. Another common characteristic is urban sprawl: the tendency to build outwards rather than upwards, resulting in an endless proliferation of small bungalows on rectangular half-acre blocks, of which architect and critic Robin Boyd has written:

> The Australian town-dweller spent a century in the acquisition of his toy: an emasculated garden, a five-roomed cottage of his very own, different from its neighbours by a minor contortion of window or porch—its difference significant to no one but himself. He skimped and saved for it, and fought two World Wars with it figuring prominently in the back of his mind. Whenever an Australian boy spoke to an Australian girl of marriage, he meant, and she understood him to mean, a life in a five-roomed house.[2]

Perhaps as a result, the memoirs and autobiographies of older Australian writers are littered with images of childhood in the suburban or semi-rural house and backyard, whether portrayed as the Western Australian idyll of T.A.G. Hungerford or as the claustrophobic environment, to be escaped from as early as possible, of George Johnston's stultifying Melbourne of the 1920s and 1930s. Often, however, the facts of sex, class and race—frequently male, working-class Catholic or middle-class Protestant, and Anglo-Celtic—have given these suburban memoirs more in common than accidental allegiance to any particular city or region. Curiously, although it is the most typical dwelling place, few Australian novels (though they may begin there) have been based entirely in the suburban milieu, unless the writer treats it as an object of satire. Australian fiction writers have always been more attracted to the polarities of Australian existence—inner city or outback life—for subject matter.

Paradoxically, the evocation of place (as compared to culture) is often stronger and more freshly and distinctively detailed in these Australian urban works than in similar writing from older English-speaking countries, perhaps because—unlike the writers who are attempting to deal anew with more highly mythologised and well-known cities such as Rome, Paris and London—the chroniclers of these young, raw Australian cities are still in the process of forming their own characteristics.

Australian urban writers seem also less concerned with producing a distinctively Australian literary mythology than 'bush', writers, even when venturing outside the city. In a recent anthology devoted to the concept of 'the bush', about half the contributors were of urban origin, or were living in cities at the time of writing. This has always been true of a large number of 'bush' writers—Henry Lawson, who lived most of his life in Sydney, and A.B. Paterson, who was a Sydney solicitor and newspaper editor, being cases in point—but the literature they produced tended to disguise this fact. Now, when contemporary urban writers write of the landscape, it is often from the point of view of city dwellers venturing into the wilderness to renew acquaintance with it. However, while urban writing is currently the strongest genre, it has always existed. There has always been an Australian inner-city cultural ethos, even when the 'inner city' consisted of a few hovels around the Rocks area in Sydney. The inner-city 'pushes' or larrikin gangs, which had their origins in convict days, exist now in remnant as a literary and intellectual affectation around certain pubs in Sydney and Melbourne.

It is significant that these two largest cities, as well as being undoubtedly the major literary and publishing centres, have produced the only regionally distinctive schools of urban writing—the 'Balmain' school of short story writers that emerged in Sydney in the early 1970s, and included writers such as Frank Moorhouse and Michael Wilding; and the inner-city comic and satiric writers of Melbourne, including Barry Dickins, Jack Hibberd and Barry Oakley. Both of these are also predominantly male groups. With their common icons of male mateship and drinking rituals, and the tendency to marginalise the female except in sexual encounters, it could be argued that these 'schools' represent the urban equivalent of the bush ethos. In the other capitals, a further recent development has been the anthology of short stories based on state or regional writing, but this may be a by-product of the recent resurgence of the Australian short story, and the perceived dominance of Sydney and Melbourne as literary centres, as much as an indicator of the development of true contemporary regionalism.

Today, a new generation of young writers—often emerging from newly developed creative writing classes in urban technical colleges and universities, often initially published by small Australian presses, and often of non-Anglo-Celtic background—write predominantly from the urban viewpoint and are influenced by international rather than local cultural theory. Until relatively recently, the regional question tended to be predicated on the simple dichotomy of city or bush; now the dichotomy is between Australia and the world: the celebration of what is uniquely Australian versus a move towards an 'internationalism' of style, although frequently this means little more than that New York and Paris have replaced London as cultural models.

Outside the literary capitals of Sydney and Melbourne, however, the picture becomes less clear. There are towns in New South Wales which are nearer to Melbourne, Brisbane and Adelaide than they are to their own capital, and which tend to look to those cities as their urban centres. Albury in the Riverina, the northern New South Wales border towns, and Broken Hill near the South Australian state line are cases in point. The picture is further complicated by the fact that the previous simple urban/rural dichotomy made little real distinction as to which state or region the 'bush' might belong, and there was little emphasis on marginal or transitional regions. Today, particularly in coastal New South Wales and Victoria, once distinct geographical regions merge into each other under a web of small centres of human habitation, with the divisions into states marked only by the arbitrary border-lines imposed in the 1850s and early 1860s when Victoria and Queensland were separated from New South Wales.

As a result, some critical commentary has identified Hal Porter's Gippsland and Les Murray's north coast of New South Wales as examples of regional writing that do not correspond to any strict urban/rural dichotomy. However, whether a single strong voice can be classed as a regional school is debatable.

On the other hand, as Hal Porter pointed out in an interview in relation to his own work:

> Of course … most writers in Australia are regional writers. The place is so big. No-one can encompass all of it. I mean, dear boy, I'm a southern Victorian—cold wind, bare apple trees, all that …[3]

As an example of the contrasting effects of place, urban or pastoral, on the developing consciousness of a writer, the Western Australian critic Bruce Bennett[4] has juxtaposed this comment with Porter's description of his arrival, at the age of six, in the Victorian rural town of Bairnsdale, 270 kilometres east of Melbourne, after beginning his life in suburban Kensington in that city.

> My first sight of Bairnsdale strikes me breathless and still and smaller.
> Space! Infinity! Light!
> In Kensington, stuck on an asphalted suburban ridge at the rim of a panorama, I had seemed taller to myself, a spy suspended above luminosity. In Bairnsdale, I feel myself let loose at the centre of an immeasurable sphere. Pure light gushes and soars away from my minuteness in every direction, upwards and ever upwards, inhabited by slicing swallows and creaking swans and stock-still hawks and pin-prick larks; outwards they arch over the northern mountains in the thick blue of which are half-forgotten, tumble-down gold-mining towns occupied by mere hill-billies as incestuous as cats; outwards and east to curve for a century of miles over the farthest eucalypts and their sumless tons of glistening morocco leaves; outwards and southwards over the river-mouths, the swan-haunted lakes, the very South itself, and the world's felloe.[5]

The question as to what extent regionalism exists in Australia depends largely on what definition of regionalism is accepted. The *Shorter Oxford* defines 'regionalism' in one sense as 'localism on a regional basis'; and, flipping back, describes 'localism' in terms of 'attachment to', 'limitation to' and 'disposition to favour'[6] one locality. By this definition, given the tendency of Australian writers to move about and not limit their subject matter to any one place, the number of Australian writers who could be described as purely regional is relatively small. Nevertheless, there are bodies of writing that seem to have characteristics in common, and that are distinctively linked to certain geographical areas. According to these criteria it is possible to describe as 'regional' the work of writers who have not been 'regional' all their lives or even for any consecutive period. These are writers who in their work have demonstrated a geographical, spiritual or psychological link with a particular place or community.

At first glance it seems unsatisfactory to link the concept of literary regionalism with the constructed or built environment—in this case the Australian capital cities. However, the difficulty arises, as the previously mentioned anthology writers seem to have found, when one tries to do anything else.

While the tags 'Western Australian writing', 'South Australian writing', 'Queensland writing', etc., have proven useful in the past, the terms 'New South Wales writing' and 'Victorian writing' do not conjure up any precise image. Geographically based terms such as 'West Coast' or 'Top End' writing, or even 'writing of the Southeastern region', may seem initially more useful, but on closer examination become somewhat vague. Is Adelaide part of the Southeastern region? Is Broome, in the north of Western Australia, part of the West or the Top End?

If we accept that regional writers and writing in Australia can be usefully linked to the capital cities rather than geographical regions, however, it is possible then to go on to attempt to define regional characteristics and exceptions to these characteristics. This sort of differentiation between cities was attempted by early visiting writers such as Beatrice Webb and Mark Twain, who delighted in discerning social (if not literary) differences between Sydney and Melbourne, a practice eventually taken up enthusiastically by the inhabitants of the cities themselves. The Sydney/Melbourne rivalry probably had its roots in the gold rush when Melbourne temporarily overtook Sydney in property and population size. Since then, Melburnians in particular have enjoyed engaging in a display of regional patriotism, and have a reputation for holding a grudge when one of their number defects to the north.

R.A. Bevan, a contributor to *Arrow*, an undergraduate magazine of the University of Queensland in the mid-1940s, in a poem called 'Observation Sociologique', summed up this sort of social preconception:

> The people of Melbourne
> Are frightfully well-born.
> Of much the same kidney
> Is the beau monde of Sydney.
> Hobart's all convicts and seamen,
> History, ruins, Van Diemen,
> Adelaide's forte is Culture —
> But in Brisbane the people insult yer
> And don't hardly know they've been rude
> They're that ignorant, common and *crude*
> — It's hardly worth
> Mentioning Perth.

More recently, in 1966, in one of the first of the illustrated 'Australiana' books that have become popular in recent decades, Melbourne-born writer George Johnston summed up the perceived differences thus:

> The stereotype beloved by most Melburnians, is that Sydney is noisy, garish, untidy, pushing, callous, glossy in life and cold-blooded in business, insincere, brash, rather delinquent, and on the whole un-trustworthy. The same Melburnians consider themselves quiet, well-mannered, industrious, perhaps a little over-reticent and reserved, but not really cold, a thoughtful people, fond of music and theatre and 'keen on' culture, great ones for the home, and with a remarkable flair for fashion, particularly among the women.[7]

More frequently, in the smaller capital cities outside the Sydney/Melbourne axis, individual prominent writers have found themselves inextricably associated with them as their spokesmen and chroniclers: Christopher Koch with Hobart, David Malouf with Brisbane, Barbara Hanrahan with Adelaide, Elizabeth Jolley with Perth. However, it would be as unfair to these writers to confine their *oeuvre*, or any part of it, to the category 'regional', as it would be to call Patrick White or even Sumner Locke Elliott a 'Sydney writer', or Henry Handel Richardson a 'Melbourne writer'. What this book attempts is to use some aspects of the work of these authors, and others, to provide a portrait of those cities and regions with which they have been at some time associated.

*   *   *

If, as has been already noted, the position of contemporary regional writing in relation to Sydney and Melbourne is difficult to define, in earlier days there were several sometimes overlapping strands of literature that could be loosely described as regional to the Victorian state. This body of writing included early pioneering novels of pastoralism and the gold diggings such as Henry Handel Richardson's *The Fortunes of Richard Mahony*; Katharine Susannah Prichard's *The Pioneers* (1915); much of the work of Martin Boyd; and a small body of semi-pastoral and lyrical writing mainly associated with the Gippsland area, such as that by Hal Porter and Eve Langley. Richardson, in *Australia Felix* (1917), the first volume of *The Fortunes of Richard Mahony*, gives an effective portrait of the growth of a town which is now the largest inland city in Victoria. In evoking Bairnsdale around the period of World War I in *The Watcher on the Cast-Iron Balcony*, the first volume of his autobiographical trilogy, Hal Porter also gives one of the fullest and most minutely detailed portraits yet of any Australian country town.

However, with the possible exception of the work of short story writer John Morrison, and more recently of Peter Carey, whose novel *Illywhacker* (1985) contains affectionate portraits of his birthplace Bacchus Marsh, the city of Geelong, and the Western District of Victoria, there remains little modern Victorian writing that could be described as distinctly regional.

In direct contrast, the isolated west of the continent, with Perth as its one major city, has produced a small but distinctive literature based on a strong sense of its own identity and its widely varying desert and coastal landscapes. In fact, Western Australian academic Bruce Bennett has identified eight Western Australian 'sub-regions' with which he claims Western Australian writers can be identified.[8] Western Australia is also unusual in that such characteristically Western Australian writers as Elizabeth Jolley and Tim Winton have remained there, while those of other regional capitals have tended to drift to the larger centres of Sydney and Melbourne. Fremantle Arts Centre Press has also recently had a role in bringing contemporary Western Australian writing before the eyes of the rest of Australia in the same way as the longer-established University of Queensland Press has done for Queensland. Western Australia is also the cradle of a significant body of Aboriginal writing in English, a development encouraged by Fremantle Arts Centre Press and further nurtured by the establishment of Magabala Books in Broome, a publishing house that produces Aboriginal literature. A symptom of this strong sense of regional identity is the Fremantle bookshop that divides its 'AusLit' section into subsections entitled 'Australian Literature', 'Recent Australian Writing', and 'Western Australian Writing'.

Queensland, also characterised by a dry hinterland and a fertile coastal strip, similarly provides one of Australia's most geographically distinct terrains. The city of Brisbane, with its diminishing amount of fine timber architecture, suburban gardens with luxuriant mango trees and tropical vegetation and its pseudo-Renaissance public buildings, seems also to have produced a more distinctive body of urban writing than other capitals, particularly in the work of David Malouf and Jessica Anderson.

However, many of its native writers, like those of Tasmania and South Australia, have subsequently chosen to live elsewhere. With, until recently, a politically conservative government, the state has tended to breed a high degree of ambivalence in its authors. In local novels, Queensland is frequently portrayed as a place from which it is necessary to escape. David Malouf's *Johnno* (1975) and Jessica Anderson's *Tirra Lirra by the River* (1978) are probably the most widely recognised novels in this genre, but the theme, along with that of the return in later life to re-examine family roots and cultural beginnings, recurs in novels such as Tom Shapcott's *Hotel Bellevue* (1986), Susan Johnson's *Flying Lessons* (1990), and also in the expatriate Janette Turner Hospital's fiction. This theme of escape, however, is countered somewhat by Queensland's image as a northern haven to escape *to*: either by restless youth drifting up the coast from surfing beach to surfing beach, or as a quiet backwater to which to retire. The Queensland coast, with its rainforest, coral reefs and tropical islands, represents to some extent the exotic 'otherworld' of the Australian psyche. Some literary commentators have remarked on a local juxtaposition of the tropical landscape and scenes of violence, particularly in the work of Thea Astley, who in her early novels is one of the most successful chroniclers of Australian small town life and the frustration and cruelty that can accrete in small communities.

Tasmania, which seems to have produced a disproportionate number of writers for the size of its population, is notable for a body of reminiscent writing preoccupied with Tasmanian history and landscape, as in the work of Christopher Koch, Peter Conrad and Carmel Bird. Once again, however, these writers have tended to gravitate north before turning their attention to their birthplace.

In Adelaide, writers of South Australian origin such as Barbara Hanrahan, Nick Jose, and Murray Bail have dealt with South Australian urban themes, and the well-known children's writer Colin Thiele has recorded the landscape of areas such as the Coorong and Coober Pedy. However, many Adelaide writers, Peter Goldsworthy being a notable example, universalise the urban environment to the extent that their native city is not particularly recognisable.

In Darwin, European regional writing is probably less robust than in other cities due to its small size. The Northern Territory, a large area with a small population, is the state of Australia where the balance between the constructed and the natural environment remains most weighted in favour of the natural, and, with the largest population of Aborigines, it remains the state where Aboriginal art and culture survive closest to their traditional forms. The central Australian town of Alice Springs has become a centre for buyers of Aboriginal art to funnel works through to galleries in Australian cities and the wider world. Darwin—with Broome and Perth—is also becoming a centre of Aboriginal publishing and writing. This cannot be judged a regional phenomenon as much as a remnant of a cultural manifestation that would once have encompassed the whole of Australia. Similarly, due to its lingering frontier ambience, a sizeable body of popular travel writing, pastoral sagas, and period adventure yarns about cattle droving, crocodile shooting, buffalo hunting and pearl fishing have been produced about the area through the years, of which only a small proportion is still read.

Canberra, also a relatively small capital, suffers some of the same disadvantages of size. However, a number of poets, often academically connected to the Australian National University, have been associated with the Canberra area for various periods, including Dorothy Auchterlonie, Bob Brissenden, David Campbell, Rosemary Dobson, Alan Gould, A.D.Hope, Les Murray, Geoff Page and Judith Wright, although their poetry has generally not been exclusively concerned with either the city or its environs. The Canberra short story writers and novelists who make up the female writers' group called 'seven writers' and who have collectively edited *Canberra Tales* (1988), a volume of stories devoted to the city, have probably made the most effort consciously to identify themselves as urban regional writers.

After the capital cities and their suburbs, the next most common place for Australians to live is in secondary industrial cities of the southeastern region, such as Liverpool, Wollongong and Newcastle; in rural cities such as Bathurst, Orange and Dubbo in New South Wales or Geelong and Ballarat in Victoria; and in the small country towns that service agricultural and mining areas throughout the states. Whatever signifies 'city' in the average mind does not widely exist outside the capitals: for many years even Brisbane, the capital of Queensland, enjoyed a reputation for retaining the atmosphere of a country town.

*   *   *

Historically, the real country towns of New South Wales and Victoria have not been treated kindly by Australian writers. With the exception of such unmitigatedly positive works as Ada Cambridge's *Thirty Years in Australia*, published in 1902, which gives a glowing description of late nineteenth-century life as a clergyman's wife in the towns of Wangaratta, Yackandandah, Ballan, Coleraine, Bendigo, Beechworth and Williamstown, the small, raw, developing urban settlements of Australia have suffered the same prejudices as the landscape and have been described largely in terms of what they lacked. Rachel Henning, arriving in Bathurst in May 1856 to join her married sister, wrote home to another sister in Cheshire and found the houses 'too far apart', the view 'un-English', and the public buildings altogether unsatisfactory.

> I cannot say I admire the 'city' of Bathurst. It stands in the midst of the Bathurst Plains without a tree or a shrub near it. It is all built of red brick, stone being unknown in the district, and it blazes away in the

sun, being the most boiling place in summer and the coldest in winter that is to be found in Australia. It looks large at a distance, because the houses are scattered about at long distances from each other, as if it had rained brick buildings. There is an immense square, or rather open space, in the middle of the town on which are scattered a church, which is called 'Norman', though I hope the Normans were never guilty of such architecture. A Romish church, with a big square tower of no particular order, and a Scotch church all over pinnacles, a Dissenting chapel, a prison with a huge brick wall round it; a bank and various shops; all one mass of red brick.[9]

Similar sentiments are reiterated by the poet Emily Manning, who published poems in the *Sydney Morning Herald* and other journals under the name 'Australie'. Her poem 'From Clyde to Braidwood', published in 1877 in her collection *The Balance of Pain and Other Poems* (1877), though a paean of praise to the surrounding landscape, describes the southern New South Wales town of Braidwood as:

> A township like
> All others, with its homes, church and school -
> Bare, bald, prosaic—no quaint wild tower,
> Nor ancient hall to add poetic touch,
> As in the dear old land—no legend old
> Adds softening beauty to the Buddawong Peak,
> Or near-home ranges with too barbarous names.
> But everything is cold, new, new, too new
> To foster poesy; and famish'd thought
> Looks back with longing to the mountain dream.[10]

Similarly, Henry Lawson seems to have had few fond memories of the small northwestern New South Wales town of Gulgong, where his family moved during the local gold rush of 1871. He makes a passing reference to it in the post-gold rush days in one of his best known 'Joe Wilson' stories, 'Water Them Geraniums', when Joe's wife Mary reminds him of the life they have traded for their small selection in the bush:

> A wretched remnant of a town on an abandoned goldfield. One street, each side of the dusty main road; three or four one-storey square brick cottages with hip-roofs of galvanised-iron that glared in the heat—four rooms and a passage—the police-station, bank-manager's and schoolmaster's cottages. Half a dozen tumble-down weather-board shanties—the three pubs, the two stores, and the Post Office. The town tailing off into weatherboard boxes with tin tops, and old bark huts—relics of the digging days—propped up by many rotting poles. The men, when at home, mostly asleep or droning over their pipes or hanging about the verandah-posts of the pubs, saying ''Ullo, Bill!' or ''Ullo, Jim!'—or sometimes drunk. The women, mostly hags, who blackened each other's and girls' characters with their tongues, and criticised the aristocracy's washing hung out on the line …—that was Gulgong.[11]

Significantly, in the same story, the wife also quietly reminds the husband of something he himself already knows: the reason they have not gone to live in Sydney instead of on a poor selection is because of Joe's drinking problem—the temptation would be too much for him in the city. And so Lawson indirectly reintroduces the great Australian dichotomy: the only place to live is 'Sydney or the Bush'. Too often in early Australian literature the small town is portrayed merely as a place to be avoided, a staging place for the itinerant rural worker to drink away his wages in (if he cannot make it to the city), or alternatively as a painful place of exile for the city dweller.

Certainly to nine-year-old Ethel Richardson, the small town of Koroit, some 280 kilometres west of Melbourne, where her mother had taken a job as a postmistress in 1878, the year before her father died insane, contained no pleasant memories. In *Ultima Thule* (1929), the last book of Henry Handel Richardson's trilogy *The Fortunes of Richard Mahony*, Koroit becomes 'Gymgurra'.

> Two wide, ludicrously wide cross-roads, at the corners of which clustered three or four shops, a Bank, an hotel, the post office, the lock-up; one and all built of an iron grey stone that was almost as dark as the earth itself. There were no footpaths, no gardens, no trees: indeed, as she soon learnt, in Gymgurra the saying ran that you must walk three miles to see a tree; which however was not quite literally true; for, on the skyline, adjoining a farm, there rose a solitary specimen …
>
> Their new home, the 'Post and Telegraph Office', with on its front the large round clock by which the township told the time, stood at one of the corners of the cross-roads. Facing it was a piece of waste ground used for the dumping of rubbish: thousands of tins lay scattered about, together with old boots, old pots, broken crockery: its next-door neighbour was the corrugated iron lock-up. Until now, it had consisted only of an office and two small living rooms. For her benefit, a three-roomed weatherboard cottage had been tacked on behind …[12]

For Patrick White as well, the small country town of the period has the redeeming features of neither city nor wilderness: in a brief, vivid passage from *The Tree of Man* (1956), he recreates a similar exile and its effects.

> 'When I put in for a transfer from Durilgai,' she said, 'after Mr Gage had taken his life, as you will remember, I was sent to Winbin.'
>
> At this cruel township in the South, ice crackled on the puddles. You could hear it. You could see the long, yellow rain come down the valley and strike the yellow grass. There was the street with the black-smith and the pub. There had been a murder once, but many years ago, and then never another. At the office, which was brown, on piles, above a lot of empty kerosene drums and broken chairs, the wind had loosened some of the woodwork. In the brown office, in which Mrs Gage stood, there was a smell of stoves and dry ink. Ink, other than the dregs of ink, frequently froze, so that you were not encouraged to put it. That frozen ink just made you feel bad. So Mrs Gage, chafing her chilblains and listening to the sound of her brown-paper sleeves, offered the public a pencil instead.[13]

In *The Twyborn Affair* (1979), which Patrick White considered one of his best novels, he presents a stark vignette of the Western Australian port of Fremantle, eighteen kilometres south of Perth on the mouth of the Swan River, and traditionally the first and last port of call on the European shipping run. This time Fremantle is encountered by a young Australian soldier returning after World War I—a soldier who, in this novel, is an *alter ego* for White himself. For White, returning to live in Australia after years of expatriatism in England and elsewhere, Fremantle, 'the first glimpse, the first whiff of a fate which can never be renounced, is enough to drive the pretensions out of any expatriate Australian'.

> Dream streets: the tiny houses in maroon or shit-colour brick. Paint-blisters on brown woodwork. Festoons of iron doilies which suggest melting caramel. Blank, suetty faces of women framed in grubby lace or muslin curtains, as they peer out in search of something to whet their interest. A little pomeranian dog, white coat with patches of pink eczema. An ageing blonde stands holding the dog to her bosom, fat dissolving on her vast arms. A gold armlet eating into a fatty bicep, the neatly folded, obsessively laundered hankie held in place by this dented gold circle.

This small-town Fremantle is another incarnation of Sarsaparilla, that fictional suburban community of conformity and self-satisfaction which White traditionally uses as a vehicle for describing the malaise he calls 'the great Australian emptiness'—which, for White, grips Australia itself.

Drank a schooner in a tiled bar. The acid smell, not quite urine, of draught beer. The 'head' forming as a red hand pulls on the joystick. The barmaid's rattling cough accompanied by a blast of morning gin.

One old professional blue-nosed soak, a finger crooked above the slops in his glass, tries to engage the interloper.

O.S.:   Owdyer findut, eh? in Fremantle.

ME: All right, I suppose. Yes, all right. [Hopeful laughter]

O.S.:   Not all the Poms do. An' I can't see why. [His turtle's neck at work as he swallows the last of the slops.]

ME: I'm not a Pom.

O.S.:   Go on! You're not? [He stands looking in need of a reassurance he does not expect to get.] What are yer, then?

ME: [because it's useless to explain.] I'm a kind of a mistake trying to correct itself.

Too much for Fremantle. The silence hits me in the small of the back, like the sheet of frosted glass with BAR engraved on a lyre of ferns.[14]

More recently, both Thea Astley and Amy Witting in novels such as Witting's *The Visit* (1977) and Astley's *Girl With a Monkey* (1958) and *A Descant for Gossips* (1960), have dealt with the unkind fate of the young single female schoolteacher in various outlying New South Wales and Queensland towns. Olga Masters, particularly in her 1985 collection of interconnected stories *A Long Time Dying*, set in the New South Wales town of Cobargo in 1935, focuses on the cruel emotional politics of a small and enclosed society. Masters, herself born in the New South Wales town of Pambula in 1919, in her brief writing career established herself as an accomplished portraitist of life in small towns. A previous novel, *Loving Daughters* (1984), explores the personalities of a newly arrived clergyman and the two sisters who vie to marry him, and *Amy's Children* (1987) examines the economic and social position of rural and urban women, particularly the deserted mother. Elizabeth Harrower, who lived until the age of eleven in the port and industrial city of Newcastle in New South Wales, describes a similar boarding-house bleakness in the city she calls Ballowra in her novel *The Long Prospect* (1958).

In those days there were no flats; a moat of steelworks and factories surrounded hills and plains of drab bungalows and shops. Cinemas, hotels, reared up from the encircled plain like small cathedrals. At night the sky glowed dusky red with industry.

For a newcomer, a single woman, accommodation meant a room in a stranger's house in whichever suburb was nearest to her place of work.[15]

Twenty years later, Peter Corris treats the same city with little more sympathy in his detective novel *White Meat* (1981), (although in his later work *Aftershock* (1991), his portrait is more benign).

Newcastle sprawls about like a drunken whore: it trickles off towards the coalfields in one direction, climbs up into the hill country in another and slides down to the sea on the east. The beach is a surprise; a fair-sized slice of white sand in front of a reasonable stretch of water for humans to swim in. It's like a reward to the city's inhabitants for putting up with so much else that is appalling. I hadn't been there for five years but the bird's eye view I got of it from the highway suggested that it was much the same, only worse. The long flat approach from the south is a ribbon of used car yards, take-away food stands and decaying wooden houses. A string of motels five miles out from town invite you to stop over, miss the city and push on to the clean country up ahead. I pulled into one of them, the Sundowner, which had a 'Vacancy' sign with the second 'a' flickering fitfully on and off.[16]

Newcastle is also the subject of William Olsen's semi-fictional memoir of a Depression childhood, *Down the Breakie* (1988): a tough but vivid account achieved with poetic simplicity.

The cattle railheads, ports and mining towns of the remote northern regions, still sparsely inhabited by Europeans, figure more frequently in adventure fiction and travel writing than elsewhere in Australian literature, although in the case of the isolated northwestern pearling port of Broome, the town became quite fashionable in literary terms for about twenty years. Broome has figured with varying degrees of accuracy in *Blue North* (1934) and *Sheba Lane* (1936), Henrietta Drake Brockman's novels of the pearling industry, and in a number of other works including J.M. Harcourt's *The Pearlers* (1933)—unpopular because of its portrayal of racial prejudice—Ion Idriess's *Forty Fathoms Deep* (1937), Katharine Susannah Prichard's romantic novel *The Moon of Desire* (1941) and Arthur Upfield's detective story *The Widows of Broome* (1951).

These areas have also frequently been described in outback histories such as Ernestine Hill's *The Great Australian Loneliness* (1937) and Mary Durack's *Kings in Grass Castles* (1959). In some of these genres, however, there is a temptation to make these settlements seem more isolated, barbarous or picturesque than they may perhaps appear to those who actually live in them. The Western Australian cattle town of Wyndham, 3500 kilometres northeast of Perth in the Kimberley region, is one such example. In Hill's *The Great Australian Loneliness*, the town figures in lurid detail.

> Surrounded by marshes, the town, which consists of a jetty, store, post office, and a double row of shack shops, mostly Chinese, hangs on [like] grim death to a narrow fringe of mangrove shore at the foot of the Bastion and Mount Albany. These two majestic ledged hills rise a thousand feet sheer from the houses in lofty disdain, as though they were about to kick them off into the sea. Crocodiles occasionally show an expectant snout along the waterfront, and over at the big State meatworks, a mile across the marshes, in sluices that are a crocodile's heaven of mud and blood, inert and ugly they lie, to use a Kimberley term, 'in mobs'.[17]

While Dora Birtles' novel *The Overlanders* (1946), from which the film of the same name was made in 1947, describes the town at night.

> Wyndham lay flat under the moonlight, its main street, its corrugated iron roofs, its mud flats by the mangrove edges, drawn into main relief, in highlight and dark shadow like the strong, rough contrast of a lion-cut, white and black. The salt pans glittered sharp as ice. It was not without beauty in its starkness. The hill behind the town loomed spine-backed and bald, the great meatworks on the other side a temple to the industry of the whole hinterland.[18]

Similarly, in a disparaging exchange in *The Merry-Go-Round in the Sea* (1965), Randolph Stow's otherwise relatively idyllic novel of a small-town and country childhood in Western Australia, the mining town of Magnet, 560 kilometres north of Perth, is mentioned with a type of laconic, derisive and racist bush humour rarely heard in Stow's own authorial voice.

> 'Oh, I thought we'd go and eat in Magnet.'
> 'The bright lights,' Alan Lamb said. 'The big smoke. Here we come.'
> 'What's at Magnet?' the boy asked.
> 'Three gins and a goat,' Alan Lamb said, 'most days. And a street about a mile wide.'
> 'They're kind of pathetic,' Rick said, 'these goldfields towns. They all started out with the idea that they were going to need the streets built for ticker-tape parades.'[19]

Frequently these more remote inland towns are portrayed fictionally with a great deal of psychological and physical violence, as in Thea Astley's *A Kindness Cup* (1974), which deals once again with the persecution of the nonconformist, this time in the matter of white/Aboriginal relations, and *An Item*

From *The Late News* (1982), which has a similar theme. Probably the most unrelenting fictional account of such an outback town is contained in Kenneth Cook's *Wake In Fright* (1961), which examines the reactions of a newcomer (once again a schoolteacher) to the more destructive aspects of the small town male mateship ritual. Here, in a scene which presages with subtle threat the events that will follow, Cook's protagonist runs up against the pride-of-place of local inhabitants.

> 'New to The Yabba?' said the taxi driver as he drove through the wide streets, lined with buildings affecting awnings supported by poles which looked as if they suffered from rickets.
> 'Yes,' said the schoolteacher.
> 'Staying long?'
> 'Just tonight.'
> 'That's hard luck. You ought to see a bit more of The Yabba than that.'
> One would have thought, reflected the teacher, that the driver was trying to sell a conducted tour, but he had noticed before that all the people of Bundanyabba seemed to be extremely patriotic.
> 'You think it's worth seeing?'
> 'I'll say I do. Everyone likes The Yabba. Best place in Australia … Try and stay a bit longer,' urged the taxi driver as the schoolteacher paid him.
> The teacher fancied he had been overcharged, but he wasn't sure.[20]

Another example of the outsider's view is contained in the short story 'For the Patriarch' (1981) in Angelo Loukakis's collection of the same name. A young Greek Orthodox priest, Dimitris, has gone to Marralin, 500 kilometres from Sydney in wheat country, population four to five thousand people, for three months to replace the local priest. The year is 1977. He writes to his parents in Greece.

> The people were kind though, and sometimes came to visit. I was invited into their homes, strange places, so different to the traditional Greek home. They are much larger than any village house, but they are cold. Too many modern things, plastic and metal.
> I lived while I was there behind the church. It was on the edge of the town away from everything. From the back you could look over some fields which were being prepared for planting while I was there. At night looking out the window there was all black emptiness. I was lonely to tell you the truth, but do not worry …
> The town is a peculiar place. It reminded me of the cowboy films you took me to see as a child, Father. The streets are wide and dusty and the shops have wooden verandahs and even horses tied up outside. But I didn't like to walk too much in the town. People would stare, not the Greeks so much, but the Australians, and the children were sometimes very bad to me, singing at me, or laughing, or throwing things. It was hard not to pay any attention, even though the shopkeepers said that was what I should do. I never saw such-like happen in Greece, not to a man of the Church.[21]

On the rare occasions that they have taken country towns as subjects, Australian poets have been kinder, although once again often from a safe distance. Kenneth Slessor, a poet generally less than bucolic by inclination, in 'Country Towns' (1944), takes the somewhat ingenuous view of the outsider passing through.

> Country towns, with your willows and squares,
> And farmers bouncing on barrel mares
> To public-houses of yellow wood
> With "1860" over their doors,
> And that mysterious race of Hogans
> Which always keeps General Stores….[22]

While Les Murray, in 'Driving through Sawmill Towns' —in this case also passing through—exercises the still sympathetic but rather more closely observant view of one who is himself a country dweller.

> You glide on through town,
> your mudguards damp with cloud.
> The houses there wear verandahs out of shyness,
> all day in calendared kitchens, women listen
> for cars on the road,
> lost children in the bush,
> a cry from the mill, a footstep—
> nothing happens.
>
> The half-heard radio sings
> its song of sidewalks.
>
> Sometimes a woman, sweeping her front step,
> or a plain young wife at a tankstand fetching water
> in a metal bucket will turn round and gaze
> at the mountains in wonderment,
> looking for a city.[23]

More recently David Malouf, in a story called 'A Traveller's Tale' from his collection *Antipodes* (1985) has benignly observed the north coast of New South Wales, but once again it is a transient view.

> Poor white country. Little makeshift settlements, their tin roofs extinguished with paint or still rawly flashing, huddle round a weatherboard spire. Spindly windmills stir the air. There are water-tanks in the yards, half-smothered under bougainvillea; sheds painted a rusty blood-colour, all their timbers awry but the old nails strongly holding, slide sideways at an alarming angle; and everywhere, scattered about on burnt-off slopes and in naked paddocks, the parts of Holdens, Chevvies, Vanguards, Pontiacs, and the engines of heavy transports, spring up like bits of industrial sculpture or the remains of highway accidents awaiting a poor man's resurrection. A tin lizzie only recently taken off the road suddenly explodes and takes wing as half a dozen chooks come squarking and flapping from the sprung interior.
> … At one point on the highway, surprisingly balanced above ground and about the same length as the Siamese Royal Barge, is the Big Banana, a representation of that fruit in garish yellow plaster. Two hundred miles further on and you come to the Big Pineapple, also in plaster, and with a gallery under the crown for viewing the surrounding hills. Between the two you are in another country. Men work in shorts in the fields and are of one colour with the earth, a fiery brick-red. Kids go barefoot, moving off the track on to the tufted bank with a studied slowness, as if they heard somewhere that there is a fortune to be made by getting struck. Little girls in faded frocks hang over gates, dispiritedly waving, or in bare yards sit dangling their legs from an elongated inner-tube that has been hoisted aloft and found new life as a swing.[24]

Signs of a possible readiness to depart from this general rule of ambivalent but largely naturalistic treatment of small towns is indicated by Rodney Hall's long fantasia of a novel, *Just Relations* (1982), the saga of a decaying New South Wales mining settlement. Once again the town is in a desolate region and many of its young have deserted it for the city.

> Underfoot the dust shifted. Before him, that single street of goldrush buildings wound grittily up the mountain. You had only to turn off between any two buildings to be instantly in the countryside, confronted by paddocks tipped crazily on edge and gymnastic cattle walking their tightrope trails or standing still with their eyes shut to recover from vertigo. The forested peaks and ridges beyond the town were, he knew, pocked with abandoned mineshafts. As for the hinterland, it was uninhabited and

commonly believed to be dangerous with the forces of Aboriginal spirits: little mountain men, the hairy men, gunjes and the like. This territory stretched a hundred miles to the west, a thousand miles to the south, and three thousand miles north where it was said islands like stepping stones connected it with the secret tribal grounds of New Guinea headhunters.

Jesus a man could hate this place[.]

Lost in the past and kept fitfully alive by the dream or nightmare of a renewed discovery of gold, the town of Whitey's Fall is presented as an enclave of elderly interrelated characters who fiercely guard their secret histories and mythologies against the encroachment of change from the outside. Hall uses the mountain landscape, a seemingly self-created and self-perpetuating region based not entirely in the real, as an allegory for the desires of its people:

> Whitey's Fall perches halfway up a mountainside, the mountain the people created. Year by year they accumulated the knowledge, the experience. They have the words so they know how to live with it. By their toughness they survived to heap up its bluffs, by scepticism they etched its creekbeds. They've lived and spoken every part of this mountain, they've dreamed it and cursed it, looked to it for salvation and penance. Its outcrops of granite are the very ones the people named, quarried and picnicked on before you could say there really was a mountain in this place at all. And the forest covering tells of its secrecy.
>
> So the mountain came to be there and indifferent to the people, those clinging whittlers in mud burrows. The mountain stood up, hunched and massive, shouldering into the rain, drenched and indistinct; just as it sprawled in the sun drifting with spirits of steam and giving out birdsong from grateful pores ... All the while, whatever the season, the mountain was busy. The life of the mountain had to go on. Water wavered in trickles of hair down to the gullies, runnels joined together with a sparkling clash. Soil broke apart to re-form itself as grass, grasstrees, fern and treefern, vines and huge eucalypts for the vines to hang from. Rock strata at the mountain's heart, with inconceivable slowness, tilted themselves, turning in millennial sleep, the seams of quartz fracturing brittle as glass, clay compacting to stone, fissures appearing and closing, soil falling away to reveal cliffs and cliffs crashing one by one down the mountainside in a flurry of gold dust. ...
>
> In those times the Aborigines of the region, the Koories, had no knowledge of any such mountain. Neither did they know nor care about gold. And but for gold, who would have stayed here at all? The people were families of diggers. Never mind that later they turned their hand to dairy farming and making cheese.[25]

It is later revealed that some of these characters have magical and semi-magical powers; Hall, however, taps into the roots of English, Irish and European history and myth, rather than Aboriginal sources, to provide a cultural depth that might otherwise be found to be lacking in a portrayal of a settlement less than a century and a half in age.

Other recent exceptions include David Foster, who in his novels *Dog Rock* (1985) and *The Pale Blue Crochet Coathanger Cover* (1988) describes life in the fictional country town that bears a playful resemblance to his home town of Bundanoon in New South Wales; and Carmel Bird, who in her novel *The Bluebird Cafe* (1990) employs futuristic fantasy in her creation of the imaginary Tasmanian mining village of Copperfield. However, while writers such as Hal Porter and Randolph Stow have written sympathetically in their semi-autobiographical fiction of childhoods spent in Victorian Bairnsdale and Western Australian Geraldton, modern Australian novels still rarely find their locations in the smaller towns and cities, and when they do, even more rarely are they treated affectionately. Perhaps, as has occurred with the Australian wilderness, sympathetic portraits may become common with the maturing of a generation who regard the country towns as their own, and the mellowing of those towns that survive an increasing tendency to centralisation.

SYDNEY

THE VIEW FROM THE BOAT:
Nineteenth-Century Sydney

… Some cities are air-cooled like antiquated aero engines (Rome, New Delhi, Adel—), others are water-cooled like the majority of four-stroke car engines (San Fran, Venice, Sydney).

The first time Shadbolt took a bus into the city the harbour appeared to be never-ending. It filled the hollows and gaps, water finding its own level, it leaked into the corners of his eyes whichever way he turned. Deep! The lapping mass glittered and penetrated, lapping at the descending layers of terracotta houses, submerging the boards of the wooden jetties, slap-slapping sullenly at rocks, a heavy mass, narrowing the main road into an isthmus. Water everywhere. It shortened the side streets into dead ends. Shadbolt noticed it right and left and straight ahead, the road climbing to escape it, and doglegged, only to return to it at the next bend; and always he felt its cooling properties, caressing his cheeks.

Murray Bail, *Holden's Performance*, 1988[1]

Sydney above all Australian cities is defined by the sea that pokes its fingers of blue water into more of its suburbs than any other. From the beginning Sydney Harbour, described by Governor Phillip in 1788 as 'one of the finest harbours in the world, in which a thousand ships of the line might ride in perfect safety'[2] has proven a magnet to writers. The site was found equally unmatched by Anthony Trollope in 1871, who thought it so 'inexpressibly lovely' that he asked himself whether it would not be worth his while to move his household gods to the eastern coast of Australia, in order that he might 'look at it as long as he could look at anything'.[3]

However, not all writers were so unequivocal in their praise.

'Is that all?' said a pretty sick girl on our steamer, who had had her chair brought close to the bulwarks, that she might not miss the first sight of Sydney Harbour. 'Oh, I don't call it much more than just pretty. Seems somehow as if the mountains had been forgotten.'

She expressed it exactly. One does feel as if the Creator had forgotten the mountains. And yet, indeed, how beautiful Sydney Harbour is, though one begins to wonder whether it is as beautiful as the harbour of Nagasaki, or of Hong Kong, or even of Algiers, or of many other places one has heard less about. There is always the want of the background.[4]

When Rosa Praed wrote this passage on returning to visit Australia in 1894 after nearly twenty years in England, in a sketch entitled 'The Old Scenes' published in London in 1899, she exemplified two aspects of much of the writing about Sydney in that era. Firstly, after mentioning the beauty of the harbour, the author almost invariably goes on to describe the city in terms of what it lacks and secondly, though Praed herself was born in Queensland, the piece was published in a collection of stories by authors described as 'Australian writers living in England'. Typically, the literary viewpoint is informed by cultural preconceptions of another country, and the narrator's position is usually that of a new arrival or a temporary visitor.

The colonial experiment in Australia, and the city of Sydney, began in January 1788 with a patch of virgin bush, a somewhat grandiose plan, and the more than somewhat unwilling efforts of convicts to carry it out. Some thousand English soldiers and their prisoners simply disembarked from boats and began to build houses on the edge of a continent on which houses had never been built before.

The first huts erected for the marines were of soft cabbage palm wood and designated only for imme-diate shelter. The convicts huts were even more slight, being composed of upright posts interwoven with light twigs plastered with clay. Stone was plentiful, but there was no lime for mortar, although to make a 'small house' for the governor, convicts were set to collecting and burning oyster shells from Aboriginal middens in the neighbouring coves. Good clay for bricks had been found nearby. Fortu-nately it was still summer, and warm enough for the flimsy shelter not to a be disadvantage.

Despite the architectural fragility of these beginnings, Governor Phillip had a vision for the future of Sydney. Early in the same year, he recorded:

> Lines are there traced out which distinguish the principal street of an intended town, to be terminated by the Governor's house, the main guard, and the criminal court. In some parts of this space temporary barracks at present stand, but no permanent buildings will be suffered to be placed, except in conform-ity with the plan laid down. Should the town be still further extended in future, the form of other streets is also traced to ensure a free circulation of air.[5]

Grants of land would be made with the provision that only one house could be built on each block, and that each block should have a frontage of 50 feet (14.8 metres) and a depth of 150 feet (44 metres). Although Phillip's plan was disregarded by subsequent governors, the tradition of the quar-ter-acre block, adorned with a five-roomed cottage, the molecular unit of urban sprawl, was set. This would become the prototype of Australian cities.

Less than two decades later, in 1819, W.C. Wentworth wrote with apparent pride and satisfaction that the town of Sydney, from the point of view of its aspect alone, must become of considerable importance.

> The views from the heights of the town are bold, varied and beautiful. The strange irregular appearance of the town itself, the numerous coves and islets both above and below it, the towering forests and pro-jecting rocks, combined with the infinite diversity of hill and dale on each side of the harbour, form altogether a *coup d'oeil*, of which it may be safely asserted that few towns can boast a parallel.[6]

Alexander Harris, in his semi-fictional novel *Settlers and Convicts* (1847), wrote of Sydney Cove in the 1820s as 'a waterside town scattered wide over upland and lowland', where 'if it would be a breezy day the merry rattling pace of its manifold windmills, perched here and there on the high points, is no unpleasing sight'. There were several 'large piles' of public buildings in sandstone, and on the steep and lofty extremity of the promontory, houses had been built on a series of terraces rather than streets. It was just under these terraces, known as the Rocks, that Harris's 'emigrant mechanic' and a friend disem-barked from a ship anchored off the King's wharf in the twilight of an early summer evening.

> As we walked down George Street we found Sydney according to custom during the first hour of a summer's night, all alive, enjoying the cool air. The street was clear of vehicles, and parties of the inhab-itants, escaped from desk and shop, were passing briskly to and fro, in full merriment, and converse. At the main barrack-gate the drums and fifes of the garrison were sounding out the last notes of the tattoo. In Sydney the barracks occupy a noble sweep of ground in the very centre of the town … Leaving the long line of barrack-wall behind us we at length reached the market-place. The fine building that now occupies the spot under the same name, was then not even in projected existence: but the settlers drove their drays into the open area amidst the old shed-like stalls that here and there stood for the occupation of dealers; and the whole was surrounded by the remains of a three-rail fence. As we wandered through the rows of drays and carts I could not but remark a striking difference between them and the contents of the carts of any general market for the produce of the land at home.

They cross George Street and turn right to reach the market wharf.

> There was not a creature on the wharf but ourselves, and the continual melancholy plash of the flooding tide, among the boats that lay moored in numbers close together, made the hour and the scene appear more lonesome still. The moon was just glinting over the dark wooded hills, so that I could plainly enough see the masses of forest on the opposite shore. What a wonderful advance had this same locality experienced when, a few years afterwards, I bade adieu to it for England! This solitary landing-place had become a street, and busy steamers at the same hour came roaring past with their teeming cargoes from the northern and southern settlements.

Harris's guide takes him to the sly-grog sellers in Gloucester Street in the Rocks.

> The noise of the carousal we began to hear when we were within about a hundred yards of the Sheer Hulk might fairly have led to the belief that there was nothing there to be concealed from the police, particularly as one old constable in his blue coat and red collar stood bâton under arm at the corner of Frazer's Lane listening to it in all the appearance of serene reflectiveness. Full a score of voices were singing each its own song in its own tune and its own time. Now there was a bellowing volley of men's voices, then sounds such as the voices of women can make only from the stimulus of intoxicating drink. It was a perfect frenzy of drunken vociferation. From the lofty terraced ledge of rock along which we walked we could see the dark blue waters of the harbour, all life, as strong tide came flooding, dancing, glittering in under the beams of the full moon. At times in Australia the moon may be seen to look perfectly globular to the naked eye, and so it was to-night. No two things ever were in greater contrast than the sounds within and the scene without.[7]

'Slap went ... the door, and down went the heavy bar; and there we stood in an atmosphere of tobacco smoke', writes Harris, before going on to describe the large, low, dilapidated, half-lit room filled with benches and greasy tables benches covered with glasses and pipes, the forerunner to another great tradition, the Australian pub.

In 1836, when a homesick Charles Darwin, aged twenty-six arrived on the *Beagle* after four years at sea, his initial reaction to Sydney was generally favourable.

> Early in the morning a light air carried us towards the entrance of Port Jackson. Instead of beholding a verdant country, interspersed with fine houses, a straight line of yellowish cliff brought to our minds the coast of Patagonia. A solitary lighthouse, built of white stone, alone told us that we were near a great and populous city. Having entered the harbour, it appears fine and spacious, with cliff-formed shores of horizontally stratified sandstone. The nearly level country is covered with thin scrubby trees, bespeaking the curse of sterility. Proceeding further inland, the country improves: beautiful villas and nice cottages are here and there scattered along the beach. In the distance stone houses, two and three stories high, and windmills standing on the edge of a bank, pointed out to us the neighbourhood of the capital of Australia.
>
> At last we anchored within Sydney Cove. We found the little basin occupied by many large ships, and surrounded by warehouses. In the evening I walked through the town, and returned full of admiration at the whole scene. It is a most magnificent testimony to the power of the British nation. Here, in a less promising country, scores of years have done many times more than an equal number of centuries have effected in South America. My first feeling was to congratulate myself that I was born an Englishman.

On closer observation, this initial euphoria did not last. Darwin was among the first visitors to comment on the embryonic beginnings of a characteristic urban sprawl, and also on signs of a relative class egalitarianism in the town.

> Upon seeing more of the town afterwards, perhaps my admiration fell a little; but yet it is a fine town.

The streets are regular, broad, clean, and kept in excellent order; the houses are of a good size, and the shops well furnished. It may be faithfully compared to the large suburbs which stretch out from London and a few other great towns in England; but not even near London or Birmingham is there an appearance of such rapid growth. The number of large houses and other buildings just finished was truly surprising; nevertheless, every one complained of the high rents and difficulty in procuring a house. Coming from South America, where in the towns every man of property is known, no one thing surprised me more than not being able to ascertain at once to whom this or that carriage belonged.[8]

Two decades later Frank Fowler, who left London in 1855 for his health and worked for several years as a journalist in Sydney, was less impressed with the town and its inhabitants.

Serpent-like gutters, choked with filth, trail before the tottering tenements, and a decayed water-butt, filled with greasy-looking rain catchings—across which indecent slime-bred flies dart and dazzle in the sun—stands and rots at the end of each court. Brazen women, hulking bullies and grimy children loll about the doorways …

However, the eastern suburbs from Rushcutters Bay to hilly, picturesque Paddington and the houses of the gentry on Darling Point, then as now, presented a more desirable aspect.

The beauty of the road down which the old 'bus goes rumbling, increases at every yard. The quiet Rushcutters Bay—deep in graduated shadows and walled in on one side with great rocks which look at a distance like a row of half-ruined castles rising from the water, —the rich foliage clustered all about the humble tenements which skirt the road; —the old cows grazing in the little narrow green lane, which runs off at the foot of the hill just beyond the cozy-looking 'White Conduit House',—the few great solitary trees standing on the brow of the heights towards Glenmore; —the white wing of a boat seen flying now and then across the bay; —the stray geese and grubby pigs luxuriating in the roadside paddocks; —all these with a thousand other picturesque items meet us in the course of our journey. I should think there are few finer villas in the world than some of those dotting the declivity facing Clark's Island on our left.[9]

Fowler, who became a member of the developing literary coterie under the patronage of the Balmain solicitor Nicholas Stenhouse, collected his impressions of the colony in *Southern Lights and Shadows* (1859), which was published when he returned to England after an unsuccessful candidature for the Legislative Assembly.

Richard Rowe, an English tutor in Sydney for several years in the 1850s and also a member of the Stenhouse circle, paints a similar picture in *Peter Possum's Portfolio*, published in 1858 after his own return to England. Riding by horse cab to the wharf to catch a steamer up the coast to the Hunter River, he encountered one of the first in a long line of Sydney cab drivers who seemed not to know his way around the city.

Up and down the steep, dimly-lighted streets that lie between Wynyard Square and the water, my cab goes blundering like a huge humble, —or, as I would rather write it, *Bumble*-bee—that beadle amongst insects. Cabmen are generally supposed to be well acquainted with the ins and outs of Sydney—some of them, unfortunately, are *too* well acquainted with the *inns*, and my driver is one of this description. In a glorious state of topographical uncertainty, hither and thither he jerks and lashes his horses; not infrequently bumping his pole against dead walls in vain attempts to find previously undiscovered passages to the wharf through *culs-de-sac*.

    I begin to fancy that I shall have to pass the night in wandering along rows of houses that seem as fast asleep as their owners (their closed shutters reminding one of eyelids sealed),—in watching dissipated cats out upon the loose, and wearing the half stealthy and ashamed, half swaggering and independent air that marks their human congeners, young gentlemen with latch-keys; homeless dogs, hungry and fierce, foraging for garbage; hulking fellows as fierce and ravenous, without the dogs' excuse of homelessness and hunger; and the slow-footed Erinnyes in shiny hats, great coats, and oilskin capes, who have *not* their eyes upon these scoundrels,—when, suddenly inspired, my jarvie pulls up at a dark archway.[10]

Marjorie Barnard and Flora Eldershaw, writing in combination as M. Barnard Eldershaw in their novel *A House is Built* (1929), an historical saga set in Sydney in the years from the 1850s to the 1880s, present a more optimistic picture than Richard Rowe. M. Barnard Eldershaw's fictional re-creation of Sydney on Christmas Eve in the 1850s, as seen by the quartermaster James Hyde's young grandsons James and Lionel on an evening shopping expedition for family presents, portrays the city as home to a generation of Australians whose memories will be rooted in Australian rather than British soil. Christmas, for James and Lionel, is now irrevocably signified by the fairytale enchantment of a warm Sydney summer evening.

> The streets were crowded. The fifties were rich, ornate years. There was plenty of money, plenty of work, plenty of hope. Life went with a fine flourish. The days were spacious, and they had need to be to hold the crinolines and the mustachios and the *bric-à-brac* that crowded them. There was plenty of colour and stir and noise in the streets, red coats of soldiers, and the blues, maroons, and silvers of the urban bucks. The embroidered waistcoats and flowing ties of the men were scarcely less brilliant and diverse than organdies, tarlatans, and muslins of the flounced and ruched crinolines that glided beside them. The air in the narrow streets was heavy with the smell of humanity, food, and dust; they resounded with the clamour of shop-keepers crying their wares and children blowing their Christmas trumpets, upon a background of laughter, conversation, and footsteps. In its upper reaches sound merged into vision, and, on the edge of the subconscious, smell merged into sound. The senses were perfectly blended, and the blend was rich.[11]

Sydney by the end of the nineteenth century, when the expatriate Rosa Praed returned from England, was a city of approaching a million people, and while it was obvious that the embryonic new society had much more to offer in the way of a healthy environment than the slums of England or the poverty-stricken potato fields of Ireland, and in economic opportunity for the entrepreneurial middle class, it was also apparent that many of the faults of the parent society had been imported with the virtues. The settlement had shaken off some of the stains of its lowly origins with the abolition of convict transportation to Australia in 1868, and was now sufficiently developed to be compared, favourably and unfavourably, with other cities.

In 1868 Sir Charles Wentworth Dilke, in his survey of Britain's colonial possessions, commented—after noting once again the beauty of the harbour—that in the upper portion of the town '... the houses are of the commonplace English ugliness, worst of all possible forms of English imbecility; and are built too, as though for English fogs; instead of semi-tropical heat and sun'.[12]

Francis Adams, radical poet, journalist and author, who came to Sydney in 1884 for health reasons at the age of twenty-one and later worked for three years as a journalist in Brisbane, notably for William Lane's radical newspaper *Boomerang*, was another who, though impressed by the Botanic Gardens, was ultimately disappointed in Sydney. In *The Australians, A Social Sketch*, published in London in 1893, he also wrote disparagingly of the extensive rows of brick terraces thrown up as working-class dwellings by developers from the 1840s on.

> Sydney is a city with charm, with the element of the ideal.
> Its sea-gardens, planted right in the centre, are as lovely as anything of their kind in the world.
> Nature has done her very best.
> The blue waters of the winding harbour are everywhere.

Sunshine, that often seems sempiternal, lights up jewelled hues in the sky and sea as tender as Athens, or Naples, or Cadiz.

The beauty of the inlets and seaside bays is almost equalled by that of the surrounding bush.

And yet the final impression is disappointment.

No European manufacturing city 'boasts' more hideous suburbs.

Places like Newtown and Enmore, Paddington and the Glebe, are simply that congerie of bare brick habitations, which is just as much an arid, desolate waste as the mid-desert.

… Something of convictism and the convict still shows itself in Sydney—in the brutality of the old slave-owning official families administering hideous and unrepealed statutes—in the hopeless criminality of the old lineal descendants of the 'lags', gathered together in Woolloomooloo, a small low-lying quarter of their own.[13]

Adams, unsuccessful in his search both for health and a better society, returned in 1890 to England, where, seriously ill, he committed suicide three years later.

In September 1898 Sidney and Beatrice Webb, Fabian socialist reformers, arrived in Sydney after touring America and New Zealand, and—in a section of her diary that was later omitted from the subsequent published work as being of 'little consequence'—the rather snobbish and intellectually superior Beatrice also gave her opinion of the city.

Sydney, in spite of its exquisite harbour and lovely Botanical gardens, is a crude chaotic place. It is seemingly inhabited by a lower-middle class population suddenly enriched; aggressive in manners and blatant in dress. The loafer, very much out at elbows and carelessly insolent in bearing, is to be seen wandering through the streets or lounging in the public Domain. In this city there is neither homeliness nor splendour; only bad taste and cold indifference.[14]

The Webbs stayed at the Australia Hotel, which they described as being 'American in its prices and pretensions', after a 'detestable voyage' on a 'dirty steamer' among 'noisy Australian commercial travellers and squalling colonial babies'. Despite being entertained by the governor Lord Hampden, the premier George Reid, the mayor and the Leader of the Opposition, they found little was to improve. Sydney, they concluded, was nothing more than 'a lower middle-class civilization suddenly got rich'.

A similarly unimpressed visiting observer was English novelist D.H. Lawrence. In a letter to his sister-in-law Dr Else Jaffe-Richthofen, written soon after his arrival in 1922, Lawrence, in a reversion to the viewpoint of many earlier commentators, finds Sydney a city of vacancies and absences. Like the Webbs, who were proponents of democracy but unenamoured of the working man on the occasions that they encountered him, Lawrence found that a surfeit of democracy did not suit him at all.

This is the most democratic place I have ever been in. And the more I see of democracy the more I dislike it. It just brings everything down to the mere vulgar level of wages and prices, electric light and water closets, and nothing else. You never knew anything so nothing, nichts, nullus, niente, as the life here. They have good wages, they wear smart boots, and the girls all have silk stockings; they fly around on ponies and in buggies—sort of one-horse traps—and in motorcars. They are always vaguely and meaninglessly on the go. And it all seems so empty, so nothing, it almost makes you sick. They are healthy, and to my thinking almost imbecile. That's what the life in a new country does to you; it makes you so material, so outward, that your real inner life and your inner self dies out, and you clatter round like so many mechanical animals … I feel if I lived in Australia for ever I should never open my mouth once to say one word that meant anything. Yet they are very trustful and kind and quite competent in their jobs. There's no need to lock your doors, nobody will come and steal. All the outside life is so easy. But there it ends. There's nothing else. The best society in the country are shopkeepers—nobody is any better than anybody else, and it really is democratic. But it all feels so slovenly, slipshod, rootless, and empty, it is like a kind of dream.[15]

The description of Sydney in Lawrence's subsequent novel *Kangaroo*, published in 1923 after he and his wife Frieda had spent more time in the city, develops the same ideas, but in a more closely observed fashion. However, one can't help suspecting that Lawrence was using Australia—and Sydney—as a metaphor for the deficiencies of his protagonist Somers, rootless in a new world in which he has no role. Having been bested by a taxi driver in a dispute about fares, Richard and Harriet Somers set off in a horse-drawn hansom cab across the city, feeling somewhat disgruntled.

In which state of mind they jogged through the city, catching a glimpse from the top of a hill of the famous harbour spreading out with its many arms and legs. Or at least they saw one bay with warships and steamers lying between the houses and the wooded, bank-like shores, and they saw the centre of the harbour, and the opposite squat cliffs—the whole low wooded tableland reddened with suburbs and interrupted by the pale spaces of the many-lobed harbour. The sky had gone grey, and the low tableland into which the harbour intrudes squatted dark-looking and monotonous and sad, as if lost on the face of the earth: the same Australian atmosphere, even here within the area of huge, restless, modern Sydney, whose million inhabitants seem to slip like fishes from one side of the harbour to another.

Murdoch Street was an old sort of suburb, little squat bungalows with corrugated iron roofs, painted red. Each little bungalow was set in its own hand-breadth of ground surrounded by a little wooden palisade fence. And there went the long street, like a child's drawing, the little square bungalows dot-dot-dot, close together and yet apart, like modern democracy, each one fenced round with a square rail fence. The street was wide, and strips of worn grass took the place of kerbstones. The stretch of macadam in the middle seemed as forsaken as a desert, as the hansom clock-clocked along it.

Fifty-one had its name painted by the door. Somers had been watching these names. He passed 'Elite', and 'Très Bon' and 'The Angels Roost' and 'The Better 'Ole'. He rather hoped for one of the Australian names, Wallamby or Wagga-Wagga. When he had looked at the house and agreed to take it for three months, it had been dusk, and he had not noticed the name. He hoped it would not be U-An'-Me, or even Stella Maris.

'Forestin—' he said, reading the flourished T as an F—'What language do you imagine that is?'

'It's T, not F,' said Harriett.

'Torestin,' he said, pronouncing it like Russian. 'Must be a native word.'

'No,' said Harriett. 'It means *To rest in.*' She didn't even laugh at him. He became painfully silent.

Sydney's northern beaches are scarcely more impressive.

They sat on the tram-car and ran for miles along a coast with ragged bush loused over with thousands of small promiscuous bungalows, built of everything from patchwork of kerosene tin up to fine red brick and stucco, like Margate. Not far off the Pacific boomed. But fifty yards inland started these bits of swamp, and endless promiscuity of 'cottages'.

The tram took them five or six miles, to the terminus. This was the end of everywhere, with new 'stores'—that is, fly-blown shops with corrugated iron roofs—and with a tram-shelter and little house-agents' booths plastered with signs—and bits of swamp or 'lagoon' where the sea had got on and couldn't get out. The happy couple had a drink of sticky aërated waters in one of the 'stores', then walked up a wide sand-road dotted on either side with small bungalows, around the backs of which lay a whole aura of rusty tin cans chucked out over the back fence. They came to the ridge of sand, and again the pure, long-rolling Pacific.

'I love the sea,' said Harriet.

'I wish,' said Somers, 'it would send a wave about fifty feet high round the whole coast of Australia.'

Later in the novel, having met the enigmatic undercover political leader 'Kangaroo' in his chambers—'a handsome apartment with handsome jarrah furniture, dark and suave, and some very beautiful rugs', where they eat with Queen Anne silver and drink from Venetian glasses—Somers and his friend Jack Callcott attempt a little social commentary.

> 'If I were but blind,' said Somers, 'I might have a shot at Australian Homerics.'
> His eyes hurt him still, with looking at Sydney.
> 'There certainly is enough of it to look at,' said Kangaroo.
> 'In acreage,' said Jack.
> 'Pity it spreads over so much ground,' said Somers.
> 'Oh, every man his little lot, and an extended tram-service.'
> 'In Rome,' said Somers, 'they piled up huge houses, vast, and stowed Romans away like grubs in a honey-comb.'
> 'Who did the stowing?' asked Jack sarcastically.
> 'We don't like to have anybody overhead here,' said Kangaroo. 'We don't even care to go upstairs, because we are then one storey higher than our true, ground-floor selves.'
> 'Prop us up on a dozen stumps, and we're cosy,' said Jack. 'Just a little above earth level, and no higher, you know. Australians in their heart of hearts hate anything but a bungalow. They feel it's rock bottom, don't you see. None of your stair-climbing shams and upstairs importances.'[16]

However, not all visiting writers of the period were as disdainful as the Webbs and D.H. Lawrence. The Polish-born English novelist Joseph Conrad, arriving as a twenty-one-year-old ordinary seaman on the wool clipper *Duke of Sutherland*, first encountered the city he would later recall as the 'town of my youthful affection' in January 1879. He subsequently made several more visits. In *The Mirror of the Sea* (1906), Conrad remembers Sydney on that first occasion as the harbour city where

> … from the heart of the fair city, down the vista of important streets, could be seen the wool-clippers lying at the Circular Quay—no walled prison house of a dock that, but the integral part of one of the finest, most beautiful, vast, and safe bays the sun ever shone upon. Now great steam-liners lie at these berths, always reserved for the aristocracy—grand and imposing enough ships, but here to-day and gone next week; whereas the general cargo, emigrant, and passenger clippers of my time, rigged with heavy spars, and built on fine lines, used to remain for months together waiting for their load of wool. Their names attained the dignity of household words. On Sundays and holidays the citizens trooped down, on visiting bent, and the lonely officer on duty solaced himself by playing the cicerone—especially to the citizenesses with engaging manners and a well-developed sense of the fun that may be got out of the inspection of a ship's cabins and state-rooms. The tinkle of more or less untuned cottage pianos floated out of open stern-ports till the gas lamps began to twinkle in the streets, and the ship's night-watchman, coming sleepily on duty after his unsatisfactory day slumbers, hauled down the flags and fastened a lighted lantern at the break of the gangway. The night closed rapidly upon the silent ships with their crews on shore. Up a short, steep ascent by the King's Head pub., patronized by the cooks and stewards of the fleet, the voice of a man crying 'Hot saveloys!' at the end of George Street, where the cheap eating-houses (sixpence a meal) were kept by Chinamen (Sun-kum-on's was not bad), is heard at regular intervals.[17]

THE VIEW FROM THE STREET:
1900 to World War II

The Cardigan Street Push, composed of twenty or thirty young men of the neighbourhood, was a social wart of a kind familiar to the streets of Sydney. Originally banded together to amuse themselves at other people's expenses, the Push found new cares and duties thrust upon them, the chief of which was chastising anyone who interfered with their pleasures. Their feats ranged from kicking an enemy senseless, and leaving him for dead, to wrecking hotel windows with blue metal, if the landlord had contrived to offend them. Another of their duties was to check ungodly pride in the rival Pushes by battering them out of shape with fists and blue metal at regular intervals.

   They were the scum of the streets. How they lived was a mystery, except to people who kept fowls, or forgot to lock their doors at night. A few were vicious idlers, sponging on their parents for a living at twenty years of age; others simply mischievous lads, with a trade at their fingers' ends, if they chose to work. A few were honest, unless temptation stared them too hard in the face. On such occasions their views were simple as A B C. 'Well, if yer lost a chance, somebody else collared it, an' w'ere were yer?'

Louis Stone, *Jonah*, 1911[1]

With the growth of an Australian-born population came the beginning of an Australian-born literature: patriotic for the most part, aggressively egalitarian and mostly concerned with life in the 'bush'. By the turn of the century the *Bulletin* had been publishing for twenty years, fifteen of them under the editorship of J.F. Archibald, and was the first Australian journal to combine political radicalism and humour with a deliberate cultivation of a nationalistic Australian readership. The *Bulletin* published only Australian authors.

   Now some of the relatively few early twentieth-century 'native' urban writers such as Louis Stone and Lennie Lower were beginning to take those aspects of Australian society considered with such intolerance by visiting English observers and to celebrate them with wit and good humour. Stone, born in Leicestershire, England, in 1871, but brought to Australia in 1884 at the age of thirteen, submitted his first work unsuccessfully to the *Bulletin*. Both Stone's novel *Jonah*, completed in 1909 and published in London in 1911 and Lower's *Here's Luck*, published in Sydney in 1930, take as their subject the larrikinism of the city and the lives of the urban battler and the inner-city push. Both books treat the city with a sometimes ironic affection.

   Stone, a schoolteacher, lived for some time in the inner-city industrial suburbs of Redfern and Waterloo, which he used as the setting for *Jonah*, probably the first Australian novel to have an entirely urban subject matter. In it Joe Jones—'Jonah'—the hunchbacked leader of an inner-city gang, progresses from layabout to successful capitalist, while the background life of the inner-city streets is closely observed and colourfully sketched in. Here, Jonah's old friend 'Chook' visits the Sydney Markets.

Chook was standing near the entrance to the market where his mates had promised to meet him, but he found that he had still half an hour to spare, as he had come down early to mark a pak-ah-pu ticket at the Chinaman's in Hay Street. So he lit a cigarette and sauntered idly through the markets to kill time.
   ... The silence of sleeping things hung over the Haymarket, and the three long, dingy arcades lay huddled and lifeless in the night, black and threatening against a cloudy sky. Presently, among the odd nocturnal sounds of a great city, the vague yelping of a dog, the scream of a locomotive, the furtive step of a prowler, the shrill cry of a feathered watchman from the roost, the ear caught a continuous rumble

in the distance that changed as it grew nearer into the bumping and jolting of a heavy cart.

... In half an hour the grimy stalls had disappeared under the piles of green vegetables, built up in orderly masses by the Chinese dealers. The rank smell of cabbages filled the air, the attendants gossiped in a strange tongue, and the arcades formed three green lanes, piled with the fruits of the earth. Here and there the long green avenues were broken with splashes of colour where piles of carrots, radishes and rhubarb, the purple bulbs of beetroot, the creamy white of cauliflowers, and the soft green of eschalots and lettuce broke the dominant green of the cabbage.

The markets were transformed; it was invasion from the East. Instead of the sharp, broken cries of the dealers on Saturday night, the shuffle of innumerable feet, the murmur of innumerable voices in a familiar tongue, there was a silence broken only by strange guttural sounds dropping into a sing-song cadence, the language of the East. Chinamen stood on guard at every stall, slant-eyed and yellow, clothed in the cheap slops of Sydney, their impassive features carved in fantastic ugliness, surveying the scene with inscrutable eyes that had opened first on rice-fields, sampans, junks, pagodas, and the barbaric trappings of the silken East.

It is almost irresistible to juxtapose this description with that of journalist Charmian Clift, in a newspaper column from the 1960s, written soon after her return to Sydney from a fourteen-year absence in London and Greece.

'Excuse me, lady! Outa da way, please! Outa da way!'

He was young and swarthy, with shoulders and arms built through generations for labour. Over his hairy Esau pelt he wore ragged shorts and a huge leather apron studded intricately with metal and hung about with wicked-looking curved picks. On his head a blue beret, worn jauntily. And as my friend Toni and I leapt for safety (or even survival) from his hurtling cart of tomato crates he grinned at us with that joyful unselfconsciousness of a Mediterranean male whose inalienable right it is to appraise every female he encounters—of whatever age or condition.

... And: 'Out of the way! Out of the way!' (or its equivalent) they yelled in urgent accents of Italian, Greek, Yugoslav, Chinese, and even Australian, as we dodged round piled trucks pulling in from places where still things grow in spite of the drought, stacks of aromatic crates, or jumped hastily from the onslaught of those lethal handcarts—dangerously stacked with fruit and vegetables or more dangerously empty—shot out from every market aperture with the obvious intent of crushing us down into the mulched cabbage leaves and the trodden mint and parsley that bore every evidence of a progress passing.

What was exciting was that we were involved in the progress, shouting in triumph as it passed, as though these huge market vaults were really Persepolis. It was terribly early in the morning, and every single leaf, blade, shoot, root or bulb had the dew or the earth on it.

I have always thought an early morning market to be a celebration.[2]

In Stone's novel, part of Jonah's personal transformation, his softening from a street fighter loyal only to his gang to a more complex character open to the influence of love, is overtly symbolised by a physical progression from the hot, mean inner-city streets to the coolness of the harbour. This linkage of the sea with the emotional life will recur frequently in subsequent Australian writing.

They had reached the end of Cremorne Point, a spur of rock running into the harbour. Clara ran forward with a cry of pleasure, her troubles forgotten as she saw the harbour lying like a map at her feet. The opposite shore curved into miniature bays, with the spires and towers of the city etched on a filmy blue sky. The mass of bricks and mortar in front was Paddington and Woollahra, leafless and dusty where they had trampled the trees and green grass beneath their feet; the streets cut like furrows in a field of brick. As the eye travelled eastward from Double Bay to South Head the red roofs became scarcer, alternating with clumps of sombre foliage. Clara looked at the scene with parted lips as she listened to music. This frank delight in scenery had amused Jonah at first. It was part of a woman's delight in the pretty and useless. But, as his eyes had become accustomed to the view, he had begun to understand. There was no scenery in Cardigan Street, and he had been too busy in later years to give more than a hasty glance at the harbour. There was no money in it.

From the vantage of Cremorne Point, the city across the water is transformed as Jonah and his lover are taken by surprise by the approaching night.

> They had talked for half an hour, intent on figures which Jonah dotted on the back of an envelope, when they were surprised by a sudden change in the light. The sun was low in the sky, dipping to the horizon, where its motion seemed more rapid, as if it had gathered speed in the descent. The sudden heat had thrown a haze over the sky, and the city with its spires and towers was transformed. The buildings floated in a liquid veil with the unreality of things seen in a dream.
>
> … And as they watched, surprised out of themselves by this magic play of light, the sun's rim dipped below the skyline, a level lake of blood, and the fantastic city melted like a dream. The pearly haze was withdrawn like a net of gossamer, and the magic city had vanished at a touch. The familiar towers and spires of Sydney reappeared, silhouetted against the amber rim of night; the hills, robbed of their pearly glamour, huddled beneath a belt of leaden cloud; the harbour waters lay flat and grey like a sheet of polished metal; light clouds were pacing in from the sea.
>
> They stared across the water, silent and thoughtful, touched for a moment with the glamour of a dream. The sound of a cornet, prolonged into a wail, reached them from the deck of a Manly steamer. At intervals the full strength of the band, cheerful and vulgar, was carried by a gust of wind to their ears.
>
> 'Oh, I would like to hear some music!' cried Clara. 'Something slow and solemn, a dirge for the dying day.'[3]

Lennie Lower, born in Dubbo, New South Wales in 1903, published his comic novel *Here's Luck* in 1930 after working for some years as a journalist, notably on *Smith's Weekly*. *Here's Luck* captures the idiomatic humour of the 1920s and 1930s while lambasting Sydney suburban life of the period, with its six o'clock closing of pubs, stifling conformity and frustrated housewives, against which—according to Lower—any self-respecting alcoholically inclined Australian male layabout must traditionally rebel. Early in the book, Jack Gudgeon is left by his wife Agatha, a not unwelcome development in his life. Agatha's mother lives in Chatswood, the epitome of the respectable northern suburb of the 1920s.

> Chatswood is one of those places that are a stone's throw from some other place, and is mainly given over to the earnestly genteel. Here, respectability stalks abroad adorned with starched linen and surrounded by mortgages. The clatter of lawn-mowers can be heard for miles on any sunny Saturday. Sunday evenings, the stillness of death descends on the place, but if one listens very attentively one may hear the scraping of hundreds of chewed pens as they travel the weary road of principal and interest and pay-off-as-rent.
>
> Agatha's mother's home tucked its lawns about its feet and withdrew somewhat from the regular line of houses in the street. It had been paid for. My mother-in-law's chief occupations were writing letters of complaint to the municipal council, and calling upon God to look at our so-called democratic government and blight it. She also laid a few baits for the neighbours' dogs, kept a strict eye on the morals of the whole street, and lopped off any branch, twig or tendril which thrust itself from the next-door garden over the fence and so trespassed on her property. What spare time she had left was used up by various communings with God about the water-rates and the only really light work she indulged in was when she seated herself behind the window-curtain and watched for small boys who might be tempted to rattle sticks along the front fence. Altogether, she was a busy woman.[4]

Half a century later, playwright and short story writer Jennifer Paynter makes a more affectionate use of the nearby north shore suburb of Pymble to satirise a similar suburban lifestyle in her 1988 story 'The Sad Heart of Ruth', one of the few subsequent successful Sydney attempts at urban comedy.

Jesus Christ was coming to dinner, and Ruth Jackson had extended the leaf in her reproduction Regency dining table and set out six Ken Done Opera House placemats and fashioned six red table napkins into bishops' mitres. She had put red candles in her Proud's candelabra and washed and dried her floral centrepiece of red bisque porcelain roses.

'Doesn't it look colourful, hey Bob?' she asked her husband who was an alderman on the Kuring-gai Council and often pretended not to hear Ruth when she asked him things. He pretended not to hear her now because he was resetting his federation Kookaburra clock, and he was cross with her about the menu. He'd told her to serve fish, but she'd gone her own sweet way and bought a leg of lamb and made a kiwi fruit pavlova.

Apart from Jesus, their other guests were the Very Reverend Keith Creevey and his wife, Sylvia, and Jesus' interpreter, a Mr Naji Jabour.

The doorbell rang. 'That'll be Him!' said Ruth, but it was only Mr and Mrs Creevey.

Bob poured out some sherry, and they all sat down to wait.

Keith Creevey was very nervous. He kept pressing his upper lip white and saying silly things. 'Isn't this incredible?' he kept saying. 'Isn't this incredible, Bob?'

Bob kept on pouring out sherries and pretending not to hear. He'd organised several little dishes of olives and ripe figs, but he didn't want to pass them around until Jesus arrived.

Sylvia Creevey started to laugh. 'Do you suppose He's lost, Ruth? Do you suppose He's gone to the wrong address? Does He know Pymble at all? Does He know how to get to Pymble, Ruth?'

Ruth made a married signal to Bob to put away the sherry. She got up and went to the kitchen and opened a tin of orange and mango juice.

Sylvia Creevey followed her, still laughing. 'Aren't you nervous though, Ruth? Aren't you worried He'll think you're too rich? I mean, you have such a divine home, Ruth. All the Venetian glass you collected on your overseas trips. And Bob's clocks. And this kitchen. What's he going to think of this Customtone kitchen, Ruth?'

Ruth was basting the lamb. 'He won't be looking at this house, Sylvia. He'll be looking into our hearts.'[5]

The expatriate Australian novelist Christina Stead was fifteen when, in 1917, a few years after her father's remarriage, she moved with her family from the inner western suburb of Bexley to Watson's Bay. Like Louis Stone she was drawn to the harbour. Stead had ten years to observe the life of the waterways before departing for England in 1928, years which she drew on closely for her first novel *Seven Poor Men of Sydney*, written in London in the early 1930s and published in 1934.

Joseph kissed his mother mechanically and went out. The gloom of the interior dropped from him. He walked smartly round the beach-path while the coral-trees along the shore, wrapped up in themselves, murmured without wind and dropped dead calices on his hat. It was low water; a transparent wave two-inches high rang its air-bells along the sand. The receding tide had left dark lines of flotsam along the beach. The poor children of the district and their mothers, with sacks in their hands, were raking through the deposit with their fingers, gathering coke, chips, and even vegetables thrown overboard in port from the vessels. Temperate sun and cool shadow divided the air. The sea-gulls paddled in and out of the water without a cry, and the fishermen pottered about sluicing and scraping their boats. During the night, the tide had risen over the path; there was a broken oar, a boathouse cradle, and part of the gates of a harbour-side bathing-pool. Miles away, south-west, between the side-drops of Bradley's Head and Shark Point, the city sat in miniature, glittering, without a trace of smoke. Blue-blooded spring was everywhere.

The ferry had not yet come in. Joseph waited outside the Italian fruit-shop at the end of the wharf, looking at a dead shark drawn up on the beach. It was responsible for the first bathing casualty of the season. It had torn off the buttocks and right leg of a bather the day before, and had been caught with a meathook on a clothes-line tied to a buoy, during the night; the bell on the buoy had rung for over an hour. The fishermen were all gathered there, with clusters of school-children and a barman from the hotel. They stood talking amiably and endlessly, like a collection of blue-bottle flies.

… The ferry whistled and Joseph had to run down to the wharf … The boat chugged into town through the glaze of the harbour on the darlingest, dazzlingest day of spring. Morning smoked on the hills, and the trees rose up to meet the sun as if to return to their primal essence and be dissolved in light. The morning was already hot; at Nielsen Park she lolled under the still leaves, the milky tide reflecting her in pools of curdled light. The *Città di Genova*, bound for Naples, rode out across the eastern channel, her masts rising higher every minute, her flanged bow emboldened by the sun, until she overtopped the little craft on the starboard bow, almost running her down. The engine-room telegraph rang furiously, the engineer shouted down the speaking-tube. They passed under the great red nostrils where the anchors hung, and the schoolboys yelled insults at the black-shirted Italian command. The cicadas skirled in the foreshore reserves, the remarks of the season-ticket holders became drowsier and foolisher, and Joseph dreamed.

Like Stone, Stead writes of loosely connected urban people linked by their common poverty. However, while Lennie Lower, and to some extent Stone, prefer to stick fairly closely to the lighter side of life and avoid the emotional complexity that might lie beneath, Stead's vivid and affectionate descriptions of the city are merely a reflection of her characters' much more closely explored inner lives. Stead's character Baruch Mendelsohn, a printer by trade and an intellectual, lives in Woolloomooloo.

Baruch lived in a room on the fourth floor back, in a side street in Woolloomooloo Flat, not far from the old public school. His window commanded the Inner Domain, the Art Gallery, the spires of St. Mary's Cathedral and the Elizabeth Street skyline. On the right hand, as he looked from his window, were the wharves of the German, Dutch, Norwegian and Cape lines. In the backyard was a wood-and-coal shed covered with creepers, pumpkins, old tires, kites'-tails, buckets and old scrubbing-brushes. There was a clothes-line across the yard, on a clothes-prop, and upon the line the tenants' garments, washed by the woman on the ground-floor, appeared in regular succession throughout the week. ... On Saturdays and Sundays the whole neighbourhood swarmed with children, and everybody was out of doors with sleeves rolled up. Tiny living-rooms with Japanese screens, fans and bead curtains, and reeking of bugs and kerosene, with bric-à-brac, vases, wilting flowers and countless rags and papers, sent out their heat and animal odours and old dust at seven in the evening when the hot day had gone down into the violet twilight, a deceitfully shady moment promising cool, but bringing in the torrid night. Everywhere couples lounged about, the waists encircled, the lips together; henna Titians, peroxide blondes, and uncoloured women faded beneath their hair still rich and young; women blowsy and painted, worn and tired, with crow's-feet and unequal powder, fanned their bursting bosoms or their empty sacks of blouses, as they slumped in rickety easy-chairs at their doors.[6]

Now had come a time when new arrivals to Sydney were no longer automatically borne into its harbour by ships from overseas, to measure the city by the yardstick of other cities of the world. As Australia began to spawn its own generations of native children, there appeared also literary images of the arrival from the hinterland, where the only yardstick was life on the isolated sheep or cattle station, small farm or rural township, now left behind by the adventurous young for the epic journey to 'the big smoke'. The city became not only an end, but a beginning.

'To come to Sydney by train is to come in the back-door of the city,' wrote Kylie Tennant in 1943.

A flurry of frowsy suburbs, grey streets, brick-yards, old iron dumps, all these go by wearing the peevish expression of a housewife who, not having time to make the beds, grumbles: 'You must take us as you find us.' For the incoming ships are the shining towers lifted, the ripples spread on the blue carpet of the harbour, the hurrah of arrival. The trains find only a commonplace of advertisement hoardings, rumble, grime and confusion.

To the native the smell, the orientation of the place, is unique, unmistakable; but to the stranger it is just city, any city, with ranked warehouses, clanging trams, cars shining and sliding, bustle of people, shops sucking in and pouring out their human food supply like so many water squirts on a reef. There is not even a swallowing movement, as the new supply of atoms pouring from the train are engulfed in the city's vast and nonchalant insentiency.

Far too soon Shannon and Beryl had started hauling down suitcases and exchanging addresses with their over-night friends, fidgeting and fretting, ready for an excited breakaway.

'No hurry until you see the mortuary station,' their red-faced friend advised placidly.

As he spoke, the mortuary station appeared over to the left, tiny, brown and graceful under the angular city buildings. It looked like a little brown Hindu contemplating its navel between the river of railway lines and the jungle roar of the city; a Gothic stone siding with pillars and arches, left over from the days when death was quiet and dignified. How many Sydney people had made that remark: 'No hurry till you see the mortuary station,' as they folded up their morning newspapers ready to leap off the train.

Beside the mortuary station waved a stumpy palm tree enforcing the idea that here was something fatalistic, Eastern, as out of key with the great buildings as the grey stone tower of the church that had flashed by just before, a dungeon deep on a rise. Here were people coming alive; but there was still a place for them if they came dead. Living or dead, the city would receive them, as it received everything, swallowing all the incongruous particles undisturbed.[7]

A similar journey is described by Marjorie Barnard in 'The Dry Spell', a story from her collection *The Persimmon Tree*, also published in 1943. Here again Sydney is synonymous with a longing for the comforting presence of the sea, and an escape from the hard and arid life of the hinterland.

But in times of drought, as the author finds as she approaches the city on foot along Anzac Parade, the desert invades even here.

I walked because there was no reason for stopping, because it was more intolerable to stay still, and because I wanted to reach the sea …

It was the third waterless summer, and the heat had come down like a steel shutter over the city. The winters between had been as bad. Dry, with a parching, unslaked cold; westerly winds that drove and drove, bringing such clarity to the air, that a hill five miles away looked near enough to touch. The drought was in everything now, penetrating and changing life like blind roots at work upon a neglected pavement. The colours and quality of the world had been altered in the long months of desiccation. The pattern of existence was pulled awry. …

The country with its endless, aching death pressed in on the city, the drought and the heat pressed on both. In the city and its environs its stamp was no less clear. The bush on the outskirts was more than half dead. Even the deep feeders, the black butts and the like, were dying. The life that was left was drawn in and banked down, muted and secret. The scrub was shabby and colourless. Fire had licked through it, leaving patches of black and sharp red-brown. Where there were houses, wide fire breaks had been cut as the only protection. Water could no longer be relied on to combat the fires. These breaks were raw scars, even on the devastated country. They looked like the trail of vengeance. Orchards were long since dead, and the trees fallen on the eroded ground. On the eastern slopes around Dural the orange trees were burnt black. The flats that used to be vegetable gardens were bare, the last dried stalks blown away. Even Chinamen could make nothing grow.

In the wealthy suburbs of the North Shore and Vaucluse a change had taken place too. It was as if the earth had been squeezed so that all the fine houses that had nestled so comfortably in the contours and the greenery, were forced up into the light. They bulged out, exposed, and the sun tore at them. The gardens that had embowered them were perished. Tinder dry, fire had been through many of them, scorching walls and blistering away any paint that remained. Most of these houses were empty or inhabited as if they were caves, by people who had come in from the stricken country. The owners had fled, not so much from present hardship, as from the nebulous threat of the future, the sense of being trapped in a doomed city. The shores of the harbour were lion-coloured or drab grey. Sandhills showed a vivid whiteness. Only the water was alive and brilliant. And it was salt.

In the crowded districts, there was less to perish, but light and air were equally abrasive, chafing all surfaces, fading and nullifying all colour. There was no pleasure of touch left anywhere, for the dust was

35

undefeatable. It pulled down pride and effort. The suburbs sagged under an intolerable burden.

I was perpetually aware of all this. It cumulated into a black wave which hung over me in threatening suspense. Nothing that I knew had escaped. From my windows I looked over the golf course and that had taken, because it was defenceless, the clearest print of all. Its silvery green hills were stripped to pale brown and tawny purple. The earth was like starved, sagging flesh on an iron skeleton. Here and there a fire had run for a few yards before it died for lack of tinder, and left a black smear with a little edging of white ash. I used to think that the desert of Arizona looked like that. Now I know that heat and drought can bring even the gentlest country to it.[8]

Ruth Park, the New Zealand-born writer who came to Sydney in 1942 and married the Australian writer D'Arcy Niland, with whom she travelled extensively in the outback before settling in an inner-city suburb of Sydney, is also intensely aware of this penetration of the desert into the city during the Australian summer. In her novel *Poor Man's Orange* (1949), a story of Irish-Australian working-class life in the urban slums of Surry Hills, Park celebrates the life-giving properties of the regular seaborne afternoon wind known as the southerly buster.

Charlie went out on the veranda, sitting on the gas-box in the corner under the ragged shadow of the vine, hiding himself from the curious stares of passers-by. The whole world seemed poised on a pivot, palpitating in the heat. From the sky like grimy glass came the smell of dust, speaking mutely of the drought-bitten hinterlands, of the cattle-skulls gaping out of the soft-sifted soil, the deserted towns half drowned in sand, the earth grinning and cracked, and the very flesh and blood of the continent whirling out in a cloud to the sea, and nobody caring a tinker's whether it did or not.

But there was something else, too, an expectancy, an awareness, which even he in his numb apathy could not help but feel.

After the unbearably hot day, the old men on the balconies were snuffing the air and saying, 'Here she comes!' The southerly buster, the genie of Sydney, flapped its coarse blusterous wing over the city, a hearty male wind with a cool and spirited breath. The women undid the fronts of their frocks, and the little children lifted up their shirts and let it blow on their sweaty bottoms. Even the dogs crawled from the oven-hot shade of parched trees and hung out their tongues like banners in the cool. Now there was movement everywhere, the trees tossing their arms upwards, the torn shop awnings undulating, and the scattered papers on the road taking flight, leaping upwards in gleeful tackings, up, across, over the roof and round the garret chimney, until like a ragged flock of cubist birds they disappeared into the rents and ravines that the southerly had torn in the high far roof of cumuli. Doors slammed, windows rattled, and Lick Jimmy's clothesline spun round like a top.

And the birds, too, exploded into the sky. There was no telling where they had come from. They pelted out of the dusky sunset, no more than black dots, as feckless, as disorderly, as swift as insects bursting from a hedge. There were starlings in a loose-flung flight, like a cast net; a rocket of sparrows, and then, far up, the strong-winged, disciplined webfoots, the ducks with necks outstretched, the heavy geese, and the wild black swans; they passed up there where the colour and the light were fading from the after-sunset, leaving behind them the eerie sound of their voices, discordant, forlorn, like distant bugles.[9]

Elizabeth Harrower, in her first novel, *Down in the City* (1957), which describes Sydney in the years after World War II, is another who uses the distinctive Sydney weather to signify a change of mood in the city.

Autumn came in slyly, in the way of all Australian seasons, with a blustery day here and there squeezed in between late summer scorchers, praised for coolness, not recognised, so far away was last autumn, for what it was: gay striped blue days when freshly-washed sheets flapped and cracked in suburban backyards, while along miles of foreshore dark green gums tossed and writhed in the sunny wind.[10]

One of the liveliest accounts of Sydney during the World War II period is contained in *Come In Spinner*

(1951), a controversial and lengthy novel by Dymphna Cusack and Florence James which was completed in 1948 and, after some editorial delay, appeared in an abridged edition in 1951. *Come In Spinner* was unusual for the period because it described wartime conditions in Sydney entirely from the point of view of women—a feat also attempted with mixed success by Xavier Herbert in his 1962 novel *Soldiers' Women*. *Come In Spinner* won the 1948 *Daily Telegraph* competition for an unpublished manuscript of a novel, but on publication it was described by the same newspaper as a 'muckraking novel fit for the literary dustbin'. This was apparently due to the fact that, despite a somewhat women's-magazine-style sentimentality, it made matter-of-fact allusions to illegal gambling clubs, liquor licensing corruption, black market activities and the exploitation of minors in prostitution.

Most of the action in the novel is based around a big city hotel resembling the Hotel Australia, but in pursuing their various lives and interests the characters range throughout wartime Sydney from the rural outskirts to the inner-city slums and the brothels of Kings Cross. The authors attempt to convey the restless mood of the wartime city and its effect on all strata of society. In this passage, once again it is a hot midsummer night, and Val, who works in the beauty parlour of the hotel, and whose husband has been posted missing shortly after their marriage, walks from the old Hyde Park barracks and among the rattling trams to Hyde Park, where a band is playing.

> She made her way across the broad intersection above St James's Church where five roads meet and the holiday traffic shuttled swiftly to and from the city. Three Indonesians loitered by disconsolately, their voices a low, liquid gabble. Two young English sailors watched her with heartsick eyes. So pitifully young, she thought. All these people so lonely, so far from home. The city is full of them, adrift on the tide of war, talking of home in alien tongues … servicemen and women, refugees, evacuees who have lost all the dear, familiar things—all those to whom this city is a refuge without sanctuary.
>
> She stood a while under the big tree overhanging the footpath near the underground station; above her in the thick foliage a colony of sparrows still chattered like men in a bar at closing time. Across the street she could see the lighted windows of the American Centre and the stream of khaki figures drifting in and out, pausing on the edge of the footpath, laughing and talking with the shoeshine boys, moving on again, crossing in twos and threes the broad street to the park.
>
> In all the restless crowd, only she was alone. There was nothing so terrible and so terrifying as being alone. All these people were trying to push loneliness away for an hour, for a night …
>
> That was why the girls stood in pairs on the steps near the station entrance, their young faces hard, their voices raucous, brazen and enticing, waiting for hungry men, some man, any man, so long as his need matched their own. Live while there is yet time, before the young men's bodies are broken and the young girls' flesh stale from too much lusting without time for love.
>
> We were lucky, Ven and I. 'We are invulnerable,' he had exulted, holding her close that last night …

Later in the same novel Helen McFarland, a shy middle-class country girl uncomfortable with the relative glamour of the city, goes out with a soldier she has met while doing volunteer work in a canteen. This enables a description of two of Sydney's more venerable institutions: Luna Park and the annual Easter agricultural show.

> They climbed the little hill from the wharf and stepped through the giant's gaping grin straight into a noisy, swirling crowd. Down both sides of the park the sideshows beckoned in brilliant electric lights; mechanical cars raced thrillingly up and down miniature hills and valleys; the mermaid lay languorously on her seashell inviting you to throw balls at her; the octopus swung squealing couples on the ends of its giant arms into the darkness above the harbour; crowds of young men in uniform pressed round the rifle ranges and coconut shies, handing back their prizes to their girlfriends with self-conscious laughter; girls carrying dolls on sticks and hideous china vases clutched the arms of servicemen and turned on them eyes

full of promise. Everywhere there was laughter and harsh lights and a moving kaleidoscope of colour.

It was a bit like Sydney Royal, was Helen's first thought. Ever since she could remember, the whole family had come to Sydney at Easter for the Royal Agricultural Show, and it had been one of her greatest thrills as a small child to visit the sideshows and slide down the slippery slide. And yet there was a difference. At first she thought it was the uniforms, but it was something more, something undefinable, a feverishness, a snatching at pleasure, rather than the way one enjoyed the Show with time to pause and laugh and saunter on. Here, everyone seemed greedy. She watched the couples coming out of the Magic Cave, flushed with their kisses in the darkness. A sailor had his arm round a slip of a girl with her brown hair fuzzed up into an exaggerated pompadour and her lipstick all smudged. He had lipstick around his lips and a hungry look in his eyes. She turned away involuntarily.[11]

If *Come In Spinner* tends to sharply divide its sympathies across the class barrier—working-class women generally have hearts of gold, while upper–middle-class socialites are briskly caricatured—then Xavier Herbert's *Soldiers' Women*, which deals with the same milieux, is gleefully malicious over the entire social range: almost all his female characters are basically self-seeking, sexually voracious, and murderous to an unlikely degree. The reader is led to suspect that a true portrait of wartime Sydney lies somewhere beyond either novel.

More recently, writers George Johnston and David Malouf have provided vividly immediate sketches of post-war Kings Cross: Johnston in the second part of his semi-autobiographical trilogy *Clean Straw For Nothing* (1969) and, twenty years later, Malouf in his novel of Australian life *The Great World* (1990). In *Clean Straw For Nothing*, Johnston's protagonist David Meredith and his new wife Cressida escape Melbourne for Sydney and find makeshift accommodation in an apartment building which, like its inhabitants, is a little shopworn from the war. Johnston's description has the authenticity of firsthand observation, however brief.

The bathroom—there was only one to each floor of the Princeton Apartments—was at the far end of a corridor angled around a euphemistic light-well from which came weird clankings and groanings and the catarrhal cleansings of other residents. The corridor was all gloomy tones of brown, and the excremental colour and smell of this dingy byway went with the bathroom, which had stains on the walls like old maps and blotches more repellent in the toilet-bowl where something seemed to have happened with Condy's Crystals. The cracked wash-basin was a mess of squeezed-out toothpaste tubes and rusted bobby-pins, and the final sordid touch was a framed printed sign screwed above the ringmarked bathtub which said GUESTS ARE POLITELY REQUESTED PLEASE NOT TO SHIT IN THE BATH. 'Politely' was the bit I liked. It was hard to believe that this had been set up in type, in Bodoni Bold, and printed.

When I went back to the musty bedroom I said hopelessly, 'We should have looked around more. We should have tried for something better than this dreadful bloody dump.'

'It's cheap,' she said, not pointing out that the Princeton had been my choice. 'It will do for a few days while we sort out. When you get a job we can look around for something else.'

'Someone said it was used as a kind of leave place for the troops, GI's mostly, a brothel, I suppose, more than a residential,' I said as if this was a mitigation. 'Put toilet-paper down on the seat first,' I warned.

Cressida went to a window so opaque with city grime that it seemed as if a wartime black-out screen was still pasted over the glass. 'Well, at least it's in the Cross,' she said, as if she could see out. 'It's a pretty street. There's a flower-stall over there. And we don't have to sit up here. We can always go out.'

For the couple, Kings Cross represents a respite from the pressing problems of the future.

Coffee in cramped dim cellar lounges and the soft play of branch- and leaf-shadow in the autumn streets, as much as the emotional and physical raptures of a love still sublime even in the sleazy room of the Princeton, have been the palliatives of this uneasy arrival time of the last few weeks. There really are flower-stalls in Kings Cross and the one run by old fat Maggie has become *our* flower-stall, and we now have *our* tobacco kiosk where Cressida has charmed an old dragon of a woman into letting her have under-the-counter cigarettes, so we no longer have to save the tobacco from our used butts to be rolled up in airmail paper or smoke those terrible South African things that smell as if they are made from wildebeeste dung. And we have our own coffee-lounge where we can hear *our* songs on the juke-box, 'Laura' and 'It Might As Well Be Spring' and 'The Breeze and I' and 'Deep Purple' and 'Night and Day'. And in the evenings we can walk through light-dazzle and cacophony under the bronzing leaves, and nobody takes any notice of us or cares who we are, although heads turn wherever Cressida walks.

Kings Cross is a little spurious and more than a little self-conscious, and its air of cosmopolitanism is an awkward masquerade, but there are misfitting foreigners about and odd eccentrics, and a raffishness has persisted, and it is better than Melbourne, and better than the rest of Sydney, which is even more war-scarred from its self-inflicted injuries. A coarser, tougher city, poised on an edge of violence. A cocky, callous place. Amid the merciless tensions of the town these past weeks have been a time of abeyance, of putting things off.[12]

In *The Great World*, Brisbane-born David Malouf has his returned soldier Digger, survivor of the Japanese prisoner-of-war camps of Thailand, encounter the same raffish suburb in the same period, but this time with a slightly broader and more detached vision. Digger, like many soldiers before him, finds comfort in the air of transience and almost dangerous vitality to be found there.

On hot nights late in Darlinghurst Road Digger found what he had always been in search of, a crowded place with the atmosphere of a fairground, but one that did not have to be knocked down and set up again night after night. It was simply there, another part of town.

It was a rowdy place, the Cross. It could be violent, sordid too at times, but it had put a spell on Digger just as Mac had told him it would.

Girls, some of them toothless and close to sixty, worked out of mean little rooms up staircases smelling of bacon-fat or sharp with disinfectant. The pubs were blood-buckets.

You would see a couple of fellows come hurtling through the door and in seconds a full-scale brawl would be going on, right there on the pavement, with passers-by ducking aside to get away from it or standing off on the sidelines to watch.

Often it was seamen; but mostly it was young blokes, louts, who had come in on motorbikes to roar about and see what was doing, keen to get a reputation and discover how tough they were. ...

Occasionally it was a woman you saw, still clutching her handbag but with her mouth bloody, one arm like a broken wing, and the man who had done it shouting right into her face, spitting out obscenities but weeping too sometimes, justifying himself. This was peacetime again.

And in between these savage episodes the delivery boys would be out and old people, or women dragging a suitcase in one hand and a reluctant child in the other, would be going about their daily affairs. Well-dressed ladies walked pug dogs. Kids sucking sherbet sticks dawdled back and forth to school. Old fellows slept it off on benches or stood with their sleeve up to the elbow in bins.

There were coffee shops, continental, with mock-cream cakes in the window, and other, darker ones downstairs where it was rumoured that satanic cults were being practised. The paintings on the walls, which were pretty bold, gave you a hint of what they might be: a woman with her legs round a shaggy male figure with horns above his ears, another in which a girl was coupling with a gigantic cat.

Then there were the milk-bars all fan-shaped mirrors and chrome, spaghetti places where men in business suits lined up for lunch, and barber shops, some with a dozen chairs; always with two or three fellows lathered up for shaving while the barber, razor in hand, harangued them while others, further down the room, would be snipping and chatting or showing a customer the back of his head in a glass, and in the doorway one of the idle assistants hung on a broom.

Barber shops, billiard saloons, dark corners in pubs—this was where the SP bookies followed their trade, using runners and a 'nit' to watch for the cops. But everyone up here had something to sell: petrol, stuff without coupons that had fallen off the back of a truck, nylons, second-hand cars, pre-war of course, and girls.[13]

Escape from Suburbia:
The 1950s

They were living in a house with a tower and a view of the Heads. They had embroidered chairs, crystal dishes that chimed when flicked with a fingernail, and a fragment of oak from Nelson's flagship in a small velvet box. At school Caro was up to the Spanish Armada and the sad heart of Ruth, when the ferry called the *Benbow* turned over in Sydney harbour and hideously sank. Grace was on a blue chair in the kindergarten and still had Miss McLeod, who had come out after the Great War and would be super-annuated at Christmas.

Miss McLeod played the organ for the school at morning prayers. 'Hush'd was the Evening Hymn,' 'For All the Saints,' and, in season, 'Once in Royal David's City.' Everyone was C of E or something like it, except Myfanwy Burns and the Cohen girl. Religion was the baby in the manger, the boy with the slingshot, the coat of many colours.

Shirley Hazzard, *The Transit of Venus*, 1980[1]

Sydney in the mid twentieth century, like the other Australian capital cities, was in a state of transition to the modern age, but it was a sluggish state of transition. Previously an almost exclusively British colonial enclave, Anglo-centric by tradition, and insulated from rapid change by its distance from Europe, the city experienced an era of conservatism and provincialism. The heavy hand of the public censor hung over the literary scene: Sumner Locke Elliott's play *Rusty Bugles*, distilled from his experience in an army camp at Mataranka in the Northern Territory in 1944, was officially banned in 1948 due to its blasphemous language, but the author, like many others, had already left for the United States. The Anglophile, conservative prime minister Sir Robert Menzies, who in his second term retained power from 1949 to 1966, personally vetoed Commonwealth Literary Fund grants to writers whom he suspected of left-wing tendencies. In the post-World-War-II years an influx of non Anglo-Celtic migrants and refugees, along with improvements in transportation and communication between Australia and the rest of the English-speaking world, would begin to make the isolation and insularity of this colonial outpost seem less great. In the meantime, however, in the years immediately before and after the 1950s, Sydney's writers treated their city with something less than generosity.

In an early section of her novel *The Transit of Venus*, set largely in England and America, Shirley Hazzard describes a certain type of suburban Sydney childhood of the 1930s: conservative North Shore middle class, redolent with Church of England hymns, European history and inappropriate English poetry. In an incident resembling the November 1927 sinking of the harbour ferry *Greycliffe*, during which some forty people died, Grace and Caro's parents are drowned and the girls' life irrevocably alters. (Others also found this incident a handy device for removing surplus characters: Sumner Locke Elliott in his semi-autobiographical novel *Careful He Might Hear You* (1963) makes use of a fictional ferry accident, and Christina Stead in *The Salzburg Tales* (1934) mentions the sinking of the Watson's Bay ferry in 1927. Coincidentally, by the time of this publication both of these authors, like Hazzard, were long-time expatriates in New York.)

For Hazzard's sisters, Grace and Caro, life in the pre-war harbour city, at least in the safe regions of the North Shore, is stable, muted, secure and, apparently, ultimately capable of causing madness. In this period, it seems only death or departure overseas has the potential to cause change.

The house to which they now moved with Dora was smaller, with camellia trees on the lawn but too many hydrangeas. At the back it was buffalo grass and spiked shrubs, and a rockery hewn from the sandstone slope. Indoors, the responsive crystal, the splinter of the true cross from H.M.S. *Victory* had become museum pieces, relics of another life. At each side of their own brief horizontal, the long streets dropped to the sea. They might almost, had they known it, have been at Rio or Valparaiso. Night followed night, nights of oceanic silence not even broken now by the screams of bandicoots in traps on the Hornimans' English lawn.

In the slit of two headlands the Pacific rolled, a blue toy between paws. The scalloped harbour was itself a country, familiar as the archipelago a child governs among the rocks: it hardly seemed the open sea could offer more. Yet, passing into that slit Pacific, ocean liners took the fortunate to England. You went to the Quay to see them off, the Broadhursts or Fifields. There was lunch on board, which Dora did not enjoy because of a small fishbone caught in her throat. Sirens were blown, and kisses; streamers and tempers snapped. And the *Strathaird*, or *Orion*, was hugely away. You could be home in time to see her go through the Heads, and Caro could read out the name on the stern or bow. Even Dora was subdued at witnessing so incontrovertible an escape.

Going to Europe, someone had written, was about as final as going to heaven. A mystical passage to another life, from which no one returned the same.

Hazzard's *The Transit of Venus*—once described by Patrick White as a 'blue-stocking romance'—was published in New York, where the writer had been living since the early 1950s. Hazzard herself, born in 1931, grew up in the Sydney suburbs of Beauty Point, Mosman, and in Balmoral, and left at the age of sixteen when her father was posted overseas in 1947 as an Australian trade commissioner.

After their parents' death, the orphaned sisters Grace and Caro are cared for by Dora, their increasingly neurotic half-sister, and their pre-Dora existence becomes a separate life, only distantly remembered. In this previous, vivid but somewhat more dangerous world of the city during the Depression years of the 1930s, the disabled from the 'Great War'—another example of Australian colonial patriotism—still beg on the streets.

Where they got down from the tram there were windows brilliant with coloured gloves and hand-bags and silk shoes, and shopping arcades lit like rainbows. The women passing along Pitt Street or Castlereagh had cooler faces and wore hats of violets or rosebuds, with little veils. Kegs of ale were nonetheless drawn on drays right past the best shops by pairs or teams of Clydesdales: chestnut necks straining in collars of sweated leather, great hooves under ruffs of streaked horsehair. And the driver collarless, frayed waistcoat open, no jacket, with his leather face and stained mop of horsehair moustache. Manure underfoot, and bruised smell of dropped cabbage trodden by blinkered ponies harnessed to vegetable carts. Along the curb, barrows of Jaffas and Navels, or Tasmanian apples. All this, raffish and rural, at the fashionable conjunction of Market and Castlereagh streets.

At the same corner they would come upon the spectres dreaded by Caro and by Grace; and, from the looking and the looking away, by all who passed there.

… Some of them stood, including those with only the one leg. The legless would be on the ground, against shop-windows. The blinded would have a sign, to that effect, around the neck—perhaps adding SUVLA or GALLIPOLI. Similarly, on the placard GASSED that hung beside pinned medals, might appear the further information, YPRES or ARRAS. Or the sign might say MESOPOTAMIA, quite simply, as you might write HELL.

The North Shore, as Grace and Caro return from school each day, represents an enclave of bourgeois respectability, embattled by the wilderness of humanity outside, in a city about to be changed forever by the events of World War II.

Caro and Grace walked home uphill in raging heat. Brick houses were symmetric with red, yellow, or purple respectability: low garden walls, wide verandas, recurrent clumps of frangipani and hibiscus, of banksia and bottlebrush; perhaps a summerhouse, perhaps a flagpole. Never a sign of washing or even of people: such evidence must be sought inside, or at the back. Caro was beginning to wonder about the inside and the back, and whether every house concealed a Dora. Whether in every life there was a *Benbow* that heeled over and sank.

You felt that the walls of such houses might topple inwards, that they would crush but not reveal.

Refinement was maintained on the razor's edge of an abyss. To appear without gloves, or in other ways suggest the flesh, to so much as show unguarded love, was to be pitchforked into brutish, bottomless Australia, all the way back to primitive man. Refinement was a frail construction continually dashed by waves of raw, reminding humanity: the six-o'clock shambles outside the pubs, men struggling in vomit and broken glass; the group of wharfies on their Smoke-O, squatting round a flipped coin near the Quay and calling out in angry lust to women passing. There were raucous families who bought on the lay-by, if at all, and whose children were bruised from blows or misshapen by rickets—this subtler threat contained in terrace houses whose sombre grime was a contagion from the British Isles, a Midlands darkness. Britain had shared its squalor readily enough with far Australia, though withholding the Abbey and the Swan of Avon.[2]

This is the same North Shore suburban existence to which, a few years later, Sumner Locke Elliott subjects his Flagg sisters, Ila and Geraldine, in *Water Under the Bridge* (1977). World War II Mosman, where tea must be served correctly and madness kept under wraps, is a social indicator in itself.

The Flagg sisters considered having their party in the garden. There was the view of Mosman Bay. But on the other hand, all that carrying of trays up and down the steps and wasn't the weather just a bit uncertain? Suppose it poured rain as they served tea. It would be just their luck. They decided on the back veranda. Scrub the wicker chairs and trust Mrs Watson could keep Dadda in his room.

'Risky,' Ila said, 'if it's one of his bad days.'

'Well, but if we never make any effort—'

Geraldine had had the idea of giving a little tea party Easter Saturday and this time not just girls. Ask some boys for heavens sake, why not? Ila brooded on this. Boys?

'Because if we never make any effort—'

Lately an anxiety had begun in Geraldine as slight as a breath of cold air, as though someone had left a door open far away at the end of the house. She could not be specific about her anxiety but it had begun to chafe that people always said 'Here come the Flagg girls' and when Mrs Watson said 'Tut, tut, *The Messiah* again and everyone else going to Christmas eve parties?' Sometimes Geraldine heard a spiteful voice whispering 'Suppose.' She had begun looking at her naked body in the bathroom mirror and appraising her flawless skin and nice breasts. It was unthinkable that some man would not, someday, lust over her. But suppose (the cold draught of air touched her) no man ever got the chance. And Ila didn't appear to mind, it was almost as if Ila had already begun to be as set as jelly into the possibility of a future with no men. If Mumma hadn't died they'd have been tactfully steered toward matrimony, that was what mummas did. But they only had mad Dadda. If only they weren't—well, so damn well-bred, so 'Mosman'. She'd overheard someone say from behind a rack of coats that those Flagg girls are nice but so *Mosman*. Mosman meant suburban rectitude, brick-fenced gardens, glass cabinets, Ovaltine before bed. Only in Mosman could one have known a miserable girl who had had the dreadful name Lesbia wished on her by her unenlightened parents, the wretched girl shortened it to Les and waited for mum and dad to die.

'If we don't lift a finger, we'll never get out of Mosman.'

'Do we want to?'

'Well, *I* do.'

Ila ran her fingernail along the seam of the gray velvet sofa, she seemed uneasy; being with other girls all the time required no guile and she was without guile.

Geraldine was not without guile.[3]

In a more recent evocation of the post-war period, *Holden's Performance* (1988), when Murray Bail's innocent picaresque hero Holden Shadbolt first arrives in Sydney in the early 1950s from sleepier Adelaide (where Bail himself was born in 1941), the city is signified by cars and crowds and water. Shadbolt's arrival coincides with the visit of the newly crowned Queen Elizabeth II of Great Britain, and the city is in a patriotic fervour. However, in streets that are named after 'British monarchs, a British prime minister, Pitt, and various inbred brothers, uncles and even fathers of British monarchs', Shadbolt witnesses a first murmuring of dissent, a first eccentric gleam of defiance against the heavy hand of conformity.

> Joining the pedestrians he allowed himself to be carried along, bumping into others, one foot in the gutter.
> They swept across Sydney Harbour Bridge.
> … Shadbolt would become a connoisseur of crowds; but not yet. This was by far the largest he'd seen. A steady hum reverberated and merged with the surrounding buildings; it tended to blur people's swaying senses. More and more people pressed from behind, and as the hour passed an anticipatory restlessness, beginning with the schoolchildren and the cripples in wheelchairs lining the front, ran back in waves like a wind or fire along grass, before stopping against solid matter, and then shifted again the other way. Standing patiently Shadbolt had no trouble looking over the heads and up the swept-clean street towards the Town Hall; and he was among the first to see the glitter of the slowly approaching black car. Almost simultaneously a murmur rushed towards him turning all heads, a murmur overlapping into a chatter of higher exclamatory voices, more like a rattle, everybody shifting forward an inch, multiplying and erupting into a clapping, a hoarse yelling and a cheering, figures swaying holding their first borns aloft, waving hankies, miniature Union Jacks or just their arms and fingers. As Shadbolt tried to remain in the one spot the torrent surged forward and back, mercury rolling across a table, pausing and stretching the elastic leading edge where policemen gritted their teeth and turned purple in the face.
> Shadbolt had consumed countless grey-and-white images of the young Queen, but as she drew level, seated well back in the open Daimler, he was hypnotised by her pinkness—she'd burn to a frazzle if she stayed in Australia—set off by the clarity of her neck, pale blue hat and raised hand. The immaculate black coachwork threw such details into relief: cunningly clever choice in duco. By then the worker-bees surrounding him wanted to cluster around their queen, their ecstatic scribbled faces and sticky hands strained forward again, and Shadbolt found himself waving frantically too, smiling desperately for the pale face to turn in his direction, and for even a fraction of a second to acknowledge his presence. As she passed, the bod in front turned with shining amazed eyes, and his nose, an unusual bulbous nose, registered to Shadbolt as one that had enveloped a ball.
> It was then he heard the voice.
> 'Sheep, merino sheep! Look at you all. Grown-up people, making fools of yourselves. What are you all here for? Tell me that.'
> The push around Shadbolt hesitated.
> 'That's right, you're all jungli, the lot of you. Wave to the Queen! Bow and scrape. She went thataway. Follow the leader. This mania for worship. Has anyone stopped to consider?'
> People began calling out and turning. It's a free country, but. Shadbolt felt the flow of the crowd dismantle into unpleasant elements. The way some grow indignant, others accept; Shadbolt glimpsed the force of the majority.[4]

However, in other parts of Sydney, it seems little has changed. In 1964 George Johnston and his wife Charmian Clift, both journalists and novelists, returned from their fourteen-year residence in London and Greece and moved into harbourside Mosman, the suburb so pilloried by Shirley Hazzard and Sumner Locke Elliott several decades before. Initially the couple lived in a rambling bungalow at Kirkoswald Avenue; they later moved to Neutral Bay, and then back to Raglan Street in Mosman.

Johnston had just won the Miles Franklin Award for his partly autobiographical novel *My Brother Jack* (1964) but, to offset this triumph, he was suffering severely from tuberculosis, a long-standing affliction that led to his death in 1970. In 1969 they were living at Raglan Street, both drinking heavily, and Johnston was progressing slowly with the final volume of the trilogy already containing *My Brother Jack* and *Clean Straw For Nothing* (1969). Charmian Clift would commit suicide the same year. *A Cartload of Clay* was finally published in 1971 in an edited but uncompleted form. In this period Johnston would sometimes force himself to take short walks along Raglan Street for exercise, and these became a structural motif in the book—and, according to his biographer Garry Kinnane, a repetition of the idea of a journey which runs through the other volumes. Johnston's *alter ego* David Meredith, in an unnamed suburban street, now rediscovers in Sydney the qualities that, in Melbourne in the 1930s, first drove him to look for a different type of life.

> He always felt intimidated by the opacity of an Australian suburban street.
> He did not have to look up to know about it. In one garden there was a power lawnmower thudding and snarling and chattering at the turf. A heavy-bellied man with grey bristly hair and a red neck and wearing a fancy cardigan clipped at the over-tidy edges of a flower-bed, big veinous hands active amid lobelia. Wallflowers, cinerarias and pink everlastings waiting their turns. In the big guesthouse named for a victory over the heathen Hindoo, a turreted edifice white and buxom as an overpowdered matron beneath the florid blush of its red tiles, anonymous and uncommunicative men and women sat in the sunshine in their separate garden chairs, sipping at their tea, reading their newspapers and magazines, knitting, just staring out. At a picket gate an elderly woman, her hair in coloured plastic rollers, lounged in an avid unexpectancy, waiting for the postman's whistle. The time, Meredith realized, that flowed along this street was measured by these things, the angry snarling of a lawnmower, snip of shears, rustle of paper, clinking of a spoon on china, a gate creaking, rasp of dry leaves, a braking bus, the postman's whistle. There were bird songs there too, but foreign to these other settled sounds; just as there was life there, a sort of life, but perhaps it was this life that created the wall of opacity, because one could never escape the feeling that if you could finger these lives there could now be no response except a sigh, and if you reached over to feel behind them there would only be emptiness. Nothing. The vacuum. It was possible that this was why the wall of opacity had to be preserved, so that nobody might ever suggest that there was nothing there ...[5]

In direct contrast Charmian Clift, in the period shortly before her death, surveys the same cityscape from the same suburb but finds little of the 'opacity' of Johnston's experience. Instead, in a series of contemporary newspaper essays that give little hint of the depression and anxiety from which she is suffering, Clift optimistically lifts her eyes towards the sky and finds a symbol of a new and healthy sense of freedom there.

> The skyline is something else though. Perhaps because it is so high, and so far away, and there is so very much of it—miles and miles and miles—and so very much sky too, and all the shapes on the skyline are delineated as precisely against this enormous sky as paper cut-outs. All that prodigality of space and light is rich bounty in a city—the tender blues and the soft greys and the cotton-wool lumps of clouds drifting about the lightning flashes and thunderheads and rainbows and dawn greens and dusk gentians. In London in winter the sky was as thick and murky as stockpot and the sun swam in it sluggishly, a squashy crimson blob that never struggled higher than the rows of grimy rooftops. I admit that I thought it incredibly romantic at the time, but I can breathe here.
> One of the charms of my skyline is its variety. It was wild country once, that's for certain. All these hills were wild. Sandstone. Scrub. Jungle. Up there, miles away, like upturned mops drying out above the domestic rooftops, there are three cabbage-palms in a row. I like them so much better than the poplars

beyond, perhaps because they have this grim little air of defiance. They've held out against the taming of these hills. And that's good, because they are a kind of symbol. Convicts wove cabbage-tree fronds into hats for protection against the fierceness of a cruel and alien sun, their children, the currency lads and lasses, adopted the fashion—defiantly too, perhaps—and by the time of the gold-rushes the cabbage-tree hat had become the cult-emblem of colonial independence, of swagger, of spiritual freedom, of anti-imperialism, of individuality. Men would pay as much as ten pounds for a cabbage-tree hat, so prized had they become. Henry Lawson writes of men in the nineties wearing cabbage hats black with age, hats that had been handed down like heirlooms.[6]

Similarly, to Hal Porter, previously a country schoolteacher and now arriving in Sydney after a stint as a librarian at the small town of Bairnsdale in Victoria, Sydney in the mid-1950s is already a lively contrast to Melbourne, Hobart and Adelaide; a city steeped in history, and a portal to the literary world he hopes to enter.

> Australia's oldest and most raffish city, rapacious and quasi-tropical, it still has, in those days twenty years ago, that galvanic quality which—so many years and cities later—you now recognise as profoundly metropolitan. It enfevers and infatuates not only visitors but its own excitable citizens as only a half-dozen of Earth's bigger hives can do. Its name's not ever, like *Calais* on that troublesome woman's heart, not to be found branded on its denizens' hearts. You sense its hurtling towards the twenty-first century is shameless and cynical, yet happy-go-lucky and vivacious. *Nichevo!* If it can no longer be as beautiful as it once was, it'll insist on being fantastic: an enormity of an opera house, all cockatoo beaks, built on gambler's losses. Who cares? *Eternity* is already everywhere screeved, copperplate, primrose-coloured chalk, on alley wall, footpath and granite plinth: the eighteenth century, that cureless influenza, cannot here be shaken off. The First Fleet still hounds about off-shore. With what skill the Hogarthian hordes swinging by wear the disguise of the 1950s, the newly fashionable Japanese thongs, the Baden-Powell shorts and rolled umbrellas, the spike heels and Audrey Hepburn hair-dos, the beehive coiffures, high and black as busbies, and the fake breasts, pointed like pine cones: the outline and gewgaws and catchwords of the moment.
>
> You're not fooled by all the *fanfarrio*.
>
> Milk-bar attendants and sales representatives they appear to be, these other-age coach-trimmers, bell-hangers, cutlers, milliners and millers; these makers of combs and candles, brushes, whips, masts, bonnets, baskets, ginger beer, wax flowers and lucifer matches; these curriers, chimney-sweeps, well-sinkers, wharfingers, lime merchants, dyers and scourers. That junior executive has a toyman's gait; that bus-driver a turnkey's scowl. Girl Friday, forsooth!—she there, passing by, could well be Madame de Rémy, French midwife, black eyes sparkling after a nip from the flask of gin in her placket.
>
> The nineteenth century? There are still boles, vermiculated quoins, marble fountain-rims, and early Victorian pillars of Hawkesbury sandstone across which have flickered the shadows of such shadows as Trollope, Conrad, Darwin, Mark Twain, Sarah Bernhardt, and Nat Gould—ah, the days when chattering and ribald shadows had famous shadows to cast!
>
> … Beatrice Davis's office, a mere attic above Angus and Robertson's bookshop at 89 Castlereagh Street, is at the altaltissimo of a perilous steep of strait wooden stairs. It excites you to know that the soles of your brogues can now freely and indiffidently press on the same creaking harmonium-pedal treads so many Australian writers have, literally and metaphorically, ascended by: Miles Franklin, Kenneth Mackenzie, Ion L. Idriess, Frank Dalby Davison, Robert D. FitzGerald, Douglas Stewart, Brian James, Eve Langley, Ethel Anderson, Xavier Herbert, Norman Lindsay, Hugh McCrae, Peter Hopegood *und so weiter*.

This was an exhilarating period for the writer, then in his mid forties. As recounted in the third part of his autobiographical trilogy, *The Extra* (1972), Porter was arriving in Sydney to meet Davis, editor at Angus and Robertson, at their office at 89 Castlereagh Street, to discuss the publication of Porter's first novel, *A Handful of Pennies* (1958). The visit also marks the beginning of his friendship with the poet Kenneth Slessor, at whose house at 18A Billyard Avenue, Elizabeth Bay, Porter stays when he is in the city.

> This is Kenneth Slessor's house, a sort of cottage *orné* in stucco, laundered shabby by weather and years, and overlooking Sydney Harbour. Here, and in an earlier nearby flat, he wrote most of his later poems, *Five Bells* among them. You're staying at 18A. The great poet is now amazingly your amazing friend.
>
> Most times, on a Sydney visit, you sleep in a room, separated from Slessor's kitchen by a little vestibule from which, through an outside door, you can descend break-neck wooden steps into a musk of rotting leaves and stale seaweed, a precipitous wilderness of centenarian camphor laurels and jacarandas once part of the long-built-on garden of Elizabeth Bay House. The aged trees soar upwards towards a hidden sky from the ever-salivating rocky bluff overrun with ground ivy, periwinkle, and primeval ferns. At the foot of this vegetable disorder, at the Harbour's edge, the remains of an Edwardian bathing pool you can people, if you will with E. Phillips Fox personages or George Lambert ones, and the murmuring sunlight of less crowded summers it would be improper not to regret. Now its tree-trunk piles are barnacle-encrusted, or sprout a gloomy sea-growth like black-green hair. No longer a place to sit by, reading *The Story of an African Farm* under a pearl-grey parasol. As for sea-bathing!—the Devil knows what senile merman or trapped ocean-horror lurks in the depths shadowed by a tangle of boughs.

Porter uses this Elizabeth Bay house of Slessor—who, he is disconcerted to find, resembles a marmalade cat—as a base from which to ramble around the area.

> Ah, now you're at the Botanic Gardens!
>
> Vagrant among Moreton Bay fig-trees, de-Europeanised European trees, acanthus and agapanthus, you come upon the dreary pre-arboreal province to be found in public pleasances from Rio de Janeiro's Botanico to the Palm House in Kew Gardens: few plants are less enlivening than the cabbage-tree and bangalow palms. Monkeys and Black Sambos and intricately decorated pythons are needed. Curious about what's next, you encounter curiosities: rustic bridge; botanist's obelisk-stoppered tomb; a Government House under a ring-a-rosy of proletarian sea-gulls; a crenellated Conservatorium, once stables, out of which issues, like plaint from oubliette, the wan trilling of a soprano. And, everywhere to be encountered, those soiled and engaging personages of marble with cobwebs in their armpits, amputated finger joints, and dirty eyes, who are the aptest inmates of gardens. You belong, of course, reactionary fellow, to a class that finds angular exercises in aluminium, welded conglomerations of scrap-iron, and Henry Moore's doughnut-bellied slobs of stone, unfit for public display though perfectly fitting for millionaire's courtyards and the patios of the nouveau riche. You're relieved to find the insufficiencies of palm garden made up for by marble people, Neptune, Winter, Commerce, Albert the Good, Agriculture, the Boy with the Thorn. You know what they mean; their sanity shows. Most engaging is the Savoyard chimney sweep, marble hands muffed in marble sleeves, frozen toes clenched, body hunched with cold in the blast-furnace high noon, the heat pulsating down from a seamless, electric-blue sky.[7]

Sumner Locke Elliott, born in Sydney in 1917 but later resident in New York, also remembers the Botanic Gardens in *Water Under the Bridge* (1977), but is even less impressed. His character Neil Atkins' childhood vision, remembered in later life as he waits for his difficult love Carrie Mazzini, lacks the exuberance of Porter's.

> When he was a little boy, long ago, he had been perplexing to people by hating the circus, shedding tears in the zoo (*Why, Neilie? You are a funny child*), and disliking outings in the Sydney Botanical Gardens. There was a sense of gloom where the giant fig trees met overhead and over-shadowed the glass-roofed kiosk where he had been reluctantly led to have ice cream at the fly-spotted marble tables. The green light made it spooky, predatory pigeons eyed him through the dirty windows.
>
> In between the stifling bamboo groves there were stagnant reeking ponds where under the water lilies sinister fins moved. The statuary was abysmal, wan stone nymphs missing noses and fingers and pugnacious boxers in fig leaves faced each other across paths, grottoes that should have enticed led implacably to rest rooms. On the hard benches the unemployed lolled, reading newspapers they retrieved from rubbish baskets and from the nearby Conservatorium of Music, the continuous chanting of scales

came to seem after a time like the wailing of the damned.

She was twenty minutes late, maybe wasn't coming. You could never be sure; she was as unpredict-able as her moods.[8]

The novelist Christopher Koch, born in Hobart in 1932 and thus like Hal Porter also an arrival from the colder south, similarly discovered Sydney by way of Kings Cross and Elizabeth Bay when he came to the city to work as a radio producer for the Australian Broadcasting Commission in the late 1950s. In his novel *The Doubleman* (1985), Koch no doubt draws on his own experience for his character Richard Miller's first ventures into Sydney life a few years later.

William Street was my entrance hall to Sydney. A boulevard linking the city and King's Cross, it carried me down into a tunnel of hotels and pawnshops and then climbed the Darlinghurst ridge. I had taken a taxi directly from the airways office, since the Cross was where most actors and show people were said to live.

Scalded by the vinyl of the seat, riding through an air that smelled like tin, I seemed to be wrapped in sacking. I had never been so hot in my life, and it was borne in on me that I had come to a foreign latitude, not very far below the Tropic of Capricorn. The taxi driver wore a T-short, thongs and indecent shorts, and was glancing sardonically at my heavy southern suit.

'You better get outa that gear, sport, or you'll turn into a grease-spot.'

Sydney's surfaces were all strange after the cool south; it was nearing sunset, and a hot, honey-thick light coated the low brick business buildings with a weird density, and was reflected in the Cross's hill-side windows up ahead, making them flash blinding messages. Strangest of all was a line of three-storey Victorian terraces at the top of the hill, which was the junction of the Cross. Rearing on the skyline, their fantastic arches and spires half colonial-Gothic, half Oriental, these buildings had a worrying, even night-mare quality; they suffered, melting and mouldering in the heat like ancient wedding-cakes, trapped in the wrong hemisphere. Painted in hideous colours, degraded in every way that simple imaginations had been able to devise, they were covered over every inch of their façades by advertisements and neon signs: new, faded and almost invisible, going back to the twenties and perhaps even earlier. One of the said: *Hasty Tasty.* Highest of all, on the pinnacle of the Cross, was a giant bottle of sherry tipping its neon liquor into a glass. It faded out when the glass was full, only to reappear as we passed, like the Cheshire Cat's smile; like a rune of the city I'd need to decipher.

Miller has come to Sydney in 1964 to work as a radio actor, and he finds himself living among the European migrants and refugees who have clustered around the inner-city suburb in an attempt to recreate in some way the cafe society they have left behind. Twenty years on, there is an echo still of the postwar Kings Cross of George Johnston and David Malouf.

We were all refugees, in the Cross.

Dormitory of Displaced Persons, New Australians and old, King's Cross was a ghetto for those on the run: from wives, from husbands, from jobs; from that past where they'd once been respectable. Roosting on its hill above the city centre, this junction where five roads met was so small it had no official right to separate existence. Mini-village in the ward of Fitzroy, gaudy patch on the slum-grey ridge of Darlinghurst, it existed through force of personality, and was more than a district. The Cross was the capital of devi-ance, and Australia's most densely-populated square mile.

Back in the twenties and thirties, in the era of mutton chops and meal tickets, of the famous Arabian coffee shop and the poets Brennan and Slessor, this had been Bohemia; a southern hemisphere Mont-martre. In the summer of 1964, when I first came there, it still wrapped itself in the tatty dressing-gown of these pretensions like one of its own landladies. But by now it was a teeming rookery of male and female prostitutes, show people, failed artists, successful criminals, and the wrecked and displaced flung to Australia from Europe after the War: people like my landlord, Bela Beaumont.

Like other Australians as Sydney became a more polyglot society, Richard Miller is confronted with the idea of living with people he has never encountered before.

> The native-born eyed these people with mingled condescension and suspicion; the pain the Displaced had emerged from, the pain they carried in them always, meaning nothing. Australians could make little of them, this tribe from Eastern and Central Europe, in the first two decades of their coming: women with formal, old-fashioned dresses and strangely braided hair; shabby men with fanatical, pale blue eyes, in the jackets of old suits teamed with sad, neat sports trousers, carrying executive briefcases in which there were probably bombs, or books that no one would want to read. The native-born dubbed them Reffos, or Balts; and in the coffee lounges, the Reffos confirmed all suspicions by endlessly reviewing their broken lives, railing against the Machiavellian leaders who had sold them out, so that now they found themselves here, in this flat, huge country of flat, emotionless people who knew no other language but English, at the end of the world. Sometimes a Balt or some other refugee would go mad, succumbing to the wartime blackness he had carried to the southern hemisphere in his brain; would shout and rave, standing alone in the street of meaningless sun, his pale eyes coldly crazy, his flood of foreign words proof of his insanity. Then the police paddy wagon would come from Darlinghurst, and the big Sydney cops in their dark blue caps would haul him away, their disgusted faces saying: *A Reffo. What else can you expect?* They would look inside his briefcase, when they got him to the station up at Darlo; they would finger his foreign books.[9]

These are the first of the new arrivals who will, by degrees, change the cultural face of the city forever.

*… the harbour appeared to be never-ending. It filled the hollows
and gaps, water finding its own level, it leaked into the corners of
his eyes whichever way he turned. Deep! The lapping mass glittered
and penetrated, lapping at the descending layers of terracotta
houses, submerging the boards of the wooden jetties,
slap-slapping sullenly at rocks …*

MURRAY BAIL, 1988 (PAGE 22)

SYDNEY FROM THE WEST, NEW SOUTH WALES. PHOTOGRAPHER: OLIVER STREWE

*Australia's oldest and most raffish city, rapacious and quasi-
tropical, it still has ... that galvanic quality which—so many
years and cities later—you now recognise
as profoundly metropolitan.*

PATRICK WHITE'S SYDNEY

… I wanted at the same time to paint a portrait of my city: wet, boiling, superficial, brash, beautiful, ugly Sydney, developing during my lifetime from a sunlit village into this present-day parvenu bastard, compound of San Francisco and Chicago. I had a lot of exploring to do. It was not so much research as re-living the windswept, gritty, or steamy moods of the streets, coaxing dead-ends, narrow lanes, and choked thoroughfares to release those voices, images, emotions of the past, which for my deplorably atypical Australian nature evoke guilt rather than pleasure.

Patrick White, *Flaws in the Glass*, 1981[1]

When Patrick White wrote these words in 1981 about his novel *The Vivisector* (1970), it was seven years since he had moved back into the city from his semi-rural home at Castle Hill on the northern edge of Sydney. *The Vivisector* was, he continued, about an artist—'composite of several I have known, welded together by the one I have in me but never became'. The words might equally well have applied to Sydney: a city composite like most cities of a series of earlier ones, and containing within it the potential for the one it might never become.

White returned with Manoly Lascaris to Australia after World War II, having studied modern languages at Cambridge, travelled extensively in Europe and the United States, and served as an RAF officer in the Middle East. No doubt, like many others before him, he saw his native city with the fresh eyes of the returning expatriate. Often, these newly perceived images were transformed into the brooding houses and suburbs that figure in his work. White himself identified the locations of several of his novels.

All the houses I have lived in have been renovated and refurnished to accommodate fictions. The original structure is there for anybody who knows: 'Lulworth' for *Voss*; 'Dogwoods' for *The Tree of Man* and *The Solid Mandala*; Martin Road for *The Eye of the Storm*; the cottage, the homestead, the sheds, the dunny at 'Bolaro' for *The Twyborn Affair*. In some cases it has not been so much architecture as atmosphere which has transferred the house to the page. The spirit of the stairs at Martin Road as I groped my way down most mornings in the dark conveyed the interior of a Paddington house I saw as Duffield's without ever having been inside. In the theatre of my imagination I should say there are three or four basic sets, all of them linked to the actual past, which can be dismantled and re-constructed to accommodate the illusion of reality life boils down to.

White was born in London in 1912 while his parents were visiting that city, and grew up in Sydney at Lulworth, a large house with an overgrown garden at 73 Roslyn Gardens, Elizabeth Bay—later a maternity hospital and now a nursing home. He was sent at the age of thirteen to Cheltenham College in England, from which he returned in 1929. He subsequently worked for two years as a jackaroo at a sheep station called Bolaro in the Monaro region of southern New South Wales, and at Barwon Vale, a family station, near Walgett, before returning to England to take a degree from Kings College, Cambridge. He stayed on in London after graduating in 1935, and his first novel, *Happy Valley*, appeared in 1939. He joined the RAF in 1941 and, after a short return visit in 1946, returned permanently to Australia with Manoly Lascaris in 1948. White was then thirty-six.

What Manoly and I needed most was somewhere to live and soon, as we had four schnauzers coming out of quarantine. One weatherboard cottage we were offered out at Kellyville set in my memory as a classic Australian image, with its dry-rotted veranda boards, the smell of mutton fat and sick lino inside, and a fig tree growing out of sinkwater. In such a house the O'Dowds would have lived. It is the house Mag Bosanquet connects instinctively with her lover Terry Legge. Eureka Steel lived there with Cuth Spurgeon in the screenplay *Last Words*. It is the house to which Manoly and I return in some of the nightmares of my old age.

In reality we passed this one up and settled for a suburban villa on six acres of paddock facing Showground Road at Castle Hill. It has been referred to as a 'lovely old home' and a 'rundown farmhouse', when it was never more than that suburban cottage in painted brick. What persuaded us to buy the place was the fact that there were pigsties which could easily be converted into kennels. We bought the house walk-in walk-out. We lived with the furniture for years, learning about those who chose it, till we could afford what pleased us better.

It was at this first home, Dogwoods, that *The Tree of Man* (1955), *Voss* (1957) and *Riders in the Chariot* (1961) were written, in an eighteen-year sojourn during which, by White's own account, they painfully and stubbornly attempted to produce milk from two dairy cows, and grow fresh vegetables and flowers to sell from buckets by the roadside. All the while, he wrote, their neighbours waited sardonically for them to be driven out by failure.

In that clay hollow, freezing in winter, breathless in summer, amongst the heavy ergot-bearing paspulum, hassocks of Cape weed, rusty rye grass, Patterson's curse, I was constantly will with asthma. We seemed almost waist-deep in weed. While rotary hoeing our wax-infested citrus orchard, Manoly wrecked his back for ever. We were this pair of amateur actors, miscast through our own determination, or pig-headedness. It was worse for Manoly in that he had attached himself to a prickly character who protested against his fate by throwing saucepans of Irish stew out of the kitchen window, cursing, and getting drunk.
… The earlier part of *The Tree of Man* I wrote at the kitchen table during sleepless nights when spasms of asthma prevented me lying down. Much of *Voss* was written in bed, and after it *Riders in the Chariot*.
It was not all asthma. There were nights of exquisite cool when the southerly arrived from Sydney in the clapping leaves of the camphor laurel which wistaria hadn't yet succeeded in choking. On such nights the mown grass prickled against naked flesh as we lay this side of the sheet of moonlit shastas.[2]

In White's memoir *Flaws in the Glass* (1981), the title itself is also a reference to the house at Castle Hill: 'The mirror in the bathroom at "Dogwoods" had a flaw in it like a faint birth mark'. White associated this bathroom with a period of intense loneliness in 1958 when Manoly Lascaris returned to Greece for a visit.

One of White's best known creations from this period is the fictional suburb of Sarsaparilla, also on the outskirts of Sydney, for which White drew on Castle Hill in the 1950s and 1960s. Sarsaparilla figures in his novels *Riders in the Chariot* and *The Solid Mandala* (1966); in two plays: *The Season at Sarsaparilla* and *A Cheery Soul,* both 1965, and in a number of his short stories.

In *The Solid Mandala* Sarsaparilla, like Parramatta, is a location that 'had a history' in the early days, but then, it appeared, 'history drew in her horns'.

There was Allwrights' store, and the post-office stuck in the side of Mrs Purves's house. There were the cow cockies and market gardeners. There were the homes of the aged, the eccentric, the labourers, the rich, though the last hardly counted, existing only spasmodically on kept lawns, amongst their shrubs, in varnished dogcarts, or, in Mrs Musto's case, behind the wind-screen of a motor car. It was really the grass that had control at Sarsaparilla, deep and steaming masses of it, lolling yellow and enervated by the end of summer. As for the roads, with the exception of the highway, they almost all petered out, first in dust, then in paddock, with dollops of brown cow manure—or grey spinners—and the brittle spires of seeded thistles.

For White, Sarsaparilla also came to be a metaphor for what he most disliked about Australian society: a perceived unquestioning acceptance of the mediocre, a materialism combined with a deep suspicion of difference, and a profound resentment of those with aspirations to anything more. Mrs Poulter, one of Sarsaparilla's inhabitants, reflects this.

> You couldn't say she wasn't comfortable. He kept the home painted up. Bill never showed his age. Lucky to still, in spite of his quirks, have his strength. Took a few jobs on the side to make the something extra. Like grass-cutting and pruning roses. For the few extra luxuries. You had to keep up with the times. They had bought the plastic awnings for the front. She had the electricity, she had the phone. Sarsaparilla wasn't on the sewer of course, and Bill wouldn't come at a septic, but she had her health, in spite of one or two aches, and what was a few steps across the yard to the dumpty. Altogether you couldn't say she wasn't comfortable. She had the radio, but no longer used it all that much, not since they got the telly, or anyway began the payments, like most else since recently. Bill said people had never in history had it so good, and well, she admitted, you couldn't say it wasn't pretty good. You couldn't complain. Not with the electric frying-pan—never used her oven now—not with the phone, and two doctors. And the telly. If she didn't have any friends without the ones she yarned with over fences, in buses, or the street, she didn't need any. She had the telly, the nice announcers, and world figures in your own lounge. She could afford to mind her own business, without Mrs Dun reminding her of it.[3]

And yet this is the city—where 'asphalt sinews ran with salt sweat' and the 'fuddled trams [tunnel] farther into the furry air', arriving at last 'under the frangipani, the breezes sucking with the mouths of sponges'; this quasi-tropical city: 'Sodom had not been softer, silkier at night than Sydney'[4]—that harbours, as well as the spiteful denizens of Sarsaparilla, those spiritually enobled misfits who will be brought together to become the 'Riders of the Chariot'.

In its more comic aspects, Sarsaparilla is in some ways the Sydney equivalent of Humoresque Street, Moonee Ponds, the Melbourne address of comedian Barry Humphries' original version of his creation Edna Everage, who with Sandy Stone of Waterloo Road, Glen Iris, shares some characteristics of the inhabitants of White's fictional suburb. White first saw Humphries perform in *A Nice Night's Entertainment* in the early 1960s. The two men became friends, and White, according to his biographer David Marr, subsequently wrote material for him.

In 1964 White and Lascaris moved to a house at 20 Martin Road, Centennial Park, where White wrote *The Vivisector* (1970) and *The Eye of the Storm* (1973), which he has described as the two novels that 'belong' to Sydney.

The Vivisector was already forming in his mind at Dogwoods, he wrote, '… but in spite of time spent in my native city during childhood and youth, and frequent visits while we were at Castle Hill, I had to feel Sydney around me, day and night, in my maturity, before I could undertake the novel'.[5]

In a number of his works set or partly set in Sydney, including *The Aunt's Story* and *The Vivisector*, White draws extensively on his experience of the city in his earlier, pre-expatriate period there. Sometimes these descriptions have the rather impressionistic texture of childhood remembered, the distant and sepia tones of a lost age, through which his visionary characters move in a dreamlike way. However, in other works where Sydney is used, this early vision is rarely completely absent, and forms a subtext to the modern city, just as his early childhood memories of Belltrees and other family stations inform his latter-day perceptions of the Australian landscape.

In early sections of *The Aunt's Story* (1948)—which White began after his demobilisation in London and completed while returning to Australia by sea in 1948—the author describes a distinctively Australian (i.e., Anglocentric) pre-World War I existence which somewhat resembled that of his own family during the same period. In this early novel there is not much explicit description of the external Australian landscape, but by sketching in details of the furnishings and architecture of his godmother's house on a property near Moss Vale (which White calls Meröe) and more particularly of Lulworth, White creates a powerful impression of a way of life. In *The Aunt's Story* White's heroine Theodora Goodman ultimately rejects this materially rich but stultifying existence—one which White himself had earlier rejected—by retaining the right to define herself instead of being defined, as were most middle-class women of the period, by house, property and husband.

Thus, Theodora, a woman apparently past marriageable age and without conventional beauty, has been invited to lunch at the house of Huntly Clarkson, an older but eligible bachelor.

> In this way Theodora Goodman went to the house of Huntly Clarkson, which stood in a blaze of laurels, a rich house, full of the glare of mahogany and lustre. The floors shone. There was an air of ease that disguised the industry which achieved this state. The servants were silent and well oiled. If they did not speak, it was because they had learnt their functions too well. They had a kind of silent contempt for anyone who did not understand what these functions were. So the servants of Huntly Clarkson looked at the shoulders of this woman helping herself to a cutlet, and condemned her as she tried to thank. Her glance was indication of her income and status. She was a woman of no account, whose clothes were not of this or any fashion, whose face was ageless in appearance, though they would have put her somewhere in the thirties.
> 'I did not ask anyone else,' said Huntly Clarkson, 'because you didn't suggest you wanted them.'
> 'How not suggest?' she said, dazed by the noises the silver made on the table. 'At the time I was a blank. You have read things off me that were never there. Really. I assure you.'
> But there was a kind of ease between them. She began to think that it might be a pleasant thing, a friendship with Huntly Clarkson, if she could resist his house, his servants, and his furniture. These were all magnificently assured. They fixed time in the present. Even the old things inherited from grandfathers and aunts, even these pandered to Huntly Clarkson and the present, as if they began and ended as part of his upholstery. She looked at the rich, shining, well-covered body of Huntly Clarkson and wondered if he would exist without his padding.

Later in the novel, at a dinner in the same house, at a table 'smouldering with red roses', Theodora discovers—in a way that it is tempting to surmise might parallel White's own experience—her reasons for rejecting Clarkson's hand.

> At the end of the dinner they brought with the dessert some very expensive crystallized fruits, which were no longer fruit but precious stones, hard, and their sweetness had a glitter. This was the apotheosis of the meal, in which the light brandished swords. You forgot the flat words in the glitter of glass and diamonds, the big crystallized stones that hung from Marion Neville's body, and the angelic straps on Elsa Boileau's brown shoulders. The whole of Huntly Clarkson's life lay there on the table, crystallized, in front of Theodora Goodman, and she knew at such moments that there was nothing more to know.

A guest at the same dinner is a Greek composer, Moraïtis, with whom, in a brief encounter, Theodora Goodman finds an analogy between White's beloved Greece and the landscape of Australia, and sums up in essence the double bind that brought White himself back to Australia to work.

> The Greek suddenly walked, as if he had made up his mind, and sat beside Theodora Goodman.
> 'They have put me in a room,' he said, 'where I cannot practise.'
> The words that he suddenly found he took out with precision. He smiled to see.
> 'It is all furniture,' he said. 'I cannot live in such a room. I require naked rooms.'
> 'Bare,' said Theodora Goodman.

'Bare?' said the Greek. 'Naked is the word for women.'

'Naked can be the word,' said Theodora.

'Bare,' smiled Moraïtis, for a fresh discovery. 'Greece, you see, is a bare country. It is all bones.'

'Like Meroë,' said Theodora.

'Please?' said Moraïtis.

'I too come from a country of bones.'

'That is good,' said Moraïtis solemnly. 'It is easier to see.'[6]

However, the Patrick White novel that is most completely a novel of Sydney is *The Vivisector*, for which White draws initially on his childhood memories of Lulworth and other Sydney houses, through to a studio flat in the back lanes of Circular Quay, the sojourn at Castle Hill, a house in Paddington in the eastern suburbs of his later existence, and finally, the milieu of rich socialites and critics of the Art Gallery Society and Vaucluse.

Hurtle Duffield, the son of a rag-and-bone man and the laundress who is employed at Sunningdale, a house owned by the wealthy Courtney family, lives in the sweltering inner-city working class suburb of Surry Hills, less than two kilometres away.

It was Sunday, and Mumma had gone next door with Lena and the little ones. Under the pepper tree in the yard Pa was sorting, counting, the empty bottles he would sell back: the bottles going clink clink as Pa stuck them on the stack. The fowls were fluffing in the dust and sun: that crook-neck white pullet Mumma said she would hit on the head if only she had the courage to; but she hadn't. (It was Mumma who killed the fowls when any of them got so old you could only eat them.) So the white crook-neck thing, white too about the wattles, stood around grabbing what and whenever it could, but sort of sideways.

'Why're the others pecking at it, Pa?'

'Because they don't like the look of it. Because it's different.'

Oh the long heavy Sunday with Pa's old empty bottles. There was an old stove beside the wash-house he had bought, he said, as a speculation. The rust came off in flakes, which you tasted to see, because there was nothing else to do.

In the early 1900s Sunningdale, at Rushcutters Bay on the harbour, is approached by the child Hurtle via the door dividing the main house from the servants' quarters.

The damp stone laundry, smelling of Lysol and yellow soap, began to horrify him. He had heard of prisons in which they tortured men in the old days. Mumma couldn't have escaped, she had the washing, she was used to it, but he who was cowardly and young, he was still also free. So he went quickly quietly out. It wasn't altogether cowardly, either, to leave Mumma with the washing and their nightmare thoughts. It was necessary for him to see the Courtneys' house again. The felted door went *pff* as he passed through.

And at once he was received by his other world: of silence and beauty. He touched the shiny porcelain shells. He stood looking up through the chandelier, holding his face almost flat, for the light to trickle and collect on it. The glass fruit tinkled slightly, the whole forest swaying, because of a draught from an open window.

He was himself again.

Now he could go on towards other private memories of the house. He could hear a pen scratching in the distance as his feet slid on the mossy carpet.

The washerwoman's son is adopted, or rather bought, by the wealthy Courtneys, and the house and the life of Sunningdale temporarily subsume him. Like others before him, White makes symbolic use of the life-enhancing harbour wind, the southerly buster, this time to cement in the secret kingdom of the garden the passionate but ambivalent relationship between his adopted sister, the Courtneys' hunchbacked daughter Rhoda, and Hurtle himself, the working-class changeling.

That evening at dusk a wind from the south threatened the suffocating warmth from the fire they kept stoked behind the nursery fender. The wind blowing through the grander rooms drove the cigar smoke ahead of it and thinned out Mrs Courtney's perfume. Edith and Lizzie, scuttling to fasten windows and doors, looked as though they only half believed they would prevent whatever they were expected to. As the long wads of ink-blotted cloud passed overhead, unravelled, then matted thicker than ever, the garden, though stationary, was slowly being poured into fresh, coldly boiling forms. It was not yet raining, but the wind in the leaves made them look a liquid black.

Rhoda said with the moist breathlessness he had begun to recognize as hers: 'Let's go into the garden. I'm not supposed to. Because of the damp.' In her feverish revolt she was almost jerking the doorknob off.

He followed her out. The wind hit them. He filled his lungs, excited by his own expanding body, his almost power over flying cloud.

Rhoda, on the other hand, he saw, was gasping. She was advancing sideways. Her hair was being lifted in little pink, steamy streamers. The birthmark looked porous on her asparagus-coloured neck.

'This is *my* garden,' she shouted

She sounded so shrill and electric he realized she too had some important part in what was happening: in tortured trees and ink-stained cloud....

Rhoda led him deeper into the darkening garden. There were stone steps, the moss so thick in places his feet felt they were trampling flesh. It disgusted him, but she couldn't see it. She was interested only in what she had to show him. Each time she spoke he could feel her moist little fingers twitching on his hand.

'Those are guavas.' She tried to make it sound like a secret.

He picked one from out of the sooty leaves, but it made his mouth shrivel up.

She was enjoying it all so much, she didn't notice.

'And custard apples. They're too green to steal. The boys can't see them amongst the leaves.'

'What boys?' he asked.

'Larrikins.'

Was she trying to show him he had changed sides?

Later in the novel Duffield, now a successful but hermitic painter, lives with Rhoda in a terrace house, which he has bought complete with stained glass and ugly knobbed furniture, in the cluttered streets of Paddington.

The masked houses had a secretive air which didn't displease him: he wasn't one of those who resented lowered eyelids, for he had usually known what lay behind them.

The back staircase to his house opened on Chubb's Lane. Here the clothes-lines and corrugated iron took over; ladies called to one another over collapsing paling fences; the go-carts were parked and serviced, and dragged out on shrieking wheels. In the evening the young girls hung around in clusters, sucking oranges, sharing fashion mags, and criticizing one another's hair as though they had been artists. There was a mingled smell of poor washing, sump oil, rotting vegetables, goatish male bodies, soggy female armpits, in Chubb's Lane.[7]

In *Flaws in the Glass* White wrote, 'The spirit of the stairs at Martin Road as I groped my way down most mornings in the dark conveyed the interior of a Paddington house I saw as Duffield's without ever having been inside'.

It is in these eastern suburbs of the early 1960s, of all his Sydney locales, that White conveys most evocatively his sense of the contemporary city, imbuing it with a lush but tawdry sensuality. At Rushcutters Bay Park, Duffield meets Nance Lightfoot, a prostitute who becomes his mistress. And at

Cooper Park in Woollahra, a steep, rock-walled gully which White calls the Gash, Duffield encounters one of White's most compelling minor characters, Cec Cutbush, the homosexual grocer.

Most evenings towards sunset, the bench the council had fastened to the pavement was fully occupied by neighbourhood acquaintances. Although they would sit staring out over the wasteland, with its deep swell of lantana and sudden chasms of ash-coloured rock, the landscape meant less to the rate-payers than their glimpses into one another's lives. Even a total stranger could be persuaded to ignore landscape while exchanging the snapshots of experience, particularly as the sun was going down. But on this occasion, as the blind sockets of the white-faced houses squeezed together on the opposite cliff, reflected their evening illusion of gold, a solitary figure had possession of the bench.[7]

Of *The Eye of the Storm*, his second novel 'belonging to Sydney'—the story of a battle of wills between an elderly woman, Elizabeth Hunter, dying alone except for a retinue of nurses, lawyers and house-keepers in a great house on Centennial Park, and her greedy expatriate children—White wrote:

The Eye of the Storm came to me crossing Kensington High Street, London, after a visit to my mother at her flat in Marloes Road where she was lying bedridden senile, almost blind, tended by a swarm of nurses and servants. I knew I would write this novel about some such old woman at the end of her life, but in a house in Sydney, because Sydney is what I have in my blood. No more than this flash of prescience at the moment of crossing Kensington High Street.[8]

Like *The Aunt's Story*, *The Eye of the Storm* does not contain extensive descriptions of external city or landscapes, but relies on an evocation of an upper-middle-class life within a grand house—this time at its fag-end, when the house is dilapidated, the jewellery of its chatelaine grubby and the cosmetic mask gruesomely overdone—that is essentially Sydney. That the house is on unfashionable Centennial Park ('Why, Elizabeth, won't you be cutting yourself off living at Centennial Park? Coming from the bush to settle, practically—in the bush!'[9]) is relatively incidental; the house itself could as well be situated in the more desirable and richer suburbs of Point Piper, Darling Point or Vaucluse, but it is a sign perhaps of White's attachment to the place, or of Elizabeth Hunter's magnificent idiosyncrasy, that it is there.

In a subsequent publication, the novella *The Night The Prowler* (1978), also set in the 1960s, White again turns a sardonic eye on Centennial Park the suburb, caught halfway between the prosperous eastern suburbs and the relative raffishness of Oxford Street and Bondi Junction.

Early in their married life Humphrey and Doris Bannister had established themselves on the edge of the park. It was a comfortable rather than a fashionable quarter: its large, undesirable houses, in Sydney Tudor, late Victorian Byzantine, Bette Davis Colonial, suggested wealth without flaunting it, just as the inhabitants seemed agreed by the smiles in their eyes never to mention money, and the odd Jaguar or Daimler silently apologized. It suited Humphrey Bannister down to the ground: solid, and only ten minutes' drive from the GPO: Doris, who might have liked to cut a dash, hedged her enthusiasm with reservations. She had married late; she had time to make up for; but settled down to solidity and quiet, and park air. On the occasions when she arranged a bridge luncheon for some of her more fashionable friends, she allowed them to turn her neighbourhood into no more than a mild, party joke; no one could accuse her of disloyalty.

And when their only child Felicity was born, the near-by park was such a blessing: to push the pram through the ragged grass around the silted lake (you couldn't expect upkeep of parks with a war on and the men away); to sit on the balding slopes under the araucarias, and look deep into her little girl's eyes; to surprise each other's cheeks with the delicious flicker of eyelashes. Exchanging the breath of laughter and contentment it was as though they were still one; in the drowsy park, there seemed no reason why they should ever be anything else.[10]

Centennial Park itself, an expanse of reclaimed swamp turned into a park by Sir Henry Parkes in 1888 to celebrate the centenary of European settlement in Australia, and now comprising some half-dozen lakes covered with reeds and wild ducks, sandhills, grassed expanses and banks of coral trees, plays an important role in Patrick White's Sydney. White, who habitually wrote from after midnight till sunrise, when the park was 'full of strange noises, odd cries and the honking of wild fowl from the lakes', once described the park at dawn as 'a Corot full of mists and flooded trees'.[11] And yet the park is something more than a peaceful retreat. It is also an indeterminate zone where all may walk, and encounters not usually contemplated may be engineered or happened upon.

In *The Night The Prowler*, later filmed in and around Centennial Park, Felicity Bannister is a respectable if somewhat repressed young Eastern Suburbs girl, seemingly destined for the safe marriage that will lead her to replicate the life of her parents. Instead, after a sexually ambiguous incident with a prowler, she turns to vandalism, the mysterious nightlife of the park itself after closing hours, and the companionship of a derelict in an empty house.

In *Memoirs of Many In One* (1986), the house on Martin Road becomes the retreat of Alex Demirjian Grey, White's last and most comically complex *alter ego*. Like White himself, who with Manoly Lascaris walked the dogs in the park every afternoon, and had his favourite bench to rest on, Alex also spends time walking in the park, where she encounters a mystic scavenging in a rubbish bin.

At the end of *Flaws in the Glass*, in a chapter called 'Episodes and Epitaphs', White reflects on the city and the time of day that he loved most.

> Early morning has always been the best time of day. In childhood, gold pouring through the slats as I got up to raid the pantry for crystallised cherries, finish the heel-taps on the supper table, and settle down to the plays of Shakespeare. Now when I wake, the naked window is washed pale. As I use the eye-drops the first bird-notes are trickling in. Down in the garden, light is a glare. I'm forced to bow my head whether I like it or not; the early mornings of old age are no setting for spiritual pride. Spiderwebs cling like stocking-masks to faces that blunder into them. Dogs point at vanished cats, follow the trail of the night's possums. At the end of her lead Eureka bays and threatens to pull me over in a cataract of light, scents, dew. We collect ourselves as far as it is ever possible.
>
> If I were to stage the end I would set it on the upper terrace, not the one moment of any morning, but all that I have ever lived, splintering and coalescing, the washed pane of a false dawn, steamy draperies of Sydney summers, blaring hibiscus trumpets as well as their exhausted phalluses, ground mist tugging at the dry grass of the Centennial steppes, brass bands practising against the heat, horses cantering in circles to an accompaniment of shouted commands, liquid calls of hidden birds, a flirt of finches, skittering of wrens, bulbuls plopping round the stone bath carved by Manoly in the early days at Castle Hill, as though in preparation for the Twyborn moment of grace.[12]

This is the time and place where White, who died close to dawn, and who had battled fiercely to save the park from developers in the early 1970s, chose to have his ashes scattered.

SUNLIGHT ON WATER:
Modern Sydney

The rain eased and had stopped by the time I turned into Glebe Point Road. The sky to the west had split open, and blue and white patches were expanding, spreading and bursting out in all directions. A beam of sunlight slanted through the tall poplars and filled the car with a pale green light. When I got out in front of my house, I could taste the cleanness of the rainswept air. The tin roof of the house on the other side of the street was glinting in the sun and steam rose from it in little spurts. I walked down the narrow space beside the house and the fence and felt the overgrown bushes and vines drip on me. What the hell, I was wet already.

Peter Corris, *O'Fear*, 1990[1]

The contemporary urban writers most closely associated in the minds of Australian readers with particular city landscapes, such as Michael Wilding, Frank Moorhouse and Helen Garner, often in fact use very little description of place. In Helen Garner's early novel *Monkey Grip* (1977), for example, which is intensely evocative of inner-city Melbourne life, there is very little geographical placement. Garner relies on the characters to create their own background: in the mind's eye the cups of tea and the conversation at the kitchen table suggest to the reader the type of kitchen, and the kitchen in turn suggests the nature of the house and suburb.

This is also true, in a different way, of contemporary Sydney writers Frank Moorhouse and Michael Wilding who, rightly or wrongly, were closely identified in the 1970s and early 1980s with the inner-Sydney suburbs of Paddington and Balmain. In Moorhouse's early short stories, the practice of naming a particular pub or restaurant—or even a particular inhabitant of either who is recognisable to his readers—fulfils the same purpose. This is literal writing: these authors are not tempted by adjectives in the creation of either personality or place. Thus, while Wilding's *Living Together* (1974) and *The Short Story Embassy* (1975) are set in Paddington and Balmain in the late 1960s and early 1970s respectively, the reader will find very little direct evocation of these historically distinct precincts.

This has led to critical accusations that the 'Balmain school' of writers, and Wilding and Moorhouse in particular, were writing for a coterie of initiates. On the other hand, perhaps this is indicatative of the old tyranny of landscape being overthrown, and a new stage being achieved in the city's literary maturity, in that the writer can expect to be granted his or her own territory without the need for lengthy scene-setting.

In Kate Grenville's most imaginative work, the phenomenon is even more tantalising. *Lilian's Story* (1985) conjures up a richly evocative portrait of Sydney life from 1900 to after World War II, based on a fictionalised life of local eccentric Bea Miles, here called Lilian Singer, who in her later years roamed the streets of Sydney giving recitals of Shakespeare at a shilling a time and terrorising taxi drivers into giving her free rides. Once again, on analysis of the text there are few purely descriptive passages of the city. Once again, using the elements of sunlight and water, Grenville lightly evokes the aura of period and location in scattered phrases throughout the work.

From the silent Victorian house of a respectable middle-class childhood at the beginning of the century, the young Lilian observes the threatening behaviour of her father from various hidden and forbidden places.

> That study of Father's was a silent and dusty place. Sun leaned in between the curtains and travelled slowly across the pile of newspapers on the floor. Dust hung in a nervous way in the beam of sun and there was not enough air, so I panted and saw the dust motes dance. Stairs creaked outside although I knew Father could not be home, something rattled somewhere, a branch scraped along stone, and I knew I should hurry out and not come here again. But the pile of newspapers on the floor was a good height to climb and sit on, swinging my legs, and I hummed into the silence in a brave way.

From the domed centre of a plumbago bush, Lilian and her brother can watch the washerwoman, Peg, stirring clothes in the copper, the waves lapping in the bay, or spy on the return of their father.

> Father liked to return home in the evenings to an embroidering wife. From under the plumbago I could see the ferry dock at the wharf and guess which dark figure was Father striding up the hill. There was a patch of the road where the trees hid him from sight and I settled deeper into the mauve dusk under the plumbago, beside the hundred empty snail shells I had collected. I sat very still and watched Father stride up the last part of the hill, jabbing his furled umbrella in front of him at each step as if poking someone along, and swing into the gate.

Lilian ends her days in a similar secluded, nameless outdoor location, which is both a return to the childhood cubby and a celebration of a basic informality of Sydney urban life: a stormwater drain in a park which she shares with an old taxi-driving friend.

> I enjoyed the way Frank rolled us into our newspapers for the night, and loved to wake up when the birds were being insistent in the trees overhead and the sun was sending yellow fingers along the wet grass. Frank snored on beside me and I heard the city wake slowly and the birds take second place as hurrying men in suits and women in high heels began to clatter along the paths to the city, and everyone got ready to die another day away … Frank took good care of me in our home by the water, tucking my feet into newspapers at night, combing my thin old hair for me by the hour as we sat like a pair of baboons in the grass, grooming each other in peaceful silence …[2]

In direct contrast, the Brisbane-born writer Jessica Anderson, in *The Impersonators* (1980), a novel of family relationships set in Sydney in the 1970s, employs a meticulously exact account of the geography of the city centre as a device to engineer a brief encounter between three of her key characters, a device that also sets up their relationships and sketches in their personalities.

Sylvia, a long-time expatriate in London, recently returned, stands on an apartment building rooftop in Macleay Street, Potts Point, to renew her memory of the city.

> She stood on the roof of a building itself on the high escarpment above Woolloomooloo, but the map of the city she had carried for two decades in her head, the attempted British gridiron of streets disrupted by the lay of the land and the intrusion of water, was visible only in Woolloomooloo and the green rise of parkland beyond. Woolloomooloo, low-lying, was slung between Potts Point, where she stood, and the green rise of the Domain, which extended itself, like a long green finger from a green fist, into the harbour where its splayed tip was called, she thought, Mrs Macquarie's Point. Beyond this green fist and its finger rose the towers of the city, some bearing construction cranes. Distance impacted them and their bases, around which lay the rest of her mental map, were hidden from her view.
> But it was easy to confirm the first part of her route. There were the McElhone Stairs, just as she remembered them leading from the escarpment of Potts Point down into Wolloomooloo. During the pared-down and frugal period following her resolution to discard fashion and save to go abroad, they were part of her way to and from her drink waitress's job in a city night club. Watching her feet in thonged rubber sandals (five shillings at Woolworths), she used to count the steps as she went.

Sylvia plans to walk from Potts Point, adjacent to Kings Cross, across the inner-city harbour suburb of Woolloomooloo and the public space of the Domain, which intervene between her present position and Wynyard railway station in the city, to catch a train to visit her family on the North Shore Line. In the meantime two other characters, Steven Fyfe, a youth counsellor in the public service, and Ted Kitching, a businessman of dubious ethics, enter the Domain from another point to jog, and the precise progress of all three is observed in a tightly-choreographed passage, a *tour de force* of descriptive writing that continues over some seven pages without a pause.

Steven Fyfe entered the inner Domain by Art Gallery Road and ran under the blackish-green figs. He wore white shorts, an old red T-shirt, and a blue headband woven by Hermione.

His gait was plain, rhythmical, and relaxed, a triumph of the biped state. As he crossed the bridge spanning the gulch of the expressway he saw down in Woolloomooloo Bay a merchant ship from mainland China, black and ochre, high out of the water. Five small men dressed in blue-grey were mounting the steps from Woolloomooloo to the Domain, and others in the same colour were radiating away from the wharf and dispersing through the streets of Woolloomooloo.

Sylvia passed a group of these Chinese on the McElhone Stairs, ascending as she descended. The sandstone steps were hollowed with wear. She counted them—a hundred and thirteen.

Steven entered the Outer Domain by the top path. On this hillocky finger of land between Woolloomooloo Bay and Farm Cove, paths and roads were traced at different levels and connected by long and short flights of steps, and on all the paths and roads and steps men and boys were jogging.

Steven, his back straight and his head erect, ran up the slope of the first hillock and advanced along its crest. At a short distance ahead he saw Ted Kitching, in black shorts and a purple T-shirt. Ted's back was not straight, his head was not erect, and his arms moved as if rocking, rather desperately, a large baby.

Ted reached the Henry Lawson statue and disappeared behind its pedestal. When Steven reached the statue, Ted was standing in half-profile, his hands on his hips, staring down at the Chinese ship.

The fluid and rhythmic urban ballet continues, setting the characters precisely within the social and sociological context of the city.

Steven, on Mrs Macquarie's Point, changed his course to avoid disturbing a group of aboriginal boys making football passes with a red practice ball. A FAR WEST coach stood nearby. He ran down the broad steps to the path that followed at water level the loop of Farm Cove. On the harbour wall sat five aboriginal children, all girls, dangling their legs. One of them jumped down as he passed and ran in his rear, in a parody of his gait, making the others laugh so much that they had to get down from the wall. Steven smiled when he heard them, but then became graver than before.

… Sylvia stood at the top of the steps from Woolloomooloo. She looked at her watch and estimated that there was not enough time to walk round Mrs Macquarie's Point and enter the Gardens by the gate at water level. This was the gate through which Steven passed at that moment into the Gardens. Sylvia crossed the road and entered by the east gate. Ahead of her, the five Chinese seamen had clustered to look at a fountain.

Ted Kitching, with downcast eyes and pondering brows, left the Henry Lawson statue and departed at a dogged pace for Mrs Macquarie's Point. He reached the aboriginal boys and ran through their game. As they made way for him, they laughed at the legend on his purple T-shirt. Ted took no notice. He ran down the broad steps to water level, and passed the aboriginal girls on the wall. The same skinny little girl jumped down and ran in his rear. Laughter rose high behind him, but Ted's thoughtful expression did not change.

Steven ran on the semi-circular path between the harbour wall and garden beds cut to conform to the loop of the cove. In these beds, shrubberies full of birds were bordered by marigolds and candytuft, or scented stock and primula. 'Of course,' Hermoine had said, taking back the newspaper, 'Ted's so ingenious, he may get out of it yet.' Crescents of sweat were creeping from beneath Steven's armpits.[3]

Anderson, who has lived most of her life in Sydney apart from some years in London, covers much the same territory in the 'Sydney stories' of her later collection, *Stories From the Warm Zone* (1987). Here the same suburbs form a sometimes painful territory of broken marriages and middle-aged women alone.

In a similar device journalist Jill Neville in her novel *Last Ferry to Manly* (1984) has her returning heroine Lillian—once again a long-time expatriate in England—take a bus from Circular Quay to Chinatown, and then walk past an archetypical Sydney pub to the Fairfax newspaper offices in Broadway, near Central railway station to apply for a job.

> When the ferry docks at Circular Quay, Lillian stares at the liner anchored opposite. Funny, she could still get that same old thrill staring at the big white liners bound for England. But it was better at night. The lights on the deck all glittering, the Harbour a wobbling tray of reflections, the stars humming with prophecies.
>
> Lillian is early for her appointment. She gets off the bus and walks slowly through Chinatown. She remembers the Chinese community as tiny, shrivelled men covered with the centuries-old habit of resignation. Now their progeny are confident, full of sap, with direct, sparkling glances, their businesses thriving, their bodies tall.
>
> Some bits of the city had not altered. Like the old tea merchant with his dusty crates and bags of tea. And the pub with its open door full of men from the market, chalking their pool sticks. And over all that old familiar Sydney smell of beer, rotting vegetables and melting tar. Men ambling along, jackets over their shoulders, shirts moist.
>
> The sweat is starting to show on her upper lip, in the wetness of hair at the temples and in the kind of vacancy of expression women adopt when they are fanning themselves.
>
> It was interesting to feel the wheels of life grinding on again. She had found a flat with a millionaire's view for fifty dollars a week and made an appointment with the editor of a daily newspaper.[4]

Robert Drewe, born in Melbourne in 1943 but later resident in Sydney after a childhood spent in Western Australia, also employs the city extensively to provide a contemporary portrait of Australian urban society, usually against a sensual backdrop of Sydney's beaches and waterways. His short fiction collections *The Bodysurfers* (1983) and *The Bay of Contented Men* (1990) are set largely in and around Sydney. In his first novel, *The Savage Crows* (1976), Drewe's character Stephen Crisp is living in a rundown flat in Lavender Bay, a northern inner harbour suburb, while writing an account of the life of George Augustus Robinson, nineteenth-century Protector of Aborigines in Tasmania, and his attempts to stave off the extermination of the last of the island's native inhabitants. Drewe's description of Crisp's flat encapsulates in a few brief paragraphs both the textures of Sydney life and a short history of the area. As in much of Drewe's work, the urban landscape tends to mirror the frame of mind of his protagonist.

> He did most of his writing at night, rising sleepless and randy in the humidity and stumbling out to the kitchen to find big dappled slugs gliding over his cutlery and dinner scraps. He prised their shrinking, mucilaginous bodies from the Formica and flushed them, shrouded in Kleenex, down the toilet. It took hot water to remove their slime from the fingers. He worked at the kitchen table among their crisscrossed silver trails. The slugs returned every night, sliding out from dark soapy cracks behind the plumbing. One night he stamped one into the seagrass matting, it infuriated him so much. Next night a leopard-printed slug was grazing on the paste of his companion. A night or two later all that remained of the squashed slug was a faint brown outline.
>
> He often slept late until the harbour and motor traffic noise woke him about noon, then meandered up the hill to the corner delicatessen for bread, fat-reduced milk, oranges and a length of cabana sausage.

A weather eye was kept open for the landlord: Crisp was behind with the rent. He also owed miscellaneous sums elsewhere to people who employed threatening collection agencies.

His flat, number 6, was one of eight in a liver-coloured brick building named *Cardigan*, after the town in Wales. *Cardigan's* narrow front yard couldn't support grass or shrubbery because of the shade and damp and surreptitious gas leaks. The local tomcats sprayed the foyer. Furled and weathered suburban newspapers gradually pulped against the letterbox. Inside, Crisp's bathroom window overlooked Lavender Bay, a small silvery piece of Sydney Harbour. The bay was attractive from that distance but at close quarters the tides could be seen converging, collecting the harbour's plastic detergent bottles, beer cans and old baby carriages. Astride the lavatory he had a panoramic vista of the boatbuilders' sheds, Luna Park's ferris wheel and Big Dipper, the railway shunting yards and the half-hourly Lavender Bay ferries. At night the view was more picturesque: fairy-lights twinkled from Luna Park, the midget train on the Big Dipper rumbled into sight every three minutes. There was a steep downward slope where girls screamed every trip. From his throne Crisp had viewed the nation's biggest celebrations: the fireworks displays for the Captain Cook Bicentenary in 1970 and the opening by the Queen of Australia, Elizabeth II, of the Sydney Opera House in 1973. Forty years earlier he would have had the same box seat for the grand opening of the Sydney Harbour Bridge.[5]

The crime writer Peter Corris, born in Victoria in 1942, has also adopted the city of Sydney as his own. Corris uses characteristic urban detail, usually ironically observed, to provide a backdrop for characters who emerge as logical products of their particular milieux. Corris's private detective Cliff Hardy, an Antipodean version of Raymond Chandler's Philip Marlowe, lives in a dilapidated terrace house in Glebe near Sydney University, a suburb that Hardy (and Corris) professes to hold in particular affection. In the extract prefacing this chapter subsection, the writer's delight in his city is once again predicated on those elements of climate—cleansing rain and sunlight—that make even a cramped inner-city suburb a tolerable place to live.

In *White Meat* (1981), a novel set in the Aboriginal-dominated world of professional boxing, Corris turns a sociologically precise eye on inner-city Redfern in the 1970s.

Redfern is like an untidily shaped ink blot to the east of downtown Sydney. It's one of those places that look worst around the edges where it's bordered by factories with stained, peeling walls and rows of old terraces with rusting wrought-iron and gap-toothed skew-whiff paling fences. A couple of high-rise monsters in the middle help to make Redfern's population density one of the highest in Australia. The taxi took me past tiny houses with flapping galvanised iron roofs, shops presenting blank, defeated faces to the streets and pubs full of Aborigines and Islanders drinking their dole money, improving their snooker and resenting Whitey like hell.[6]

One might compare this observation with another view of life in the same suburb, again by a writer born elsewhere, but this time Aboriginal herself. Ruby Langford, born on Box Ridge Mission, Coraki, on the north coast of New South Wales, moved to Sydney with her family in 1949, when Ruby was fifteen. In her autobiography *Don't Take Your Love to Town* (1988), she describes her arrival in the city.

Hundreds of smoking chimney stacks; rows of houses squeezed together; a neon sign showing a man in a little aeroplane and some musical notes and then the words lighting up, 'I like Aeroplane Jelly'. We ran from one carriage to the next, looking out the windows. We seemed to be travelling miles and miles to get to Sydney, and didn't realise we were already on the outskirts.

Dad and Mum Joyce had a one-bedroom flat at 22 Great Buckingham Street in Redfern. Gwen and I shared a three-quarter bed on the balcony and at the other end was a kitchenette. The main room had a fold-out table, the big bed, double bunks for the boys, and the wardrobe where Dad kept his money.

I went to work as an apprentice machinist at Brachs' clothing factory in Elizabeth Street. It was a long room with a bench running down the middle and about twenty-five women sitting at their machines on either side of the bench, facing each other. The owners were Jewish people and their daughter Topsy Brachs, who was a model, was sent to show me the job.

… After lunch Gwen and I went to the matinee at the Empire Picture Theatre in Cleveland Street. A figure in a scarlet cloak and a black mask rode around saving people from trouble, he was the Scarlet Horseman. In 'The Perils of Pauline' the heroine was on the railway track and the baddies almost won out.

Back at home we gave our signal, a three-note whistle, and the key would be thrown down from the balcony.[7]

Some newcomers to Sydney come from even further away. Dimitris, a young Orthodox priest arriving in Sydney from Greece in 1976, figures in Angelo Loukakis's collection *For The Patriarch* (1981), in the story of the same name.

As the only other large city I have seen is Athens it is hard to make judgements about the standing of Sydney in comparison to the other cities of the world. To compare it to Athens though, I would have to say that the first difference that meets the eye is the size of the buildings. In the central district they are much taller than those of Athens. I don't remember anything over fifteen floors in that city whereas here there are some that seem to be at least fifty. The plazas in Sydney are square usually, those I knew were smaller with more trees and more shelter. One they call the Martin Plaza in the centre of the city is really a street which has been closed to traffic. It has very cold and plastic seats. The gardens, however, are large and beautiful, some much better than even the King's Gardens.

There are plenty of shops, but they don't seem very enticing to me, and many department and large stores. People are as materialistic here as elsewhere it seems. One thing is very different. The cars. Here they are very new-looking compared to the beaten up ones you see in Athens—even the taxis! They are driven better too and not so fast.

Everyone dresses so casually here. Even in the city you see people wearing rubber sandals as if they were going to the seaside. There are very few sensibly dressed men, or women, to be seen. Suits and dresses are no longer in fashion apparently, instead there is flaunting of the body. The young children often do not wear shoes and I don't think the reason is poverty, for at the same time they have enough clothes.

Sometimes they stare at me in the street here. I suppose I must be an unusual sight in my black costume and beard, very different to the fairhaired and fairskinned locals. Nevertheless, it is a little embarrassing to be looked at this way. After all, the English priests are not so dissimilar in their appearance.[8]

Loukakis himself was born of Greek parents in Sydney in 1951.

Anna Couani—born in Sydney in 1948 of Greek and Polish extraction—observes the streets of Surry Hills with the eye of an intellect informed by diverse cultures, and yet which is, ultimately, Australian in essence. In 'Parramatta Sestina' she writes:

We could smell the salt in the air at Parramatta, that's where the city began in those days. Then everything had a kind of sameness until we hit the city, everything seemed old and dirty, running beside the tram tracks. Newtown seemed particularly old and Redfern not at all red or fern-like. This was one idea of old, but not like the mountains which were ancient.

And in between this old and this ancient was European old and the ancient of the Mediterranean and Asia. But I didn't know that, it was a blank for me like Parramatta was till recently. And I never thought I'd find people I could like, not in the enduring way I loved the mountains and the city, where days counted for nothing against 'forever'. Bush love like bush tracks and city love I could trace back to Sydney's birth as a city.

My memories are my grandma's memories of the city and my mother's talks looking at the mountains, talking about The Ancient, about the beginning of the world like the 2001 movie track but more serious. And Dad feeling alien anywhere west of Parramatta or Broadway even. I felt his sense of relief on the days we came down to the city and he showed me what his Sydney was like.

Where we saw salami and olives in shops I now realize were just like ones in Greece and definitely unlike the big Franklins in the city which sold DEVON (a word my parents pronounced like POISON). There were different city days with Mum, more anglo. Sitting in the Cahills coffee shop looking at the ancient Egyptian motifs etched on the amber mirrored walls. Stopping at Parramatta for a sandwich and having a talk about the Great Western Highway when it was just a track.

Just as Mum knew the mountain tracks, Dad knew the city tracks. Not just the steps and pathways around the Cross for example, but he had a mental picture like a map. The shortcuts all the way from the coast to Parramatta. Which makes me think of Sydney as like a middle eastern city, multi-layered and only really knowable by people with that ancient knowledge which is still applicable in the cleaned-up version of Sydney these days.

I had a dream of finding parts of Sydney I'd forgotten and rediscovered on summer days in the dust and heat. Suddenly finding a lane like a track leading between some buildings. But that's ancient history to me now, that personal approach to writing. Now I like to write about the things happening around me not to me. About the city. And I want to start from the centre I know and work out past Parramatta.

Even trying to avoid nostalgia, my childhood days seem ancient and thinking about them is like archaeology. Tracking down connections and making them till they stand out as strongly and clearly as the arterial roads between the city, Parramatta and the mountains.[9]

Judith Beveridge, in 'Streets of Chippendale', maps the same inner-city terrain with the same observant eye to shades of cultural variance and the odd nomenclature of the city. The poem begins:

Streets named Ivy, Vine, Rose and Myrtle—
now lack a single tree. They could have been
the homes of kindly aunts in quiet suburbs
before factories like terrible relations

moved in, changed the place.
And Abercrombie (sounds like the eccentric,
unmarried third cousin)—you expect a place
where residents dressed in slacks and turtle necks

are walking pedigree dogs;
but Abercrombie's different—
hits the bottle with a dozen pubs,
grumbles like a drunk parent

to 'bloody well watch the road' as it crosses
to Hugo, Louis—they could be
respected gentlemen strolling past terraces
to call on the nearby Aunts ...[10]

One might compare Beveridge's 'Streets named Ivy, Vine, Rose and Myrtle—now lack a single tree' with Couani's 'Newtown seemed particularly old and Redfern not at all red or fern-like'—although it is reasonably well known that the suburb was named for William Redfern, a surgeon-convict transported in 1797, who subsequently became a leading Sydney figure and had his house in the area that later bore his name. Both these examples show a concern among younger writers with semantics and semiotics not so apparent in earlier fiction about Sydney, a desire to go beyond description into analysis and deconstruction.

The younger writer Obelia Modjeska, born in Sydney in 1974, whose piece 'Observations from Rozelle', included in the anthology *Inner Cities: Australian Women's Memories of Place* (1989), strikes a similar note in her description of the culture of school children at Balmain High school. And yet, the reader is tempted to think, despite the semi-satiric use of modern sociological jargon and the ethnic diversity, Louis Stone—whose own observations of street culture in *Jonah* were based around Regent Street, Redfern—would not find the milieu foreign.

Monday morning. I walk through the school gates and there's that familiar scent—hairspray and Impulse deodorant. This is where every Soula, Voula, Roula and Toula occupy their lunchtime talking about every Vince, Con and Christian, or maybe a new pair of white leather boots from Marketown, or Pseudo Echo's new single. Occasionally you'll meet Soula, Voula, Roula or Toula on Saturday morning at the delicatessen working behind the counter or stocking up on olive oil and feta cheese for dinner with Grandma, Grandpa, Uncle Con and the cousins.

It makes you wonder how much lifestyle has to do with a suburban environment. Most of the ethnic families live in the Leichhardt-Newtown area, but if they came from wealthy backgrounds and owned houses in East Balmain, would they, like the East Balmainites, be taking commerce instead of textiles as school subjects? Or wearing Reeboks rather than white leather cowboy boots? Cashmere sweaters instead of K-mart sloppy joes? Who knows.

Certainly kids in cashmere and Reeboks who live in East Balmain are accepted as part of the 'trendies' culture. They read *Stiletto* magazine. Designer labels or Esprit all the way—and the yuppie generation becomes obvious even among teenagers.

There isn't much to be said about the trendies. A much more interesting group are the rejects, also known as dregs, outcasts and dociles. Nobody really devotes much thinking time to them. Either they're too stupid, strange, toffee-nosed, or intelligent to fit into the other social groups. So they collect in various pockets around the school together and do their own thing. The rejects come from everywhere in the inner-west. Part of what makes them interesting is that they've cut off completely from consumerist culture, which is what dictates position within the peer group. If we never stopped off at Marketown after school, we wouldn't know about white boots. And if we never laid eyes on *Stiletto*, we wouldn't think that 'converse all stars' was skateboard talk for gym boots.

While the rejects are disliked in Year 9, they may be happier people than their parallel culture companions by the time Year 10 comes along. Struck down with tags like yuppie, greasy Greeks, and pretentious uni student, the popular kids won't be any better off than the rejects, who, if they continue as they are as children, will be free of categorisation, and probably the only people who are really in touch with the world.

So does it matter if I live Rozelle, seven minutes from the city? Rozelle kids come from every background and don't really fit anywhere either.

Friday night in the city. Wandering down George Street on a Friday night is another time you're likely to come across more examples of culture and suburban environment. Dawdling away from Hoyts after a late movie, some friends and I bumped into a mob of Gothics. The Gothics are a very distinctive subculture. They parade about in black clothes, black eye-makeup and wear huge silver crucifixes around their necks. Where do they all come from? I dodge past the group and then it all becomes obvious. Darlinghurst. Surry Hills. Paddington. The Havens of the Alternative Lifestyle freaks. A quick detour down Oxford Street will tell you that this is where 'different' people come to live. So what's the attraction? The inner-east area is fashionable and expensive, which probably makes it perfect for wealthy eccentrics and fun-loving weirdos. With all the trendy boutiques and rainbow dye-job hairdressers, it's inevitable that they flock to these suburbs—even those who don't have the money.

On returning home to Rozelle, I walk in through the front door to discover the house full of my mother's friends. They talk about work, anthropology, post-modernism etc., as they sip their coffee. I try to figure out where they come from. Their appearance gives no clue. I suppose with adults, categorisation doesn't work the same way. Perhaps as you grow older people can only slot you into a group when they know how you think. I wonder what people will call me when I grow up and I'm still an existentialist.[11]

George Papaellinas's story collection *Ikons* (1986), which follows the progress of the Cypriot Mavromatis family, who have migrated to Sydney thirty years before, explores similar territory on a

If I were to stage the end I would set it on the upper terrace, not the
one moment of any morning, but all that I have ever lived, splintering
and coalescing, the washed pane of a false dawn, steamy draperies
of Sydney summers … ground mist tugging at the dry grass of the
Centennial steppes, brass bands practising against the heat, horses
cantering in circles to an accompaniment of shouted commands …
as though in preparation for the Twyborn moment of grace.

PATRICK WHITE, 1987 (PAGE 56)

CENTENNIAL PARK AT DAWN. PHOTOGRAPHER: PHILIP QUIRK

*Like that of Tierra del Fuego, the extremity of Van Diemen's land
presents a rugged and determined front to the icy
regions of the south pole …*

LIEUTENANT GOVERNOR DAVID COLLINS, 1840 (PAGE 70)

QUEENSTOWN, TASMANIA. PHOTOGRAPHER: CAROLYN JOHNS

more gritty level. Papaellinas, himself born and brought up in Sydney, has worked as a factory hand, storeman, waiter and taxi driver while pursuing a career as a writer. In the story 'Around the Crate', the Mavromatis's Australian-born son Peter finds that even outside the family, wracked by its own conflicts and internal lack of allegiances, migrants are frequently united only by society's view of them as outsiders.

The boy gripped the hammer, one hand clumsily aligning a length of wood with another. His unhealed blisters stung with the bite of the oil seeped into the handle. The Lebanese answered his wince with a grunt and his taunting child's grin and Peter looked up and away again. Abou's eyes were yellowed, like his work clothes, by the oil that hovered everywhere in a thin mist. Peter's hammer clattered on the pitted concrete. He got up off his knees, composed now, and picked at the damp, sticky patches on his jeans. Conscious of Abou's silence, he looked away across the warehouse, up to its high, vaulted ceiling, at the racks of heavy tools, at the piles of lumber, at piles of rubbish, at scatters of crumpled, rusting car panels, at troughs full of oily parts, across to the tall, bare crate frames. Men worked in gangs around each crate, clustering defensively it seemed to Peter. Each crate would be built in stages, filled layer by layer with large car parts, the chassis, axles and panels, and between these anything that lay in the troughs. Each gang worked smoothly, each man's movements timed to at least one other's, all motion paced by the woosh-thump of a staple gun.

The boy looked back down at the Lebanese who frowned and wiped an arm across his face, at the sheen across his bristled cheeks, across his forehead. Peter stared back at him, at his curling mouth, and stroked the hair falling in ringlets down his shoulder.

'When am I going to work on one of the crates?'

And the Lebanese grunted again, speechless, and held up one, two, three fingers, 'three … three …' and he pointed to Peter's unfinished crate.

'Three …' and then he pointed to the watch strapped to his thick wrist, 'three … then eat.'

'Yes … but when am I going to work on the big crates?'

Peter turned away from Abou's arm-waving and the puff of his dark cheeks and squared his sharp, thin shoulders. He flicked back his hair and looked towards the nearest gang. Greeks, all of them, he thought. He had heard them talking at lunch. They moved slowly around the crate, their work clothes creased neatly, almost crisp, despite the indelible laundered grease. And there, on top of the crate, the old man crouched, pointing here, over there, and the others would move, shouldering a fender or on either side of a bonnet or a staple gun. The old man's authority was a mute one, unlike Abou's. His settled features, his broad hands, the grizzled, crinkly hair, his rounded belly were its caste marks. Peter stared and chuckled self-consciously. His low laughter brought the Lebanese lumbering to his feet. Abou grabbed Peter's shirt sleeve. The boy shrank. The size of him would brood over the boy. Pete sank back down to his knees, his face lowered, testing a sneer....

During the lunch break, the boy sat by himself on the slippery concrete, his back against a finished crate. The Lebanese had disappeared right on the bell. He didn't know where. He sat a little distance away from the Greeks who were sprawled in knots around a tarnished tea urn. Other gangs lounged elsewhere. He made no attempt to join them, to intrude. He hadn't been invited to do so. Some sat on sagging vinyl car seats, some on the floor, like himself, while others slouched about the urn, in noisy attendance. For a fortnight he had waited, sitting in the same spot, mouthing his sandwich and squirming on the cold concrete. He watched, eyes slitted in concentration, alert to their loud voices and their laughter, sharing it secretively, waiting to understand their relationships, the hierarchy. He couldn't see one today. The old Greek was not there. He waited until the bell rang, only occasionally looking away.[12]

David Ireland likewise takes industrial Sydney as his subject in his biting satire *The Unknown Industrial Prisoner* (1971). Set at Botany Bay, Cook's landing place and now an area of factories and refineries, the novel describes the experience of a group of workers employed by the 'Puroil' multinational oil company. This is industrialisation at its worst: the workers—known only by their nicknames—are portrayed as exploited by a cynical government, leading a dehumanised existence in a complex resembling a technological termites' nest, and Ireland uses the language of technology to further depersonalise the Orwellian work environment. Ireland, born in Lakemba, New South Wales in 1927, himself worked for a long period in factories and an oil refinery before becoming a full-time writer in 1973. One thing that has not changed since colonial times, according to Ireland, is that power and profit still reside overseas.

Puroil's land included a stretch of what had once been parkland. Residents' petitions, questions in Parliament, real estate developers' organized, agonized pleas, no amount of democratic pressure was able to beat a foreign oil company. A few words were altered on a piece of paper somewhere, the parkland was declared industrial land and Puroil had a foothold in New South Wales. The total of 350 acres included, on the river side, some of the swampiest land this side of Botany Bay, but mangroves were cleared, swamp flats partitioned and drained and filled until only a few dozen acres on the river bank were left in their natural state. Another hundred acres of mangroves still stood on the other side of Eel River, just down from the gasoline depot of a pretend rival of Puroil: Puroil supplied them from a nice fat silver pipeline that nuzzled into the slime of the river bed and came up again out of the ground handy to their shiny white tanks.

Puroil supplied the depot of another company too, with a line that ran half a mile under cleared clay. Wagons of rival companies that ran out of their own brand, simply called in and gulped down a load of Puroil, went out and sold it as their own. Even Puroil sent out grey unmarked wagons—they had brother companies with different names. The rival companies fixed the price between themselves in the first place, the Government approved their figure then made a big deal of getting them to reduce half a cent a gallon when crude went down a cent. Then they all advertised like mad and called it competition.

At Puroil the largest vessels of the new cracking plant were in position and complicated mazes of pipelines were being lagged with glass wool and aluminium sheet. Turbines, pumps, compressors, heaters, coolers, columns were assembled from many parts of the world; there were even a few girders and pipelines from Australia. Puroil never gave out the usual unctuous bumph about the refinery belonging to the Australian people; it was very clear that whatever faceless people owned it were a long way off. They were clever faceless people.

A little further on in the novel, which hovers between social realism and speculative fiction, Ireland uses one of his characters, called the Great White Chief, to describe their condition to another, called the Glass Canoe. (The Glass Canoe is also a term in Ireland's fiction for a schooner of beer.) The workers are prisoners, but now they are prisoners of capitalism rather than colonialism, although they are reduced to the same dehumanised state.

When he had his boots on, he went to wipe some dried mud on to a pile of rags in the corner, but stopped himself in time. The Glass Canoe didn't, and was busy rubbing his feet on the rags before the Great White Father tapped his shoulder.

'Humdinger,' he said. The Glass Canoe looked down. The rags stirred and stretched, yawned and looked up.

'Is that what you think of your fellow workers?'

'Christ, I'm sorry, mate,' said the Glass Canoe and everyone gaped. Perhaps he was getting sick again.

On the job, events moved slowly. On the drawing board in the Admin block though, for eight hours a day, the pace was frantic until four, when the white-shirted multitude suddenly went home. Their effort might have been more widely spread over the twenty-four hours to take advantage of the quiet of the dark hours, but white-collar men don't yet do shifts.

The tall man had another word for him when he was dressed for work. 'No one enters those blue gates only to make gasoline, bitumen or ethylene from crude. Oil *and* excreta, that's what they fractionate here. Us and oil. With foremen, controllers, section heads, superintendents, managers and all the rest, there's maybe forty grades. Forty grades of shit. That's all any of us are. White shirts, brown shirts, over-alls, boiler suits, the lot. Shit. The place is a correction centre. The purpose of giving you a job is to keep you off the streets. It's still a penal colony. All the thousands of companies are penal sub-contractors to the Government'.[13]

Gabrielle Lord, one of the few writers to base a work of fiction in the largely ignored west of Sydney, sets her crime thriller *Jumbo* (1986) in a nameless new housing development on the city's inland fringe. Lord, born in Sydney in 1946, lived in this area while working as a Commonwealth Employment officer prior to writing the novel.

Richard and Verity's house was one in an estate in Sydney's western suburbs, bulldozed from straggly bush several years before. Sydney, like life itself, had grown from the sea, the Pacific and the mouth of the Parramatta River. Even the hottest days in Sydney are refreshed by the sea-borne easterlies. The river valley goes back from the coast, through a plain for about forty miles, until it meets the walls of the Great Dividing Range. These are ancient mountains deceptively worn, so that they appear young and rounded and exceptionally blue. They form the end and rim of the hot Cumberland Plain. A few miles from the sea the easterly breeze is no longer felt and in the summer the heat increases. Twenty-five miles from the coast the air is still and oppressive, fumed with car exhausts and the spouts of industry. Forty miles inland, under the shimmering rim of the mountains, the heat can be desperate. The mountains throb in the background, gathering up and absorbing the sun's power, radiating it back on to the plain where the houses of the outer western suburbs cluster. Land is cheaper out here, forty miles from the Harbour. The householders plant narrow pencil pines and wall their homes with glass. Huge picture windows glare out on to the black tar, and slow bubbles swell on the roads. The heat ricochets from the mountains and the black roads straight back through the picture windows. Night falls, but the houses stay hot for hours as the mountains radiate the enormous stored heat of the day. Bush fires burn for days up there, so that you might almost think that the sandstone of the ridges has burst into spontaneous flame. There are a couple of cooler hours till sunrise starts the process again—people can finally get to sleep, waking drained a few hours later to have tepid showers, forced breakfasts and thence go to work.

Jumbo deals with the dilemma of a schoolteacher, Verity, who receives a letter from an adolescent ex-pupil she does not remember. The girl, unemployed in a society plagued by increasing unemployment, plans to kidnap her baby-sitting charges to 'save' them from the hopelessness of a fate similar to her own.

Richard had insisted on air-conditioning, so that their living room and bedroom, even in midsummer and at great expense, was a similar temperature to Sydney at night. It was a great relief to walk into the house after teaching all afternoon in a portable classroom steaming with thirty-four sticky, gritty and cranky kids. It was pleasant to sit in the lounge with a drink, feet on the cool glass-topped table, or to sprawl on the cream woollen carpet—the sort of carpet only childless couples have.

It could be worse, Verity thought, I could be living at Avondale where this morning's letter had come from; an area infested with wreckers' yards, used car lots decorated with dirty bunting. On the one- and two-acre lots in Avondale, the people had built their weatherboard homes and battled with the burrs and clay to make gardens. Dusty fowls scratched in the dirt under abandoned car bodies. Even the eucalypts out there were stunted, scorched by the westerly. As well as the heat and the ugliness, she thought, the outer west was bursting with children and adolescents, and adolescents having more children. Fat girls in tent dresses pushed fat babies in prams through the air-conditioned glitz of the shopping complex.[14]

Speculation about the future of Sydney does not figure strongly in fiction by the city's writers, but when it does, the works are portentous with warnings of evils to come. M. Barnard Eldershaw's novel *Tomorrow and Tomorrow and Tomorrow*, first published in 1947, is set in Australia in the twenty-fourth century and looks in retrospect at the nation's history. Life in Australia at the time—post-World War III—is seen as materially secure but spiritually enervated, with wealth concentrated in the hands of a few, and government in the hands of a tyrannical foreign elite. Sydney has been destroyed from within and without by bombs and fire, pastoral land is farmed by communes, and any attempt at regaining liberty in this authoritarian Utopia is defeated not by force but by apathy.

Gabrielle Lord's science fiction novel *Salt* (1990) is also set in a post-war future, and once again there is a corrupt and Orwellian power structure against which resistance is difficult. Now, however, the nation's ills are largely ecological, and global warming has turned the fertile coastal strip into a terrain resembling the salt lakes of the dry inland.

They flew in wrathful silence for a while, crossing what used to be the western suburbs, where once tens of thousands of people lived and died, built and manufactured, played and swam and had babies and grew in ever-widening rings to the south-west. They cleared and built more houses, until the satellite cities were as packed as Sydney itself. Fast trains linked them with the coast. But now there wasn't much left at all. The Civil War of twenty-five years ago, the ravages of the violent new climate that afflicted most of the world, had rendered these once teeming places into desert. Occasional tombstones of chimneys still stood to cast long black shadows in the dawn, but most of the masonry lay in spills of brick and tin. Any salvageable material had been reused in the rebuilding of the city and the walls after the war, or collected by the shanty town dwellers for the rough shelters where they lived; not many of them, and not very happily. Sando took the controls and flew over a ragged collection of lean-tos and shacks, one of the smallest 'towns' several kilometres from the city, wondering how anyone could survive there, day after day, living in barbarity and temperatures of fifty-seven degrees celsius.

He'd made the mistake once of landing down there, moved by a plea for help spelled out in the dust. But when he'd cautiously climbed out of his craft, looking to see where the mother and her sickly baby might be, his sixth sense made him jump back in and lift the helicopter off the ground again. As he'd swung away, he'd seen the ambush group who were to have murdered him, lying hidden behind a low wall. He had seen their twisted, sunburnt faces as they cursed after him. These desperate people lived in rough, stone bunkers close to the walls of the city. They were *personae non gratae*. Some of them were felons, escaped from the huge western prison. Others had fled from the city, wanted by City Security, or on the run from fellow citizens. But none of them wanted to be too far from the city walls.

Sydney itself is in ruins and barely habitable and its population—except for an elite few who have walled themselves in—are crazed and poverty-stricken.

It was worse, further west. The shanty dwellings were thickest in the south, because the walls formed a windbreak against the blazing westerly that blew almost constantly now from the red-hot heart of Australia. The huge walls cast long shadows, protecting the outdwellers from the deadly sun for part of the day. This was corrosion country, where each outlaw did what he could to survive. It didn't take long for the sun and driving wind to destroy most fabrics and metals these days. Not like us, Sando thought bitterly, staring at the chartreuse tone of the distant Pacific wrinkling under the westerly. Not like some of us, safe in the city, surviving with a modicum of taste and style. There was no doubt that the pilots of Western Security were the elite of Sydney. Most people had very little.

Near the coast, the sea was quite brown, as if the heat of the coastline had scorched it. But the dark areas were teeming with rusty algae and weed. Desalination plants and algae farming factories dotted the coastline. Here and there, large black patches in the brown sea indicated the frightening algae dieback. The black patches were growing bigger every year, and recent research indicated that certain of the algaes could no longer live in the high salt levels and growing temperatures of the coastal sea.[15]

However, these prophesying voices of doom remain rare in Australia's first and most hedonistic city.

68

HOBART

A NATURAL PENITENTIARY:
'Bringing Hell and Heaven Together'

In this land vague malevolence
hung upon the skin, olive grey, a gas …
Evil in the lungs, humming in the nostrils, flies;
world no one wanted to live in,
an open pit—
chained thieves peopled the camps of new Ararat.
Sadism was twisted into the root

and snuff for floggin's—
for the militia ladies, rum for amnesia,
boys for the laddies.
Fleisch und Blut.

Eucalyptus too hard to cut, nothing sharper than a shell,
no animal of service, one edible fern,
one fungus.
Land without a fruit-bearing tree,
as though Truth were exiled and the Lie remained.

Here women would carry on their teats
small black skulls;
last of the race, the scourged Eve—
Truganini, would die in a boarding-house.

This old, bloodied island,
with their dark shadows swept off the shore and the movement
in the bush—
there is the smell of Poland
despite the sea breeze and the sheep dung.
Parklands root in mass graves,
the bush is like a derelict house …

Alan Afterman, from *Van Diemen's Land*, 1980[1]

Of all Australian states, the island of Tasmania seems most prey to its geography and history in the minds of its writers. Lying 200 kilometres south of the mainland across Bass Strait, facing the frigid wastes of Antarctica to the south, the land mass was called Van Diemen's Land in 1642 by Abel Tasman in honour of his superior in the Dutch East India Company, Antony Van Diemen, and was renamed Tasmania in 1853 by the British when convict transportation was abolished. When Lieutenant-Governor David Collins arrived from Sydney in 1804 to establish a new convict colony there, he found himself in one of the most desolate and isolated outposts of the known world.

Like that of Tierra del Fuego, the extremity of Van Diemen's land presents a rugged and determined front to the icy regions of the south pole; and, like it, seems once to have extended further south than it does at present. To a very unusual elevation is added an irregularity of form, that justly entitles it to rank among the foremost of the grand and wildly magnificent scenes of nature. It abounds with peaks and ridges, gaps and fissures, that only disdain the smallest uniformity of figure, but are ever changing shape, as the point of view shifts. Beneath this strange confusion, the western part of this waving coast-line observes a regularity equally remarkable as the wild disorder which prevails above. Lofty ridges of mountain, bounded by tremendous cliffs, project from two to four miles into the sea, at nearly equal distances from each other, with a breadth varying from two miles to two and a half. The bights or bays lying between them are backed by sandy beaches. These vast buttresses appear to be the southern extremities of the mountains of Van Diemen's land; which, it can hardly be doubted, have once projected into the sea far beyond their present abrupt termination, and have been united with the now detached land, De Witt's Isles.[2]

Hobart, Australia's southernmost capital city, was established on its present site at Sullivan's Cove on the Derwent River estuary in Tasmania in February 1804 by Collins after he had rejected Port Phillip (now Melbourne) on the mainland as a suitable site for a convict colony. This new settlement, replacing one made the year before by Lieutenant John Bowen to forestall French claims to the island, was named, like its predecessor, after Robert Hobart, the Earl of Buckinghamshire and secretary of state for war and the colonies. Like New South Wales, Hobart was 'a Place of banishment for the Outcasts of Society'. David Collins died in Hobart six years later, frustrated in his attempts to make the settlement flourish in the face of an unfavourable winter climate, a barren landscape and official neglect from his superiors.

When business entrepreneur and farmer Charles Rowcroft arrived in 1821, he found a town of 'straggling, irregular appearance', interspersed with a 'pretty good house here and there, and the intervening spaces either unbuilt on or occupied by mean little dwellings, little better than rude huts ...'[3]

However, when the convict Henry Savery described his arrival in Hobart on the *Medway* in December 1825 in a series of sketches published in the town's *Colonial Times* between June and December 1829, he wrote that he was surprised to find a well-established town.

> It was a remarkably fine clear day when I landed from the ship on the Wharf. What was my surprise, to observe the large handsome stone buildings, into which, porters were busily engaged rolling casks and other packages, and at several civil looking well dressed young men, who were standing with pens behind their ears, and memorandum books in their hands, paying the most diligent attention to what was going on. A number of other persons formed little knots or circles; and the hallooing of ferrymen, the cracking of whips, and the vociferation of carters, struck me as creating altogether, a scene of bustle and activity, which indeed I had little expected. For the moment it occurred to me, that our Captain, in the hurry and confusion which the quarrels on board had occasioned, has missed his reckoning and had made a wrong port; and accordingly seeing a fat, portly, sleek-looking, apparently good-humoured Gentleman approaching, I enquired of him, with an apology, in what place I was? — Judging from my manner and appearance that I must be a stranger, he very civilly replied, that I was in Hobart Town, the capital of Van Diemen's Land.

Savery, having discovered a 'magnificent straight line of street' extending for more than a mile, an impressive governor's residence flanked by well-laid out gardens and shrubberies, a Chamber of Commerce, and the existence of no fewer than three newspapers, continued his exploration until he encountered a 'handsome brick church'.

> The church door happening to be open, I took the opportunity of judging of its interior, and I could almost have fancied myself in one of the modern churches of the metropolis of the world. Such regular well-arranged pews, so beautifully a finished pulpit and reading desk, made of wood, which at first I thought was Spanish mahogany, quite astonished me ... While I was thus employing myself, a Gentleman wearing

a Clerical hat, approached, and with much affability of manner, addressed me as a stranger, and gave me some general information respecting the religious institutions of the place. He had a lisp in his speech, which was by no means disagreeable, and his well cased ribs bore evident marks that, whatever other doctrines he might preach, that of fasting was not one upon which he laid much stress, at least in its practice. He acquainted me, that independent of the congregations belonging to this large Church, a Presbyterian Chapel, a Roman Catholic Chapel, and a Wesleyan Meeting House, were each well attended every Sunday, and it gave me great pleasure afterwards to be told of this Gentleman, as he himself had beautifully expressed of his brother labourers in the vineyard, that in their lives and conduct the religion they all professed received its brightest ornament—that they made a well formed cornerstone of the superstructure they supported. Oh! thought I, this must be the effect of a virtuous and industrious population.[4]

The well-educated but rather unctuous Henry Savery, transported to Van Diemen's Land for forgery and on his arrival given a job as a clerk in the office of the colonial secretary, soon found himself in trouble when his privileged position led to accusations that he had received overly favourable treatment from Lieutenant-Governor Arthur. In 1828 his wife Elizabeth, encouraged by his glowing reports of his prospects, arrived to join him, but found him in financial difficulties and on the brink of suicide. When she left Hobart early the next year, Savery was in prison for debt, and it was probably in this fifteen-month period that he wrote his 'sketches', which were collected and republished as *The Hermit In Van Diemen's Land* in 1830. A libel suit was brought over these sketches in May of the same year.

It is tempting to think that Savery's work might have been read by Marcus Clarke, who, in *For the Term of His Natural Life* (1874), presents a portrait of Hobart as it might have been seen by his preposterously unChristian clergyman, Mr Meekin, a mere eight years later.

'Society in Hobart Town, in this year of grace, 1838, is, my dear lord, composed of very curious elements.' So ran a passage in the sparkling letter which the Rev Mr Meekin, newly-appointed chaplain, and seven-days' resident in Van Diemen's Land, was carrying to the post-office, for the delectation of his patron in England. As the reverend gentleman tripped daintily down the summer street that lay between the blue river and the purple mountain, he cast his mild eyes hither and thither upon human nature, and the sentence he had just penned recurred to him with pleasurable appositeness. Elbowed by well-dressed officers of garrison, bowing sweetly to well-dressed ladies, shrinking from ill-dressed, ill-odoured ticket-of-leave men, or hastening across a street to avoid being run down by the hand-carts that, driven by little gangs of grey-clothed convicts, rattled and jangled at him unexpectedly from behind corners, he certainly felt that the society through which he moved was composed of curious elements. Now passed, with haughty nose in the air, a newly-imported government official, relaxing for an instant his rigidity of demeanour to smile languidly at the chaplain whom Governor Sir John Franklin delighted to honour; now swaggered, with a coarse defiance of gentility and patronage, a wealthy ex-prisoner, grown fat on profits of rum. The population that was abroad on that sunny December afternoon had certainly an incongruous appearance to a dapper clergyman lately arrived from London, and missing, for the first time in his sleek, easy-going life, those social screens which in London civilisation decorously conceal the frailties and vices of human nature. Clad in glossy black, of the most fashionable clerical cut, with dandy boots, and gloves of lightest lavender—a white silk overcoat hinting that its wearer was not wholly free from sensitiveness to sun and heat—the Reverend Meekin tripped daintily to the post-office, and deposited his letter.

However, behind this cosy picture of apparent English rectitude and the solid material achievement of a 'virtuous and industrious population' lay an entirely different reality.

'Colonel Arthur reported to the Home Government that the spot which bore his name was "a natural penitentiary"',[9] wrote Marcus Clarke in the early 1870s in *For the Term of His Natural Life*, referring to Port Arthur, named after Sir George Arthur, Lieutenant Governor of Van Diemen's Land from 1824 to

1836. Arthur, a strict disciplinarian, had in 1830 established this secondary convict depot 104 kilometres southeast of Hobart for offenders reconvicted after their arrival—a third level of banishment after New South Wales and Hobart itself.

Clarke had been sent to Tasmania in 1870 by the Melbourne *Argus* to research the island's convict past for a series of historical articles, and he reacted to the landscape in much the same way as had David Collins. 'The south east coast of Van Diemen's Land,' he wrote, 'resembles a biscuit at which rats have been nibbling. Eaten away by the continual action of the ocean which, pouring round by east and west, has divided the peninsula from the mainland … the shoreline is broken and ragged.

'Viewed from the map, the fantastic fragments of island and promontory … are like the curious forms assumed by melted lead spilt into water. If the supposition were not too extravagant, one might imagine that when the Australian continent was fused, a careless giant upset the crucible and spilt Van Diemen's Land in the ocean.'

'Tasman's Peninsula is … in the form of an earring with a double drop', wrote Clarke of the site of Port Arthur, and then outlined in detail the 'wild and terrible coastline, into whose bowels the terrible sea had bored strange caverns'. He described the burial ground of 'The island of the Dead', the 'Devil's Blow-Hole', the chained guard dogs, the shark-infested waters, the wild and mountainous wilderness, the system of semaphored signals which could alert the guards to any escape, which were characteristic of Arthur's 'natural penitentiary'. Clarke guessed that the 'worthy disciplinarian' Sir George Arthur probably took as a personal compliment 'the polite forethought of the Almighty, in thus considerately providing for the carrying out of the celebrated "Regulations for Convict discipline"'—of which Arthur himself was author.

In March 1871, when Marcus Clarke's newspaper serial *His Natural Life* began appearing in the *Australian Journal*, the events it was describing were only forty years old, but would have been largely unfamiliar to Clarke's reading public. The worst cases of brutality to convicts, like the worst outrages against Aborigines, took place on the fringes of settlement—in this case in a distant island penal colony, far from public scrutiny, to which the worst offenders were sent. The location itself was regarded as a natural punishment.

Most of Marcus Clarke's images of nature and the landscape in Tasmania form a suitably vivid background to his brutal but romantic convict drama, with its tortured heroes, sadistic official villains and melodramatic plot twists involving lost inheritances and mistaken identities. However, little of Clarke's rendering of convict suffering was historically inaccurate: convicts who managed to escape from the settlements often preferred cannibalism or death from starvation to recapture. Tasmania was a trackless wilderness in which men could and did perish from cold and hunger, surrounded by a reef-filled and imprisoning sea that had already claimed ships and would continue to do so. The natural environment, which in parts many later free settlers would find as gentle and beautiful as Ireland, was for them hostile, strange and hazardous.

'It would seem,' wrote Clarke, 'as though nature, jealous of the beauties of her silver Derwent, had made the approach to it as dangerous as possible … [but] the voyage up the river is delightful.' Once past the natural barriers of the coast, Van Diemen's Land was '… fertile, fair and rich … [and t]he climate … is one of the loveliest in the world.'[5]

This is the same country that the Congregational minister John West, who arrived in Tasmania in 1838 and published his *History of Tasmania* in 1852, described as 'mountainous, with numerous beautiful valleys, rendered fertile by numerous streams descending from the hills, and watering, in their course to the sea, large tracts of country. The south-western coast, washed by the Southern Ocean, is high and cold, but the climate of the northern and inland districts is one of the finest in the temperate zone, and produces in abundance and variety all the fruits which are found in the same latitude in Europe.'[6]

A century later, in his poem 'Tasmania', the poet Vivian Smith would celebrate his birthplace as:

> Water colour country. Here the hills
> rot like rugs beneath enormous skies,
> and all day long the shadows of the clouds
> stain the paddocks with their running dyes …[7]

Marcus Clarke's rapturous description is reiterated by Mark Twain in *Following the Equator* (1895): 'How beautiful is the whole region, for form, and grouping and opulence, and freshness of foliage, and variety of colour, and grace and shapeliness of the hills, the capes, the promontories; and then, the splendour of the sunlight, the dim, rich distances, the charm of the water-glimpses …'

Mark Twain, however, goes on to note, 'And it was in this paradise that the yellow-liveried convicts were landed, and the Corps-bandits quartered, and the wanton slaughter of the kangaroo-chasing black innocents consummated on that Autumn day in May, in the brutish old time. It was all out of keeping with the place, a sort of bringing heaven and hell together.' This overwhelming sense of a shadowy past is reflected in the ambivalence displayed in the works of many later writers, both visitors and natives.

Hal Porter, arriving in 1946 to take up a position as a schoolteacher in Hobart, wrote, 'I first see Tasmania, just after sunset, as a landscape desperate, assailed, and sinister.' He goes on with a description of a landscape that echoes the descriptions of Marcus Clarke one hundred years before and of Lieutenant-Governor David Collins fifty before that, except that now his point of observation is a rain-obscured aeroplane window instead of the heaving deck of a sailing ship.

> The aeroplane, jolting and dropping into chasms of nothing above the hell-broth of storm-clouds over Bass Strait and Tasmania, is not able to land at the Hobart aerodrome until a long time after it should. We must wait in the attics of air, wheeling farther south to where at last, through slots in the streaming veils of mist, there is Tasmania, the gothic littoral of a thousand marine paintings, precipices of bitter grey basalt, insolent tusks of rock, iron-coloured waves smashing forward an enormity of water to run like skeins of smoke up the fissured scarps. Nothing is missing except the tormented windjammer, jagged masts, shreds of sail, careening to its doom. These ramparts at the edge of civilization, ceaselessly murdering water to spume and vapour, are not of the twentieth century.[8]

The colony of Van Diemen's Land was separated from New South Wales in 1825 and renamed Tasmania in 1853 to mark the abolition of convict transportation. However, by that time, the scars of its history as a penal settlement had worn deep.

In Patrick White's novel *A Fringe of Leaves* (1976), based on historical events of the last century, Mr and Mrs Austin Roxborough arrive in Hobart in November 1835 from England to visit Garnett Roxborough, Austin's brother. Ellen Roxborough walks on a chilly afternoon through Hobart, 'buffeted by wind, threatened by a great cumulus of cloud, between the mountain which presided over man's presumptuous attempt at a town, and the shirred waters of the grey river …'.

Earlier, Garnett has asked Ellen if she approves:

> 'Of what?'
> 'Of our neat little town.'
> 'It is that,' she said. 'And English. I have difficulty in believing I am being driven through a famous penal colony of the antipodes.'

He laughed. 'You will soon believe, but need not fear, or feel embarrassed … The authorities keep the wretches suitably employed, and on the whole, subdued.'

While the brothers reminisce, Mrs Roxborough is able to enjoy her view of the 'unassuming, while often charming houses, their general effect of modest substance sometimes spoilt by the intrusion of an over-opulent facade. Hens were allowed the freedom of the streets, and an ambling cow almost grazed a wheel of the buggy with her ribs …'

But the town is 'morally infected', according to Austin Roxborough, and so—with her unconscious sensuality—is Ellen herself, according to Garnett. 'You and I would enter hell the glorious way if you could overcome your prudery,' he tells her.

This 'moral infection' which finds its metaphor in Hobart and its convict system represents to White a symbol of Australian history and national characteristic. It is the more earthy Ellen, however, rather than Austin Roxborough, who becomes 'Australian' through her subsequent experience of the landscape. Austin, the English gentleman, cannot survive it.[9]

Hal Porter reflects a similar raffish sense of 'moral infection' in his vision of Hobart in 1837, in his poem 'Hobart Town', published in 1968, which goes in part:

Mike Howe's head with frozen frown
Is on display in Hobart Town.

By Wapping Stairs the alley whale-oils blaze;
The tap-room skittle-grounds are stews of din
Where tripeman, shepherd, fence and whitesmith daze
Pock-pitted doxies with the Sky Blue gin. …

The gibbet chandeliers sag down—
Glass-frosted thieves of Hobart Town. …

Through Russia tweed and kerseymere and smock
And red shell jackets and Valencia vests
Cold, like a watchman's cutlass, drives its shock;
The apple-woman shawls her stony breasts.

In Geneva-stinking gown
Venus paces Hobart Town.[10]

(Michael Howe was a notorious bushranger, transported for highway robbery in 1811, who escaped in 1813 and terrorised settlers until his violent death in 1818.)

Arriving over a century after these events by aeroplane during a storm, Hal Porter finds that Hobart does not belong to the twentieth century any more than it did then.

It takes me a long time to find accommodation. I try to tell myself it is mere fatigue and the icy rain that make me feel I am an Ishmael in the London of Dickens, the narrow steep streets, the drains gulping and retching, the cobbled mews, the names of the hatchet-faced hotels I am turned down at—The Man on the Wheel, The Ocean Child, Sir William Don, The Ship, the thin and acrid beer, tepid from the wooden barrel on the counter of a waterfront tavern where the taxi-driver and I drink while asking advice.

Where, you pock-marked barman, where in the rain, is a bed?

Anywhere will do—A Night Refuge, Fagin's den, underneath the Arches.

We are directed to a boarding-house which has pretensions. It calls itself the Astor Private Hotel. The reception desk is in a hall haunted by the smell of galoshes, and cumbered with Benares-tray-topped tables, a cast-iron umbrella-stand, alabaster pedestals topped by equestrian statuettes of bronze, and throne-like Chinese chairs of ebony inlaid with mother-of-pearl peonies. Antlered stag-heads protrude from the panelled walls.

There is one room vacant for one night only, a penthouse room on the flat roof, a shuddering un-lined box, probably once a caretaker's. The wind scourges it. The rain falls and falls.

Perhaps fancifully, he admits, he allows himself to be ruled by his preconceptions.

Before I sleep I fight an instinct warning me that I am on an aggrieved island, that the water gobbling in the roof-gutters, and pouring off the asphalted roof can never wash away some taint of plague sensed everywhere. There has been a hell here to which Hell is a crèche. I tell myself again that I am the victim of fatigue, irritating weather, and a mish-mash of impressions—the end-of-the-world precipices; a city momentarily like a setting for Jack the Ripper; grisly convict tales of cannibalism, sodomy and the triangle; the Australian legend that Tasmania is an island of incest and haunted architectural follies; and the sailors' stories heard years ago in Madame's Williamstown Wine Depot that Hobart is one of the most immoral ports in the world.[11]

The Brisbane-born poet Gwen Harwood, who moved to Tasmania in 1960, also links landscape with history in her poem 'Oyster Cove'.

Dreams drip to stone. Barracks and salt marsh blaze
opal beneath a crackling glaze of frost.
Boot-black, in graceless christian rags, a lost race
breathes out cold.
Parting the milky haze
on mudflats, seabirds, clean and separate, wade.[12]

And so does Christopher Koch, who, in the voice of his character Richard Miller in *The Doubleman* (1985) reiterates the themes of Marcus Clarke and Mark Twain.

The other island's name had been Van Diemen's Land. It was a name that had rung and chimed in Cockney and Irish songs of hate; the name of a British penal colony that had once been synonymous with fear throughout the Anglo Saxon world. At the place called Hell's Gates, on the savage West coast, a penal settlement so terrible had been created that convicts had murdered each other to secure the release of hanging, or had fled without hope into the icy rain-forest to leave their bones there, and sometimes to turn cannibal. Van Diemen's Land had also removed a whole race: the few aboriginals the colonists had found when they came, and whose last remnants, deported to a smaller island in Bass Strait, had pined away, staring across the water to their lost home.

All this was a hundred years gone; and that century was best forgotten. Gone, the sad aborigines, who had once lurked in the bush like dark, accusing wraiths, threatening lonely farms. Gone, the old convicts from London and Cork, in their mustard suits of shame; ancestors whom nobody wanted to own. When transportation ended, the native-born colonists changed the island's name to obliterate the dread; to make it normal. As clean young Tasmania, it would start anew, the horrors forgotten.

But were they? I would sometimes wonder about this, reading about Hell's Gates and Port Arthur. How long ago was a hundred years?

Sometimes it seemed to me that the fusty odour of fear, the stench of the prison ships, was still in Hobart; and a tragic, heavy air, an air of unbearable sorrow, even in sunshine, hung over the ruined, sandstone penitentiary and the dark blue bay at Port Arthur, south of Hobart, where the tourists went. Was it possible that the spirits of the convicts were silently clustered in that air, weighting it like sacking? Were the floggings and the shackles still invisibly here, hanging above the dark green bush? Still somehow repeated, for eternity?

Miller has a convict ancestor, a fact his mother tries to conceal.

> A certain look of distaste came into Tasmanian faces when you mentioned the convicts; the look of respectability threatened. It was similar to fear, and even the jokes had fear in them. No one wanted to admit having a convict ancestor; because the truth was that long ago was not long ago: not long enough.
>
> Once I had asked my mother whether the family had a convict, in that past, and her face took on the look of threatened respectability I'd come to recognize; a prim yet evasive distaste.
>
> No, she said, of course not; how could I ask such a thing?[13]

Vivian Smith, writing of the 1930s and 40s, repeats the same sentiments in an essay entitled 'Growing up in Hobart', first published in *Island* magazine in 1988.

> In those years statements were often made that the past had to be forgotten, erased, eliminated; that the strain or shadow over Tasmania had to be wiped away. The sense of class distinctions inherited from England was very strong; there was a lot of snobbery, and social pretensions; some people were afraid that their convict ancestry might be uncovered. There were forlorn pockets of gentility and much emphasis on family background and stock. People stressed their family connections in a way that made it clear that they were completely free of the taint of convict or proletarian origins. I remember the Archbishop of Hobart … saying that Port Arthur should be demolished and thrown stone by stone into the water.[14]

Even the Sydney-based writer Robert Drewe, in *The Savage Crows* (1976), his novel of the slaughter of the Tasmanian Aborigines, is not immune. For Drewe's angst-ridden hero Stephen Crisp, visiting Tasmania on a personal quest but ostensibly researching his thesis on George Augustus Robinson, the Protector of Aborigines in Tasmania in the 1830s, the past also intrudes into the present in a disturbing way. Robinson was made responsible for conciliating the Aborigines of Bruny Island near Hobart and persuading the surviving mainland Aborigines to join a new settlement on Flinders Island in the north, an event which led to their subsequent deaths from disease and attrition. Among the Bruny Island people was Truganini, long considered the last of the full-blooded Tasmanian Aborigines, who died in 1876. The second last survivor, William Lanney or 'King Billy', had died of cholera seven years before in a room of a guesthouse, the Dog and Partridge, located—according to Drewe—on the corner of Goulburn and Barrack Streets, Hobart. Shortly after his death, Lanney's skull was stolen and replaced by a white man's, his hands and feet were severed, and other body parts were removed to satisfy scientific and medical curiosity. Crisp decides to visit Flinders Island.

> He considered the fact that within the day he had reached one of the world's most remote places—an island of extinct people. He had escaped to the arse-end of the Earth. No one who knew him knew he was there. Strangely, this didn't afford him the relief he had expected and, alone in the darkening room, the restlessness of the curtains making shadows on the walls, he cried for several minutes.
>
> A knocking came on the bedroom door. 'Tea's on the table,' called the Gaebler woman. 'Don't let it get cold.'
>
> The thudding outside was kangaroos, he discovered. At dusk they appeared out of the bushes in their hundreds, Gaebler told him, right up to the front porch, ate the bran and pollard from the hens' bins. He took a shotgun to them periodically, for dogs' meat. They came in handy as bait for his craypots set just offshore over the hill.
>
> Slaughter everywhere, Crisp thought.
>
> 'One thing's for sure,' Gaebler said, 'the old Abos didn't become extinct out of starvation. A century later, farms, guns, all that and the bloody island's still plagued with roos.'
>
> Gaebler's old mother sucked her soup, broke some bread into it. 'They died of broken hearts,' she

said. 'Pined away for their homes like animals in a zoo. My grandad had them working for him in Burnie. Lovely crochet work some of the women did, fine eye for detail.'[15]

It is as if, in the face of this denial by past generations of their roots, these must be at least acknowledged before the Tasmanian-born writer can move on to other things.

THE LOST HEMISPHERE

We were an offshore island off the shore of an offshore continent, victims of a twofold alienation. Australia was chosen for settlement because it was remote, and because it could scarcely sustain life. It served as a place of ultimate banishment and estrangement. Yet beneath its southeastern tip was an even more remote version of itself, a site of internal exile. This was our little serrated triangle of rock: Van Diemen's Land, where a penal colony was founded in 1803 at Hobart. Australia is tempted to forget Tasmania, just as I had wanted in 1968 to walk away from the small past I had accumulated. At school we were given plastic maps to trace, with dotted lines standing for those invisible fences drawn across scrub and desert to divide the states. Tasmania, disconnected by Bass Strait, was an embarrassment to this cut-out continent; it was therefore simply left off. We passed the time outlining the contours of a country we didn't belong to.

Peter Conrad, *Down Home: Revisiting Tasmania*, 1988[1]

Perhaps it is just this preoccupation with submerged history that leads to the sense of alienation evident in the work of some modern writers born in Tasmania. Two at least, Christopher Koch and Peter Conrad (both expatriates, Koch to mainland Australia and Conrad to London), have described in their fiction and non-fiction a sensation of disconnection from both Australia and the rest of the world, especially in childhood. Both seem preoccupied with the question of identity and belonging. In *The Doubleman* (1985), Koch's protagonist Richard Miller reflects on his childhood.

The whole of dry, Time-flattened Australia lies north of latitude forty, its climate Mediterranean and then sub-tropical. But small, mountainous Tasmania, filled with lakes and rivers, is south of latitude forty; and this makes it different. Politically, it is part of the Commonwealth of Australia; physically it is not.

The island lies in the track of the Roaring Forties, the westerly winds that blow from Cape Horn. In the upside-down frame of the Antipodes, it duplicates the Atlantic coast of Europe; and Brian Brady and I were children of a green, marine landscape: subjects of the stern winter cold. Our spirits were conditioned by the blood-thrilling Westerlies; snow fell in our mid-winters; we walked to school through London fogs.

'Seven o'clock,' said the morning radio announcer, 'time for a Capstan, the Empire's favourite cigarette.' Our seasons were the seasons of English storybooks, and of the films we saw on Saturday nights, brought from the northern hemisphere. Our great-grandfathers had put together a lost, unknown home in landscapes that made it all perfectly natural: Georgian houses with classical porticoes; hop fields and orchards; chimney pots rising on gentle hillslopes, in the subtle, muted lights of East Anglia. Banners of cloud hung low across dark blue hills, straight from *The Hound of the Baskervilles*; Norman and Gothic churches had appeared; discreet brothels; banks; Salvation Army hostels. Repertory Societies were run by artistic men in tweeds; trams ran on wet tramlines; and in the midlands, the gentry mulled their claret and rode to hounds. Tasmanians, I suppose, were rather like the prisoners in Plato's cave; to guess what the centre of the world was like—that centre we knew to be twelve thousand miles away—we must study shadows on the wall: *Bitter Sweet* at the Hobart Repertory; *Kind Hearts and Coronets* at the Avalon Cinema; the novels of A. J. Cronin and J. B. Priestley and Graham Greene; shadows, all shadows, clues to the other hemisphere we might someday discover. We were living, when I grew up, in the half-light of that Empire the ultimate end of whose bridge of boats was Hobart.[2]

In Peter Conrad's *Down Home: Revisiting Tasmania*, a non-fiction account of his childhood experiences and eventual return visit to his birthplace, there is a similar attempt at definition. Conrad also remembers his early sources of imagery of 'Australia', of which the island of Tasmania does not seem to be part.

> Nor did Tasmania look like that burnished, seething, ancient mass above us which we knew as the mainland; the main land, to which we were an appendix, an afterthought. At my primary school—a conurbation of sheds across the highway from our house in the northern suburbs of Hobart—the only art-work on the premises hung in a corridor: a print of Ayers Rock as painted by Albert Namatjira. He had prettified it, rubbing it down to a violet, velvety mound and surrounding it with white, wriggling eucalypts; still, if this was Australia—an omphalos of burning stone which sacredly marked a centre in the unmapped sand—then where were we? Out the school window, contradicting that hot, planetary clump on the wall, was our local monolith, Mount Wellington. It was blue, not red; above it were bundled sleety storm clouds, not the metallic scorching sky of the Northern Territory; instead of reddening in the sun, it wore for half the year a toupee of snow. It was a source of chilblains, or frozen toes on the way to school, and of those mists of condensed white life you exhaled on the bitterest mornings with every breath. The Australian landscape had omitted us. Instead of an overcooked desert, we lived in a cool dripping jungle.

For both writers, this sense of being on the edge of things, of not being part of the 'main land', is translated into a longing for an imaginary world. For Conrad, who left Tasmania as a Rhodes scholar in 1968, it is the mythical England of his early imagination, which he will replace with the actuality of expatriatism at the earliest opportunity.

> About halfway through my life—at least I hope my sums are right—I began to wonder about what I had lost. At the time, I hadn't so much lost it as thrown it away, with a negligence I considered cool. When I left home at the age of twenty, it was without a backward glance. First went two tea chests, shipping my books to England. What remained I carried, in endless relays, out to the rusted incinerator in our yard, between the fragrant compost heap and the stilted avenues of beans. There scrutinised by neighbours pausing at their kitchen windows as they did the washing-up, I set fire to all the leavings of my life so far—diaries and exercise books, bundles of letters tied by string; anything that might incriminate me by attaching an identity to me. The cruel cleanness of it now amazes me, but at that age it's easy to cremate a portion of your life: you don't believe in memory, and so can have no affections. Thus I loaded the funeral pyre with the attributes of a self I was to regard from now on—I thought—as irrelevant, the discarded first draft of a person. The two tea chests contained all I wanted to salvage. Their cargo was academic; being good at exams had earned me my second chance. Home was where you started from, not where you stayed. It was more than ten years before I saw it again.
>
> Only now, when it is irretrievable, has it acquired value for me. The thing I had lost was childhood, youth: the decades before you control your own life, when you are tussled and twisted into the being which you helplessly remain. And for me, that time was a place. I was born in Tasmania.[3]

For Koch's fictional protagonist in his early novel *The Boys in the Island* (1958) it is a dream world, an 'otherland', beyond the ridgeline of the suburb in which he lives, which is perceived most clearly in childhood and lost as he grows up. The message of the electric wires that cross the landscape is implicit anticipation of access to something unknown.

> The little boy stood looking at the Soons. *Soon! Soon! Soon!* they hummed; and it was not words the way people spoke words, it was a long humming song, going on and on. But he knew that it said *Soon*, and that he was meant to listen to it. He stood and listened.
>
> No people came in sight: there was only himself here, looking at the Soons. The brown-grassed country was very flat, stretching into the distance. It was not daytime or night-time here, it was both: the air was dark around some concrete buildings nearby, but around the Soons it was bland and bright as water. …

Sometimes it seemed to me that the fusty odour of fear, the
stench of the prison ships, was still in Hobart; and a tragic,
heavy air, an air of unbearable sorrow, even in sunshine, hung
over the ruined, sandstone penitentiary and the dark blue bay
at Port Arthur, south of Hobart, where the tourists went.

C.J. KOCH, 1985 (PAGE 76)

PORT ARTHUR, TASMANIA. PHOTOGRAPHER: PHILIP QUIRK

Water colour country. Here the hills
rot like rugs beneath enormous skies,
and all day long the shadows of the clouds
stain the paddocks with their running dyes ...

Vivian Smith, 1985 (Page 74)

Flinders Island, Tasmania. Photographer: Peter Solness

Each one of them, although they seemed to march, really stood quite still. The farthest of them were very tiny, but the one directly in front of him was very big and tall. Its face was a round silver frown. It stood like a giant spoon on its latticed handle, making the noise of power: *Soon! Soon! Soon!*

Not long ago the little boy had known what the Soons were; but just now he could not quite remember.

Koch also gives an early account of the island's history, which he interposes in an adult voice, perhaps to explain it to an audience unfamiliar with the island. But does this audience consist of other Australians? Outsiders from foreign countries? There is an implicit suggestion that to Tasmanians, everyone else is an outsider.

Tasmania is an island of hills, a fragment separated from the parent continent by a wide stretch of sea. It is different from the hot Australian mainland; it has a different weather and a different soul, knowing as it does the sharp breath of the south, facing the Antarctic. No wars, no disturbances have ever reached the island: no horrors at all, since the last convict transport made the long run from England, when the island's dreaded name was Van Diemen's Land, in that bad, smelly old century of rum and the lash. It lies now like a suburb in the sea, eventless and snug.

It has two small cities, Hobart, the capital, in the south, and Launceston in the north. At their centres are the solid, ornate stone buildings of Georgian and Victorian England; in the suburbs, small swarms of twentieth century bungalows, with galvanised iron roofs and neat front lawns, spread to the edges of the country. There are prosperous farming districts and occasional little townships; and the rest is bush, its metallic green tides ebbing and flowing mile upon mile across the hills and gullies.

And the bush is silent. There is a silence in the island, outside the towns, which the prosperous life of today cannot break: the silence of a land outside history, almost outside time. It is so far south: on the edge of the blank wastes of ice.

The island's people have not been there long: not two centuries. The dark, stone-age people the colonists found when they came have all been wiped out; they are a lost race. But they are still a reproachful memory in the island's silence. Only stone knives on the floors of gullies, and the middens of sea shells from their camps remain; the red, live eyes of their campfires went out long ago. Yet the places of the bush seem to wait: they wait, the dead-quiet eucalyptus gullies, the damp bracken hollows, the dark-haired groves of she-oaks whose grey bark is like the mummified flesh of that race; and a single roadside mailbox, far out of sight of its house, can look forlorn as a lost child. Forlorn, all marks of men, in the lonely places of the island which still doesn't quite belong to them, nor they to it.[4]

In Peter Conrad's *Down Home* there is a similar sense of not being able to take the landscape for granted. Early in the book an equivalent descriptive passage occurs, probably aimed in this case at Conrad's English readers.

Before the engineers thought of diverting rivers and shearing hills, transforming a plateau into a pond and crushing boulders into roads, all civilisation could do in Tasmania was stay indoors and look through its windows at an arboreal screen representing England. To protect their fiction of rural gentility, the colonists transplanted the landscape they had left. Along the Derwent, banks of willows attempted a facsimile of the Thames. The peeling, bedraggled local trees suffered social exclusion: there's a field outside Ouse in which haughty poplars join ranks on three sides against a single unkempt, stringy eucalypt. Oaks were honoured as imperial ambassadors, leafy flagpoles for a remote home. At Lilydale Falls, acorns despatched from Windsor were ceremoniously planted on Coronation Day in 1936. Tasmania still quotes from this distant, indistinct prototype. In the midlands, Georgian country houses shimmer through incongruous dusty heat, and little girls with riding hats and jodhpurs trot their ponies past the red-whiskered mountain men who are mending the road.

But once the view from the landowner's window stops, the truth of the country reasserts itself—dry and windy plains like a desert of spiky grasses, hilltops turned into dunes by the glare; and to confound the oaks with their umbrellas of shade, the presences which dominate everywhere are dead gums, twisted in arthritic agonies but unbowed. Travelling round Tasmania, I came to remember individual specimens as vividly as if they were human casualties I had met.[5]

Christopher Koch, who left Tasmania at the age of twenty-two, describes the same cultural and historical dilemma in his 1987 essay 'The Lost Hemisphere'.

But one felt the odd twinge of doubt. Who *were* we, marooned at 42 degrees south? Why were we *here*, and not *there*? And how perfect was the duplication really?

It was only on the surface, of course. The island's landscapes had a troubling strangeness, if you looked behind the stage sets we had erected. And beyond Port Davey's last little lights of settlement, in the extreme southwest, all normality ended. Beyond Port Davey there was nothing—there was Antarctica. In the Gothic wilderness of the southwest, 5000 square miles of impenetrable, cool-temperate rainforest lay entirely unsettled: a place where men had walked in and never walked out, and where rivers ran underground. A writer who is a native of such an island comes quite soon to the problem of trying to match its spirit with the spirit of the ancestral land in his head—the lost northern hemisphere. It's possible to love both, but matching them up isn't easy: the task of a lifetime, in fact.

What I am asserting is that this situation, to a great or lesser degree, is typical of many Australian writers, and that it's this tension—produced by the consciousness of another lost landscape and society—that produces a quality in Australian literature that is peculiar to it. It also produces a pathos of absence; so that the essential Australian experience emerges as one where a European consciousness, with European ancestral memories, is confronted by the mask of a strange land, and by a society still not certain of its style.[6]

Suburbia vs The Wilderness

Hobart can forget the bush. It is a city, but only just; the fragrant foreign breath of the country can move in summer down Elizabeth Street, catching at the people on the pavements of midday. The country's yellow grass reaches stray fingers into the suburbs; and standing in almost any street, you can glance up and see the near rusty-green and farther dream-blue ranges, their tree-serrated tops poignant against the sky: a reminder of farness.

So the island has two horizons, two barriers against the world: there are the ranges of hills, and there is the coast: in the north and east, green-laced beaches, facing the world; and in the south, black primeval capes of rock: final walls facing the Antarctic, pitted and rowelled by the wild southern storms of winter.

The little city of Hobart, like all cities, is divided into two worlds: the old districts, and the new districts. And the boy very early named the old districts as bad, and the new districts as good.

C. J. Koch, *The Boys in the Island*, 1958[1]

Christopher Koch, who was born in 1932 and spent the years from 1935 to 1946 at 1A Bay Road in the Hobart suburb of New Town (which he calls Elimatta in *The Boys in the Island*) and in Lenah Valley, provides a mellow portrait of the Tasmanian capital in that novel. Hobart is moving out of the shadow of its history, out of the neo-English village atmosphere of the small town, and into the modern, suburban age—to which, as in all Australian cities, its writers are reacting with characteristic ambivalence.

However, by his fourth novel *The Doubleman* (1985) Koch is recreating the city with a more sinister aspect. Here it is seen from the point of view of a crippled schoolboy taught by the Christian Brothers in the 1940s and 1950s.

The bruise-coloured steeple of St Augustine's was visible for miles around on the hill of South Hobart: a watch-tower over a camp of fear. When I go back to my native town and look up Harrigan Street to that tower, I still feel the old nausea, the old dread. My case is scarcely unique; as all too many others have testified, a Christian Brothers' education in the 1950s had a fine pitch of dread, unlikely to be matched again.

Fear's height was in the early mornings, when I climbed towards the steeple and its cross up the asphalt hill of Harrigan Street, the steepest in hilly Hobart. The cars of those days would stall on Harrigan Street and then turn back; but I heaved myself upwards with rhythmic jerks, helped by my single crutch, my thin left leg aching and trembling. My thoughts were small and beast-like with effort, and I was glad of this; it kept my mind from Brother Kinsella. Today was bright and frosty, and I was twelve years old.

Upwards: jerk and heave. I counted the landmarks that brought me nearer to the end of my small ordeal, and so to the beginning of the day's larger one. I knew by heart every mean colonial cottage of ochre brick; every ribbon of weedy garden; every holystoned front step; every picket on every fence. Left below as I climbed was one of the town's few small slums; a place of mean, two-storey tenements, shabby shops and small factories: Hobart's miniature Gorbals. A tall factory chimney there was lettered with a vertical message, repeating itself to my misery as I climbed, winter and summer: UP TO DATE.

The last section of hill brought me to the walls and cypresses of the Archbishop's Palace: the zone of the Church, high above slumdom and the town. Where these walls and Harrigan Street ended Byrne Street would be reached, running across the hill's brow. Here, opposite the Palace, stood St Augustine's; and the reward for my climb was Brother Kinsella, waiting at the top. Just inside the red brick gateposts grew a bare, crooked little thorn tree. I never knew its species; but each morning as I passed it, the notion came to me that this was the tree on which Judas hanged himself.

Koch, who attended several schools in the course of his education, finds in the same novel a parallel between the town's convict past and its youthful victims of the poliomyelitis epidemics of the period. At the same time the onset of World War II makes its own mark on the town in a similar way to that described by David Malouf and George Johnston in Brisbane and Melbourne respectively.

I had known I would catch Paralysis long before it happened. When it finally struck, I felt almost foolish for having hoped to be spared.

For a long time, even into adolescence, I continued to think of Paralysis as a creature—a being who belonged to childhood like my clockwork train; like chilblains. This entity had come to me in the days before Salk vaccine, when whole populations of children faced it without any talisman. Its full title then was not poliomyelitis but 'infantile paralysis', and I would never be able to think of it by any other name. It took its time about coming: it waited until I was nine years old. This was at the end of World War Two, in 1946, when the worst of the epidemics were over, and even my mother thought I was safely among the spared. But Paralysis came, after all; it had merely been toying with its lists.

Its most thorough recruiting campaigns on the island of my birth had been carried out in the 1930s, and during the war years. Poor Van Diemen's Land! The leg-irons and the lash of a hundred years before still hung near, like bad dreams; now, suburban and respectable under your new name, you found your children in irons once more, tormented by pains more searching than the lash. Through the streets of Hobart in the 1940s, the children claimed by the epidemic were wheeled by in chairs, or lurched on their crutches. They horrified and fascinated me before I became one of their number, in those years of the War. Shopping in Hobart with my mother, I would study the crippled children with fascination; a fascination that was only rivalled by my interest in the American troops on leave.

The Yanks swaggered and rioted through our staid little city in the world's utmost south, where we knew the War couldn't come; and nothing like them had been seen in Hobart before. Elephant-grey shapes of Liberty ships loomed in the Derwent estuary, dwarfing the wharf sheds below the Post Office; troopships filled with GIs who had fought in New Guinea with the Australians …

Meanwhile, GIs in well-cut uniforms and white-capped sailors called gobs passed on the narrow footpath in shoals; laughing strange laughs, shouting, whistling at girls, drinking from beer bottles and vomiting into the gutter in unbelievable fountains. But the Yanks were helping to save us from the Japanese, and there was only a small resentment in the town. My mother's cheerful, fleshy face was blank and disapproving under her brown felt hat and her stare was fixed; she kept close to the windows of familiar department stores, one white-gloved hand firmly gripping mine. And from time to time, we were passed by the crippled children who were no longer ordinary children.

Their parents pushed them in chairs or in crude, specially-constructed prams like huge trays which interested me horribly. Pasty-faced, monstrous babies of nine or ten years old, wrapped in tartan rugs, they stared sadly at me; or sometimes, inexplicably, they smiled, their legs stuck out stiffly in front of them like those of dolls, imprisoned in the paralysis irons.

'Will I get Paralysis?' I asked.

'Of course not,' my mother said 'Don't talk like that.'

But no use: Paralysis would come for me in three years' time, at six o'clock in the evening, just before dinner.[2]

For Koch's contemporary, the poet Vivian Smith, born in Tasmania in 1933 and who like Koch and Peter Conrad left the island in his early twenties, Hobart is also defined by Koch's 'two horizons, two barriers against the world'.

When I think of growing up in Hobart, two images spring to mind: the mountain at the top of the street; ships at the bottom of the road. The mountain wasn't always there, of course; there were days when it disappeared, when it was covered with mist, and a white blank filled the top of the street where the mountain should have been. And of course there were days when there were no ships at the bottom of the road. But these were the exceptions and for me Hobart remains framed between the mountain and the sea.

However, unlike Koch's protagonist in *The Boys in the Island*, Smith in his essay 'Growing Up in Hobart' displays a nostalgia for the older parts of Hobart, despite his strong desire at the time to leave.

Most of my childhood was spent at Battery Point, which still looks very much as it did then, though large sections around the edges and many houses, have been demolished, and the wharves have been rebuilt. In my childhood the piers were huge oil and tar-stained timber constructions. Ocean Pier was burnt down in a spectacular fire in the 1940s. Salamanca Place was a timber, coal and junk yard and factory area. Its transformation into art galleries and boutiques was unimaginable. The heritage industry was unheard of; the culture industry had not started. In those days immense chimneys behind the factories and warehouses billowed forth smoke …

When I think of my childhood, I think first of the sea—of the harbour seen from upstairs windows in Hampden Road or from the shore at Secheron which was not far from where I lived and where I used to go almost every day when I was very small to play shops on the shore with my cousins. We used to take with us a few old sheets of newspaper for wrapping and walk along picking up beer bottles, apples, seaweed, flotsam and jetsam and set up our shop on the rocks. In those days Clarke Avenue was almost entirely open paddocks, with only a scattered house or two. Port Huon Fruit Juices (now demolished) had its factory on the Derwent there, but when the war came and the refugees started appearing, part of it was transformed into Van Diemen Wines. On long hot summer days the slightly overripe sickly smell of sugars and fruits filled the air. It smelt as if one had eaten too many lollies.

The sea meant shops on the shore. It also meant in those days fairy penguins, seagulls, fish, schools of dolphin which often entered the Derwent; on at least one occasion a washed up sea-lion; and sometimes a whale spouting halfway between Hobart and Bellerive. I remember being told often as a child that next to one or two South American places, Hobart had the finest deep-sea harbour in the world. It certainly provided a huge mental and visual space for a child to grow up in. And then just around the corner and down the hill were the wharves and Salamanca Place and there we were likely to see Japanese, Indian, Norwegian or French sailors, and to hear a variety of languages. Various overseas liners called there to pick up passengers; soon after the war began there were warships, and then the Americans started to appear in 1942, bringing wildness and colour and violence into the streets and a murder or two in the parks …

The port brought intimations of the excitement and allure of overseas, of foreign places. I wanted to see the world. At one stage in my early adolescence I was so keen to get away that I used to haunt the shipping agencies asking if I could work my passage on a ship.[3]

By the 1950s, Hobart is becoming more suburban—at least to Conrad, who was born there in 1948.

My father's job was building Tasmania. He worked for the housing department; the new suburbs of the 1950s, bulldozed into the bush or run up overnight on squelching fields of mud, were his handicraft. He began by painting the lookalike crates, walled with weatherboard and roofed with lids of corrugated iron, and ended by overseeing others—mostly gangs of 'new Australians'—who did the same. He was proprietorial about the shaky, provisional estates. On Sunday afternoon we'd often go on tours of his building sites, to look at the timber struts and asbestos ceilings and cement paths: Chigwell clinging uncertainly to its hillside, Risdon Vale cowering under the walls of the pink prison. The paling fences seemed too frail to keep uncolonised space out; the wrinkled rooftops with their skins of colour were no security against the enormous sky. The thought of living where no one had ever lived before alarmed me. But to my father, these wooden containers signified safety. They dealt so negligently with the landscape—felling trees, creating a waste and calling it a vale—because they were his escape from it. Here where all was new the old miseries of a poor rural childhood could be forgotten. Once the bush was abolished, you could cultivate a garden.

My mother's job was tending that garden. The man builds Tasmania, the woman decorates it. At first my parents grew vegetables out the front: rows of cabbages and trellises of beans, trenches of onions and potatoes. Behind the house, they planted three fruit trees—apricot, nectarine and peach. On their little lot, they instinctively recreated in microcosm the farms they had quit.

There were even hens, cackling and crapping in their coop beside the back fence. Gradually they did away with these recollections of the agricultural past. The vegetables went into hiding beyond the wood-shed, and the front yard was grassed. Between the paths, my mother planted thickets of flowers. The house retired behind a red gum and a willow tree, which clutched at the water-pipes and had to be cut down for its greed. My father took to cultivating cacti in a hut of glass. The move from vegetables to flowers was a historical victory, completing that first conquest of the wild when the houses were set in their foundations. 'They make a lovely show, don't they?' my mother would say of her blowsy hydrangeas and flaunting gladioli. This was her creation, enticed from the dry black soil; her art, and her own floral barricade against the world.

My father built the house, she made the home.

Returning in later years, Conrad still experiences a marked distaste for the banality of the new and modern.

The most deceptive of roadside myths is that of self-renovation. The Trim Shops invites you to come in and consult its style books: since identity depends on hair, new selves are on sale within. A mile or so further on, Lyke-Nu dry cleaners renovate the language at the same time as laundering your clothes. For my parents, the idyllic offer of the suburbs was newness. They always sniffed at old things, and were dismayed when I came home from the Salamanca market this year with a pair of smeared and tattered army pants which had cost me three dollars; my father objects in principle to watching black and white films on television, and says they're out-of-date. Yet for me, in this newly-made world—colonising the cleared bush in a house built weeks before, re-enacting that first venture of settlement in a place without a history—the immediate necessity was to unearth a past. In our house, what past there was had been unregretfully packed away: a mahogany box of old photographs on the mantelpiece, a trunk in my parents' bedroom with souvenirs of the war. When starting out along the road, I was travelling in search of ownable pasts.[4]

It is in just one of these newly completed, spic and span Tasmanian suburbs so disliked by Conrad that Helen Hodgman places her protagonist in her novel *Blue Skies* (1976), set in a nameless beachside Tasmanian town or city in the 1960s.

I'd watched it from the beginning.

Before she came, our house had been the last in the road: a tatty full-stop to a long line of prosperous weather-board bungalows. It stood out a bit, as it wasn't painted in a lurid pastel shade like the others—because I could never make up my mind what colour to do it. Dead colour-selection cards littered the house.

On the far side was a small patch of scrubby bush straggling to the beach, the one remaining unsold block. For days on end I could forget that I lived in a suburb just by looking out of the right windows.

Then the land was sold and cleared. Disturbed spiders took refuge with me. Trenches were dug. Men built the house.

The woman who had bought the block came each day to the site to oversee them. I eavesdropped behind my blinds as she whined at them to get on. The large sun-reddened men were unmoved. They took their time, pausing at regular intervals to brew billy-tea, smoke and grin shyly at her through large mouthfuls of meat-pie.

The work was quickly finished, and the house balanced on an uneven area of raw reddish earth. The men left. It was a wet time and afterwards rain water stood round it in slick, sky-reflecting puddles. The sun glinted and flashed on those pools, surrounding the house with a fence of reflected metallic shards.

The woman hired another gang of large soft men, who levelled the earth and drained it. They dug it and primed it to receive the sackfuls of domesticated grass seeds.

These the woman tended herself. A square of spiky grass blades stood before the house, a vivid and unreal green. Impressive at a distance, but close to it looked pretty sad. The blades were far apart. The dusty earth, growing dustier as summer passed, showed through the gaps like mange and defied her daily watering.

The native grasses rustled and swayed at the edge of this pampered patch. Occasionally it would stake its aboriginal claim to the usurped homeland by launching a seed to fertilise and reclaim a centimetre. Tough though it was, it could not take the almost daily shaving.[5]

Hodgman, born in Scotland in 1945, lived in England until her family migrated to Hobart in 1958. During a subsequent ten-year period of expatriatism in England and Canada, she wrote her novels *Blue Skies* and *Jack and Jill* (1978), both set in Australia. As with Peter Conrad, the disaffected eye of the expatriate (or the returning immigrant) is a sharp one.

In *Blue Skies*, stranded as she is by an unwanted pregnancy, even the beach (based, according to Hodgman, on Tarooma, near Sandy Bay outside Hobart[6]) that initially seems to promise an attractive life to Hodgman's protagonist reveals itself as a trap.

Surprised early in the morning, it was a marvellous beach—a holiday-brochure cover of a beach. On each side it stretched away, pale yellow and perfect. Startling black rocks jutted up in contrast at either end, the sea bluest blue with silver-lamé glitters. It was an absurd extravagance of beauty so early in the day, a whirl of colours that I associated with midnight hours. I sat down in the dust among the empty beer cans and wept.

I was stranded up on that beach like the poor dumb female turtle I once saw in a film. It had just laid a load of eggs in great distress and difficulty and hadn't a hope of making it back to the sea, but, exhausted, was going to die.

Walking back up the road I saw a house for sale.

We borrowed money, bought it and moved in. My husband worked hard; he needed to repay the loan and prepare for his financial fatherhood. I sat back like the turtle and waited to die.

It didn't happen. The days passed and I began to doubt it would—numberless days when the clock always said three in the afternoon, no matter what you did to it. You could try turning it upside down. You could try catching it out, peeking suddenly round the door and taking it by surprise. No matter what you tried, the day ran out then, and there was nothing left to fill it with.

All the other women in that nature-reserve for females managed to invent something to fill their time in decorative and reassuring ways suggested by the women's magazines—those placebos prescribed to sugar-coat time and keep half the population quiet and useful. But such schemes required spirit, an urge to fill days acceptably. I had none.

Now safely town-dwellers, with the wilderness apparently tamed, the inhabitants of Tasmania turn their gaze once more to the landscape. However, in Hodgman's novel even this escape from the urban trap offers little relief.

From town I took another bus, a country-bound bus, square and slow. Full of mail bags and chicken crates, it went out through the surrounding bush townships in the early morning and came back in the late evening. …

We moved off, creaky and overloaded, crawling through the suburbs to where the thin stream of weatherboard houses trickled out into a pool of rusting car bodies, rotting mattresses and ragged-edged beer cans. The telegraph poles continued, pulling themselves out of the tangled mess of the town into a taut straight line and marching purposefully ahead from horizon to horizon, ignoring geography and natural obstacles and playing tricks with perspective.

First off were the lady bowlers. They disembarked just past the airport and disappeared into a little wooden clubhouse by the roadside. We drove on, round scrubby hills, blue-green and smooth at a distance, coarse-grassed and rocky up close. Dotted on these hills were little trees with rounded tops:

toffee-apple trees from nursery wallpaper. Overhead the high bright blue sky was stretched tight and shiny between pink-tinged clouds. The road ahead was a shoelace of white dust. The colours were primary, hard-edged, acrylic-clear. I scraped myself, in my bus shell, across the perfect clarity and colour of that day—a bag of white skin full of passionate reds and purples and boiling yellow-green jealousies. If the bag split, those colours would spill out and spoil the scenery. But it didn't. There was no bursting with happiness. Or anything else.[7]

Now it seems as if the Tasmanian landscape, once so dominant, is under threat before its European inhabitants have properly come to terms with it. James McQueen's *Hook's Mountain* (1982), one of a number of novels to have dealt with the theme of conservation in the last decade, presents a critical view of local forest management. In this passage McQueen's two main characters, an ex-military man, Hook, and an apparently simpleminded hermit, Arthur, discuss the threat to the local mountainside.

On one of these bright cold mornings they squatted beside their tiny fire, enamel mugs hot between their palms, a small chill breeze rippling the feathery tips of the wattles behind them and swaying the tall silver-stippled trunks a little. Before them the meshing spurs of foothills, hills, low mountains stretched away for fifty miles. In the thin crystal air the rock faces of the mountains showed in clear etched detail, the snow on the peaks so white that the shadows seemed almost indigo. The trees of the distant forests were not hazed, but a delicate smoke-coloured filigree. The dark and sullen carpets of the pine planta-tions thrust their tailored edges sharply over ridge-crests and down into shadowy valleys. On the lower slopes of Blue Hill the clear-felled strips lay like ugly wounds.

'Why are there so many pines?' asked Hook.

Arthur was silent for a few moments, marshalling his thoughts. No one had ever asked his opinion on the pines before. Or on much else, except as a jocular courtesy.

'What they do,' he said at last, carefully, 'they cut down the gums, the best ones, for the timber mills. Then they take the second best for chipping, then they just bulldoze what's left, push it up and burn it. See … along the bottom of Blue Hill?' He pointed across the valley at the broad band of grey shaley earth and windrowed trees that circled the base of the hill.

'But why the pines?' asked Hook. 'Why all the pines?'

'They grow quick. Thirty years and they've got another crop to market …'

'Crop?'

'That's the way they talk.'[8]

McQueen, born in Tasmania in 1934, worked in a variety of jobs and occupations including ship's cook, fruit picker and art student in Sydney before returning to Tasmania in 1974 to grow orchids, at which point he also started to write. Of the writers mentioned so far in this chapter, he is one of the few who still lives there.

Carmel Bird, born in Tasmania in 1940, also displays a preoccupation with man's destructiveness of the wilderness and also, inevitably, with Tasmanian history. In an essay written in 1989 for the New York literary journal *Grand Street*, entitled 'Getting My Mother's Sewing Machine Across Bass Strait', Bird once again attempts to explain the fact of being Tasmanian. For Bird, as for Koch and Conrad, there is the problem of Tasmania's 'non-existence'.

Sometimes on the map of Australia the waters of Bass Strait simply merge with the waters of the Southern Ocean and Tasmania does not appear, has taken on a completely secret existence. Because of this, when I was a child Tasmania often seemed to be nonexistent, even though I knew I lived there. Tasmanians of my generation often speak of the memory of Tasmania's nonexistence. In doing so we seek to explain and relieve feelings of past frustration. Sadly for us, the idea of leaving the island off the map has become, on the mainland, just another Tasmanian joke. So the frustration felt in the past is now compounded. I used to collect references to Tasmania in literature, rejoicing to discover for instance that Virginia Woolf wrote in *Between the Acts*, 'She had been born, but it was only gossip said so, in Tasmania.' Such references reassured me that Tasmania was real. I think, because of its isolated beauty, coupled with its dark and recent history, Tasmania was a strange place to grow up in. Perhaps all places are strange places to grow up in. This one was.[9]

And yet it is more recent writers such as Carmel Bird who show signs of ultimately being able to come to terms with the island's history in a more optimistic way. Bird, who grew up in Launceston and lived in Tasmania for twenty-three years, then studied in California, France and Spain before returning to Australia to teach in Melbourne, has described her attitude to the Tasmanian landscape as 'awe, affection and tenderness' for 'a place of grandeur and sweetness and mystery'.[10]

In 'Kay Petman's Coloured Pencils', from her collection *The Woodpecker Toy Fact and Other Stories* (1987), Bird draws on the same type of suburban detail found in the work of other observers of the Tasmanian scene.

The gods saw two rows of houses on either side of a gravel street. This was after the Second World War, before television, before the pill. Some houses had cellars where boxes of faded photographs and broken kitchen chairs gathered mould; some had little attics where trunks of crumpled ball-gowns collected dust. Between the attics and the cellars were rooms full of everyday life, busy with pop-up toasters, vacuum cleaners, pressure cookers, and radios that told the story. Each house was surrounded by a garden. In the front there were lawns with flower borders; at the back there was a lemon tree and a circular clothes-line. Next to most back fences was a little hen house, and beside some of the hen houses were the trap-doors leading to the trenches which the fathers had dug for the safety of the families in the event of an invasion of Japanese.[11]

However, in the title story from the collection Bird is able to transform and enrich these same suburbs with a good humoured affectionate evocation of family life.

My mother was a magger.*
A paling fence divided our garden from the garden next-door, and over the back fence lived Mrs Back-Fence. My mother and Mrs Back-Fence might have been posing for a cartoonist as they stood on either side of the fence, magging. Behind each woman was a rotary clothesline. We had striped tea-towels, white sheets, woollen singlets, pink pants, and knitted socks all hanging from dolly pegs. Some things were patched and darned, the mending being more obvious when the clothes were wet. It was unsafe to hang anything damaged but unmended on the line, for this would be noted by the other maggers as a sign of degeneration in the family. And once, when a torn, unmended nightdress had got through the washing as far as the line, our rabbit attacked it and shredded it so that it had to be thrown out. My mother and Mrs Back-Fence had floral aprons, and often their hair was set with metal butterfly wavers, covered by a chiffon scarf knotted at the front. They did not wear fluffy slippers. Instead they nearly always wore thick stockings and brown lace-up shoes, like nurses.
Over the back fence, these maggers passed hot scones wrapped in tea-towels, cups of sugar, bowls of stewed plums, and a continuous ribbon of talk. They sifted through the details of everything they heard and saw and thought, and arranged them into art. Children under the age of ten, considered to lack the ability to understand the narrative, were allowed to listen, provided they were still and quite. (Today, magging usually takes place on the telephone, I think, and so a child listener becomes restless because there is only one side to the conversation.) The Crusaders took from the Arabian desert the seeds of the wild flowers which later became the glory of English gardens. The maggers scoured the lives of their relations and neighbours, and sometimes the lives of famous people, to shake out the seeds from which would grow undulating plains of exotic grasses and flowers giving colour and perfume.[12]

* The magpie is the scandalmonger of the woods. The verb 'to mag' meaning 'to gossip' derives from the magpie.

And in another story, 'A Taste of Earth', death and history become suburban and intimate.

> I remember when my mother used to take me to the cemetery.
> When there were trams in Launceston, the line ended at the cemetery gates, at the top of a gentle slope. On each side of the tram-line there was a row of pine trees that formed a sombre tunnel through which the tram would glide. The cemetery was always called just 'Carr Villa' and the words stood for something terrible. My mother and I would go to Carr Villa to put fresh flowers on the family grave. I thought of it as my grandmother's grave, although in it were buried my grandfather and my uncle, both of whom had died before I was born.
> Behind the grave there was a row of pine trees, and low clipped hedges of rosemary rambled along the edges of the gravel paths. The smell of pine and rosemary, and the smell of corruption are therefore linked.
> At the end of the hedge near my grandmother's grave there was a tap. My mother and I would empty the vase, pouring the stale, smelling water into the drain under the tap. Strands of brown, slimy, translucent leaves caught on the grating and slid off as the water drained away. The vase that we were emptying was made of pottery. It was tall, glazed with splashes of prussian blue and yellow ochre. There was always the chance that it would be stolen from the grave, or that one day it would be knocked over and broken. But it was a matter of pride that there must be a real vase on the grave, not a jam jar. My mother turned on the tap hard. The sound of the water rushing into the vase was to me like a picture, and a statement going straight from the tap to my heart. In my mother there was always a strange loneliness. As the water gushed into the vase from the tarnished tap by the sinister smell of the drain and the leaves of the rosemary hedge, I knew and partly understood that loneliness.[13]

In a later work, *The Bluebird Cafe* (1990), a fabulist novel combining history, myth and speculative fiction and set in Launceston and small towns in the northwest of Tasmania, the image of the cemetery recurs frequently, but now it is an adjunct to a young girl's flights of the imagination. Bird's character Virginia sits in the ruins of a cemetery of early Scottish settlers composing a romance.

> Virginia sat in the doorway of a sepulchre with her notebook and pen. She used green ink. Before she began writing she spent some time looking down into the valley where a mist was still hanging. She could see the Chinese market-gardener working in his garden, moving up and down the rows of vegetables, watering and weeding. The man was so small in the distance that he resembled, in his basket hat, a figure in a porcelain statue. Virginia imagined she was a princess sitting on a soft green hillside, watching her Chinese slave working in her flower garden. She is wearing her blue silk dress, and on her knee she has a letter from her lover who lives beyond the mountains. He will come to her at midnight on the hillside; he has sent her bunches of white violets which lie carelessly at her feet. At her feet, the violets. Hidden in the hill behind her, rats.
> Virginia was writing a historical romance called *Savage Paradise*. The hero was David Macintosh, the heroine Elizabeth Scott. They both came from Scotland, he from Glasgow, she from the Isle of Skye. They met in Van Diemen's Land. After a long and troubled romance, they married and became the owners of whole mountains, taming the waters, and fighting off marauding bushrangers and blacks.[14]

The full Tasmanian literary circle is completed with a return to historical preoccupations. But now these are benign, turned into the stuff of myth and historical fantasy.

BRISBANE

'RAMSHACKLE HOUSES BUILT ON STILTS, AND AN AMIABLE, EXTROVERT RACE ...'

There was no sense of space, of enclosure: they lived right on the street but the Queensland sky was clear, limitless. And then the late summer clouds built up, all compressed like a dark mattress, and the springs inside would groan and thunder, and finally collapse under the weight of so much rain. Down it would come, in such a hurry he barely had time to lift his small body from the gutter—cars hurrying, trams rattling by—to obey his mother's demands to come inside. ... 'Here it comes,' someone always sang, and there it was—a deluge that put the street under inches of pounding water and trams flared a wave off their bows and cars stalled, and people ran helplessly for shelter and the heavens cracked; electric lights flickered and a great floodstream passed the clear brown eyes of Harry Tekaros, aged four or five, who even as a grown man would never cease to be excited by this most routine of weather rituals, the coming of the thunderstorm. And sometimes the lights did go. For this Theo always had a box of candles ready; they were placed in those squat, fat milk bottles and lit around the darkened shop, and this was like their family religion, in that shop which gave food and life to themselves and others and which controlled their destiny, from day to day, from sunshine to rain. And sometimes outside, the trams would lose their poles on the wires overhead, adding to lightning and thunder the immediate sparks of electricity, the driver leaping out in his Foreign Legion cap and fighting with the rope, the delinquent pole in the rain; and then the silver tram looked like a great stranded whale, the metal tracks hidden underwater and other trams, other whales, were banking up behind it from the city. Theo always complained when he saw the rain coming; the afternoon's business was ruined. But when the storm hit, he was as childlike as Harry at the door: watching with renewed energy this mad display of the sub-tropical elements, so different from the winter rains of Greece, and knowing above all it would bring a fine day in the morning.

Tony Maniaty, *Smyrna*, 1990[1]

In September 1823, nineteen years after the establishment of Hobart in Tasmania, Sir Thomas Brisbane, the governor of New South Wales, sent surveyor-general John Oxley north along the coast in the *Mermaid* to find a suitable place for a third penal settlement. In November, about 800 kilometres north of Sydney, Oxley entered Morton's (later misspelled as Moreton) Bay, named by Captain Cook in 1770, and encountered three shipwrecked convict escapees who told him of a large river flowing into the sea there. This river Oxley subsequently named Brisbane after his superior officer.

The Moreton Bay penal settlement was established at Red Cliff Point on the shores of this bay when Governor Brisbane sent Lieutenant Miller with forty convicts and their guards north in the brig *Amity* in 1824. The following year the first settlement was abandoned due to lack of fresh water and the hostility of local Aborigines, and the little colony moved thirty kilometres upstream from the mouth of the Brisbane River. On a site known to the local Aborigines as Meanjin, this settlement became known as Brisbane Town, although officially it was designated Edinglassie. The third commandant, the Scots-born Captain Patrick Logan, who governed from late 1825, was known particularly for his rigidity and harshness: he was killed either by local Aborigines or convicts in 1830 (or possibly by one with the encouragement of the other). The settlement figured in a number of convict ballads, one of which, 'A Convict's Lament on the Death of Captain Logan', bids the tyrant an ironic and less than regretful farewell. Moreton Bay, like Hobart, was isolated from other settlements except by sea, and designed for the worst offenders among the convict population. However, while Hobart was cold and mountainous and austere, sub-tropical Brisbane was hilly, hot, wet and fecund. One hundred and sixty years after settlement, Brisbane-born writer David Malouf described its terrain as follows:

> The key colour is green, and of a particular density: the green of mangroves along the riverbanks, of Moreton Bay figs, of the big trees that are natives of this corner of Queensland, the shapely hoop-pines and bunyas that still dominate the skyline along every ridge. The Australian landscape here is not blue-grey, or grey-green or buff, as in so much of southern Australia; and the light isn't blond or even blue. It is a rich golden pink, and in the late afternoon the western hills and the great flat expanse of water that is the Bay create an effect I have seen in other places only before or after a storm. Everything glows from within. The greens become darkly luminous. The sky produces effects of light and cloud that are, to more sober eyes, almost vulgarly picturesque. But then, these are the sub-tropics. You are soon made aware here of a kind of moisture in the air that makes nature a force that isn't easily domesticated—everything grows too fast, too tall, it gets quickly out of control. Vegetation doesn't complement the man-made, it fiercely competes with it; gardens are always on the point of turning themselves into wilderness, hauling down fences, pushing sheds and outhouses over, making things look ramshackle and halfway to ruin. The weather, harsh sunlight, hard rain, adds to the process, stripping houses of their pliant, rotting timber, making the dwellings altogether less solid and substantial, on their high stumps, than the great native trees that surround them.[2]

A number of Brisbane's writers have claimed in the years before and since that just this climate and geography have made Queenslanders into a different sort of people from other Australians. For Thea Astley, also born in Queensland, it is Brisbane's distinctive wooden houses on stilts, described in her 1978 Blaiklock Lecture entitled 'Being a Queenslander: A Form of Literary and Geographical Conceit', which contribute to this uniqueness.

> The human race places great store on the outward trappings of conventional behaviour—or conformist behaviour. Almost from the first, Queenslanders made no attempt to reduplicate the architecture of their southern neighbours. Houses perched on stilts like teetering swamp birds, held stiff skirts all round, pulled a hat brim low over the eyes; and with the inroads of white-ants not only teetered but eventually flew away. And then, we tend to build houses so that we can live underneath them. Perhaps those stilts made southerners think of us as bayside-dwelling Papuans. Our dress too, has always been more casual. Our manners indifferent, laconic, in temperatures that can run at over ninety for weeks on end.[3]

And for David Malouf, Brisbane's topography is also distinctive.

> The first thing you notice about this city is the unevenness of the ground. Brisbane is hilly. Walk two hundred metres in almost any direction outside the central city (which has been levelled) and you get a view—a new view. It is all gullies and sudden vistas. Not long views down a street to the horizon—and I am thinking now of cities like Melbourne and Adelaide, or Manchester or Milan, those great flat cities where you look away down endless vistas and the mind is drawn to distance. Wherever the eye turns here it learns restlessness, and variety and possibility, as the body learns effort. Brisbane is a city that tires the legs and demands a certain sort of breath. It is not a city, I would want to say, that provokes contemplation, in which the mind moves out and loses itself in space. What it might provoke is drama, and a kind of intellectual play that delights in new and shifting views, and this because each new vista as it presents itself here is so intensely colourful.
> ... Now what you abstract from such a landscape, from its greenness, its fierce and damply sinister growth, its power compared with the flimsiness of the domestic architecture, its grandeur of colour and effect, its openness upwards to the sky—another consequence of all those hills—is something other, I would suggest, than what is abstracted from the wide, dry landscapes of Southern Australia that we sometimes think of as 'typical'. It offers a different notion of what the land might be, and relates it to all the daily business of life in a quite different way. It shapes in those who grow up there a different sensibility, a different cast of mind, creates a different sort of Australian.[4]

Perhaps it is partly this sense of geographical difference that has led a number of Australian writers, including Malouf, Jessica Anderson, Patrick White, Brian Penton, Rosa Praed and Thea Astley, to base historical novels on Brisbane and its environs.

*   *   *

This is the settlement, in its embryonic stages, that the Brisbane-born Anderson recreates in her historical novel *The Commandant* (1975), in which she describes Moreton Bay as it appears to Frances O'Beirne, Captain Logan's fictional young Irish sister-in-law as she arrives via Dunwich, on the mouth of the Brisbane River, by sailing ship from Sydney. The year is 1830, five years after the settlement was established, and Frances listens to the fussy attempts of an officious returning native, Amelia Bulwer, to prepare her for it.

> 'But Dunwich is only a depot. Don't judge us by a depot. Wait, my dear Miss O'Beirne, wait till we get to the settlement. Which with this wind—' Mrs Bulwer drew a hand from her muff and held it into the wind—'will be some five hours more. It's quite a pretty little place, I assure you. And healthy besides. None of us has gone to our graveyard. Not one. And only one soldier. Quite a contrast with the India stations. At least with Madras. I am sure the rumour that the fifty-seventh is to go to Madras is quite unfounded. Agra. It will be Agra. Agra is delightful. Very little fever at Agra. On the settlement we do have the fever, but not the India sort, not the sort to carry one off. And we have the ophthalmia and the dysentery, though neither is so prevalent with *us*. But of course we have nothing so dreadful as the cholera. The cholera! Do you know how bad it was at home last year? Why, of course you do, you were there.'
>
> Frances said, well, she had been in Ireland.

As they approach, amid vine-infested gigantic trees, the young girl feels an instinctive pity and horror at the convict system.

> Then a sudden bend in the river disclosed another kind of country: on one bank pleasant wooded hills, and on the other low fields swarming with men in yellow hoeing between rows of very young wheat. They were so close that Frances could hear the unrhythmic sounds of their shifting irons and the collapsing links of chain. Overseers, carrying heavy sticks, lumbered over the unsettled soil among them, and on the perimeter of the field moved red-coated soldiers, crosses of white webbing stark against their breasts, and bayonets shining and precise against field and sky. All the passengers on the cutter were watching them. Captain Clunie exclaimed at the area under cultivation, and Amelia Bulwer told him in a fast pleased voice of other fields, other crops. In the streets of Sydney Frances had seen iron gangs coming and going from barracks, but she had never before seen so many at once, and nor had she seen them at work. It was their great number perhaps or the clumsiness of their fettered movements that made them appear sub-human, like animals adapted to men's work or goblins from under the hill. She hated herself for her aversion from them, for the recoil of her spirit and the agitation of her heart. She was still standing between Louisa and Henry Cowper. 'Dear God,' she whispered, 'why must they look like that?'
>
> Both turned to look at her. She felt in their attitudes a kind of caution. 'Like what?' asked Henry Cowper.
>
> To spare them, or herself, she temporised. 'They are so—small.'

But to Anderson's imagination the settlement—which in 1830 incorporated some thousand convicts, one hundred soldiers, a surgeon and a chaplain and a school for thirty-three children—offers some unexpected paradoxes.

> On the land a loud bell began to ring. Frances, who knew the botanical gardens to be cultivated by

prisoners, and was prepared to find them hateful, was surprised by their peacefulness, their boskiness and glow. It was the first stage of dusk, when shadows deepen but the light grows for a while more intense. The banana trees, the citrus and figs, the grapevine and cane, all in separate plantations, covered the whole of the sloping bank. The mustered prisoners, half-hidden by foliage, were visible only as an undulating ridge of yellow, a colour that glowed as innocently, against the violet shadows and vivid greens, as the shaddocks and lemons nearer the shore. The glimpses of red—soldiers' coats—could have been the flowers of Rio, and when she saw, on the crest of the hill, a small octagonal cottage with a pointed roof, she gave a cry of pleasure, and Amelia Bulwer, watching her across the deck, nodded in vindication. The bell had stopped ringing.

Again they were rounding a long point defined by the windings of the river. The gardens lay on its eastern side, and when they left them behind, and came within sight of the western bank, Frances, like Amelia herself, brought up her hands and clapped them. The row of houses set in gardens, the smoking chimneys, the tall flagstaff and spirited fluttering flag, the barge crossing the river, the windmill on the hill, the cluster of people on the wharf, all this seemed to her the essence of homeliness and familiarity. A long line of yellow, flanked by spots of red, was moving in low billows of dust down the hill from the windmill, and more yellow, this time an irregular block, could be seen on the opposite side of the river, which the barge had now almost reached. But did not labourers all over the world walk home at dusk, and wait for ferries on river banks? To fortify Frances's impression that the place was much better than she had lately supposed came memories of Sligo, came her knowledge that free men and women, and their children too, could die of hunger while these men ate. For the man lying in the wheatfield the peasant in the Sligo ditch offered himself as counterpoise.[5]

Similarly, it is at Moreton Bay in this period that the exhausted and naked Ellen Roxborough arrives on foot after an ordeal by shipwreck, capture by Aborigines and rescue by an escaped convict in Patrick White's novel *A Fringe of Leaves* (1976). White, who visited Brisbane on the coastal steamer *Manoora* in 1961 to research the novel, which he then put aside for more than a decade, found Brisbane 'full of interesting ramshackle houses built on stilts' in which 'the people, just to meet them on the street, [seem] an amiable and extrovert race'.[6] In the novel, however, White chooses to convey the atmosphere of the town more by a brief description of its interior surfaces, one of which is contained in the house of Governor Logan (here called Lovell) and his family.

He received her standing in the centre of a room which might have impressed had she been more impressionable, and had she not suffered the same fate as the furniture, of covering great distances and ending up battered, scratched and dusty, though still with a hint of having enjoyed more pretentious circumstances. There was a smell of must from a worn, dust-impregnated carpet mingling with the scents of citrus and guava which strayed in from beyond the verandah.

A little later in the book it seems to White—or to Mrs Roxborough—that, while Hobart is 'morally infected' by convictism, here

... this bend in the brown river, with its steamy citrus plantation, garden beds too primly embroidered with marigolds and phlox, and beyond a hedge, cucurbits of giant proportions writhing on mattresses of silt, was designed for revelations of evil, as the low-built, rambling, deceptively hospitable official residence presided over by the fecund Mrs Lovell and her authoritarian spouse.[7]

A Fringe of Leaves is loosely based on the story of Mrs Eliza Fraser and the wreck of the *Stirling Castle* off the Swain River near Rockhampton in 1836: Captain Fraser and some other survivors reached Great Sandy Island (later Fraser Island) by longboat, but were almost all killed by Aborigines. Eliza Fraser and the remaining survivors were rescued after six weeks by soldiers of the 14th Regiment sent from Moreton Bay. White, who first heard the story from the painter Sidney Nolan, who had hitch-hiked to Fraser Island in 1947 and subsequently executed a series of paintings on the subject, might have also been influenced by the story of James Murrel, the sole survivor of the wreck of the *Peruvian*, lost near Port Denison in 1846. Murrel was found near the Burdekin River in 1893, and later published *Sketch of A Residence Among the Aboriginals of Northern Queensland For Seventeen Years* (1863).

By the early 1840s, when the pastoral hinterland of the Darling Downs was being opened up, the prison settlement, which in the previous few years had been used mainly as a depot for reconvicted women, was disbanded. The journalist Brian Penton in his novel of pastoral expansion *Landtakers* (1934) fictionally recreates the site of Brisbane in 1842, although this would seem to place it after the convicts had been withdrawn. Here the English newcomer Derek Cabell, intending to make his fortune on the land before returning to his native Dorset, surveys the 'ramshackle buildings' of the Moreton Bay penal settlement in the dry season.

> Red earth and blue sky met in the jagged line of a near horizon. In the middle of this vault stood the settlement—a prison within a prison. Shanties built of black bark twisted by the fierce sun, with crazy-shaped doors and glassless windows. Jail and barracks of stone. A yellow stone windmill. A long, dusty, empty street. Sheep, a few cows, pigs, wide patches of yellow Indian corn. At one side of the valley a river shimmered in the sunlight; at each end of the valley the bush. Into illimitable blue distance it faded, across unexplored mountains and plains, grey, motionless and silent.

The convicts present a discouraging sight, as do their military guards.

> A detachment of soldiers and yellow-clad convicts approached from the other end of the street as though upon air. Only the rattle of a chain here and there was to be heard, for the dust was inches thick and soft as powder. It rose in clouds from their feet and cast a smoky shadow on the ground.
>
> With undisguised contempt Cabell watched the detachment go by. There were men of all sizes, in every stage of decrepitude. Shuffling feet, round shoulders, faces prematurely aged by sun, hard work and under-nourishment. The soldiers' uniforms were unbuttoned and dirty. Dust and sweat mixed in the lines of their withered faces. Of the convicts few were unmarked by disease or mishap. The scarlet rash of poisoned blood covered their arms like long gloves. Black stumps of teeth showed through their lax mouths. Legs dragged heavily that had been broken and badly set. Hands lacked fingers. And bitten deeply into all, convicts and soldiers alike, was the pockmark of spirits desolated by ennui and despair.

But when Cabell returns in 1850 after some years in the hinterland, progress has been made.

> Settlement was rapidly becoming inadequate to describe broad, well-made streets, two-storeyed hostelries, houses built of dressed timber and painted, shops that sold perfumes, pomades, stays and wedding-rings in a mysterious annex from the saddlery, horse medicine and barrelled rum department. A boat arrived fortnightly from Sydney with news of the great world, and there was a *Moreton Bay Courier* to mirror the affairs of the little world held in the claws of two bends of the river, but busy with the comings and goings of a vast hinterland.
>
> Gone, for years now, the red coats and the canary jackets. A notice on the wall of the Colonial Stores to the effect that the latest composition of that distinguished and elevated authoress, Miss Maria Edgeworth, having been received hot from the press, perusal by members of the Moreton Bay Reading Circle would commence at eight p.m. sharp on the twenty-first instant at the commodious residence of Mrs Gribble, near the Colonial Stores, attested to the dawn of a new era of civilising influences, which revealed themselves also in the offer of betrousered abos to recite the 'Lorsprer' for a consideration.[8]

Born in Ascot, Brisbane, in 1904, Penton worked briefly on the *Brisbane Courier* before leaving for

*Our great-grandfathers had put together a lost,
unknown home in landscapes that made it all perfectly
natural: Georgian houses with classical porticoes; hop fields
and orchards; chimney pots rising on gentle hillslopes,
in the subtle, muted lights of East Anglia.*

C.J. KOCH, 1985 (PAGE 79)

RICHMOND BRIDGE, TASMANIA. PHOTOGRAPHER: DAVID MOORE

*Now what you abstract from such a landscape, from its
greenness, its fierce and damply sinister growth, its power
compared with the flimsiness of the domestic architecture, its
grandeur of colour and effect, its openness upwards to the
sky—another consequence of all those hills—is something
other, I would suggest, than what is abstracted from the
wide, dry landscapes of Southern Australia that
we sometimes think of as typical.*

DAVID MALOUF, 1984 (PAGE 93)

BRISBANE, QUEENSLAND. PHOTOGRAPHER: DAVID MOORE

England, after which he worked in Sydney as a political journalist on the *Sydney Morning Herald* until 1927, while the new Federal parliament was being established. After another sojourn in England he joined the *Daily Telegraph* in Sydney, editing it for the ten years prior to his death in 1951. His novels *Landtakers* (1934) and its sequel *The Inheritors* (1936), the first two parts of a never-completed trilogy based on contemporary accounts and journals, give a graphic but cynical account of pioneering life in Queensland from the early 1840s to the 1890s. *Landtakers*, like *The Fortunes of Richard Mahony*, has the theme of the emigrant who comes temporarily to Australia to make his fortune, but who is so irrevocably transformed by the landscape and his experience of it that return is impossible.

Edmond Marie Marin La Meslée, private secretary to the French consul-general, who had settled in Australia in 1876 and travelled extensively on the east coast in the 1880s, described Brisbane as it appeared to him in 1883 in his non-fiction account *L'Australie Nouvelle*, published in France in 1883.

On the third day out of Sydney, wrote La Meslée, the *Alexandra* steamed into Brisbane River through low-lying country, thickly wooded with mangroves and lianas, where in areas of cleared land the occasional free selector's hut stood surrounded by banana trees.

> The river wound along through the rich alluvial flats in which it had dug its bed: not until its junction with Breakfast Creek did the first signs of rising ground appear. On a little knoll where the two streams joined stood a charming country house, built in the colonial style with wide verandas on all four sides. Magnificent green lawns stretched right down to the water's edge and a pleasure boat was moored to a little stone pier. Two small children, a boy and a girl, gambolled about on the grass or let themselves tumble down it together to the foot of the hill. From here on to Brisbane the river's course became more tortuous: as it was only two hundred yards wide at the most, our experienced officer needed all his skill to pilot our 800 ton steamer safely to the Company's wharf, where we tied up at about eleven o'clock in the morning. A cab took us to the Imperial Hotel in George Street, where a good English-style dinner awaited us.

There were good reasons why Brisbane would never become as large and prosperous as Sydney and Melbourne, La Meslée concluded.

> Standing on a bend in the river that bears its name, about thirty miles from the sea, it cannot be reached by large modern ships. The depth and breadth of the channel are not great enough to accommodate vessels of more than about twelve hundred tons. Consequently port charges are high, for sailing-ships must be towed up the river from Moreton Bay and big steamships have to unload their cargo at the river's mouth, whence it is carried to Brisbane by lighters.

In the intervening years Queensland, taking its name from Queen Victoria, had in 1859 become a separate colony from New South Wales, and in 1864 the area bounded by Queen, George, Elizabeth and Albert streets were destroyed by fire. Now the city contained some 35,000 people, with suburbs stretching into the surrounding hills, and a thriving business centre (in which, La Meslée noted, as in other Australian cities of the period, no one lived).

> The city's main thoroughfare, Queen Street, is already becoming rather crowded with business activity, and the shops are gradually invading the parallel streets nearby. In years to come the business quarter of the city will cover the whole areas between the Botanical Gardens and the foot of Spring Hill. Some people are already predicting that the beautiful Gardens will have to be destroyed, in order to widen the river and build wharves in their place. It is to be hoped that the economic benefits to Brisbane, from such an act of vandalism, would be small enough to prevent its being committed. During our stay of nearly six weeks in Brisbane we often visited the Gardens, and it was always a fresh delight to walk along

the great avenue of *bunya-bunyas* that borders the river. There too I met again those old acquaintances from Mauritius, flame-trees, with their exotic scarlet flowers that for a whole season of the year replace the leaves, and create such a beautiful effect among the surrounding greenery.

Unfortunately, the Frenchman was ultimately unimpressed with the entertainments of the city, and left by an American-style paddle steamer called the *Emu* to explore the landscape further upriver.

In any city, except Paris, Rome or some other great metropolis with innumerable attractions, a visitor with no great business to attend to cannot fail to be bored after quite a short time. Brisbane is quite a pretty town, it is true, but it bores you to death.[9]

By the early 1890s, the period in which Rosa Praed set her romantic novel *Lady Bridget in the Never Never Land* (1915), the city was well established. In the first pages of the novel Praed recreates Brisbane as seen from the suburb of Kangaroo Point on the south of the river, where Praed herself once lived. Praed refers to Brisbane as 'Leich[h]ardt's Town' after the explorer who led an expedition from here to Port Essington (or Darwin) in 1844-5, and subsequently vanished on a second expedition to the Western Australian coast.

Mrs Gildea had settled early to her morning's work in what she called the veranda-study of her cottage in Leichardt's Town. It was a primitive cottage of the old style, standing in a garden and built on the cliff— the Emu Point side—overlooking the broad Leichardt River. The veranda, quite twelve feet wide, ran— Australian fashion—along the front of the cottage, except for the two closed-in ends forming, one a bathroom and the other a kind of store closet. Being raised a few feet above the ground, the veranda was enclosed by a wooden railing, and this and the supporting posts were twined with creepers that must have been planted at least thirty years. One of these, a stephanotis, showed masses of white bloom, which Joan Gildea casually reflected would have fetched a pretty sum in Covent Garden, and, joining in with a fine-growing asparagus fern, formed an arch over the entrance steps. The end of the veranda, where Mrs Gildea had established herself with her type-writer and paraphenalia of literary work, was screened by a thick-stemmed grape-vine, which made a dapple of shadow and sunshine upon the boarded floor. Some bunches of late grapes—it was the very beginning of March—hung upon the vine, and, at the other end of the veranda, grew a passion creeper, its great purple fruit looking like huge plums amidst its vivid green leaves.... The heat-haze over the town and the brilliant sun-sparkles on the river suggested a cruel glare outside the shady veranda and overgrown old garden.[10]

In February 1893 the Brisbane River, which was prone to flooding, rose nine and a half metres twice in three weeks and caused much loss of life and damage to property. This was an event that, coming after the great fire of 1864, no doubt contributed to the unflattering description of the city by the travelling socialist reformers Beatrice and Sidney Webb. The Webbs, who visited in 1898 and stayed at the pleasantly stately Bellevue Hotel, since demolished by developers, found Brisbane 'a pretentious little place: blocks of buildings run up in good times, now half-tenanted and badly kept …'.

In January 1974 a similar flood made thousands homeless, and probably inspired David Malouf's account of a flood in the same period in his early and most famous novel of Brisbane, *Johnno* (1975).

But each night after work, with the bridge lamps casting their yellow glare far into the sky, crowds gathered on the footwalks of the bridge and on the high embankment along North Quay to see the river come swirling down between the iron pylons of the bridge and to point out to one another the strange cargo it carried: huge tree-trunks that strained and splintered where they struck, chicken-coops, water tanks, butter-boxes, even sometimes an odd piece of furniture, a genoa velvet lounge-chair, for example, that bucked about on the surface of the water like the Tilt-a-Whirl at the National Show. And other things even more wonderful to city eyes: dead cattle with their feet in the air; great islands of waterlilies where the field creatures, bushrats and lizards, swarmed as on a raft. And leaving the river swollen and brown for days afterwards, whole acres of rich Brisbane Valley topsoil. A farmer standing here might have seen two or three of his best paddocks go past. The river, usually placid enough with its rainbow-slick of oil and its bubbles of ferment popping in the heat, boiled up now into lighted peaks like the sea, and its roar could be heard from tramstops half a block away. Twice daily, with the tides, it rose up through the drains into low-lying suburbs and left its ripple mark on the walls. People went out in rowing boats to see a dressing-table drawer full of stinking mud or a dozen catfish gasping in a bath. It was a month of wonders.[11]

Added to idiosyncrasies of climate, vegetation and topography, a certain precariousness to their existence might also be said to be a formative influence in the lives of Queenslanders.

THE CITY IN THE HOUSE

The houses are of timber, that is the essence of the thing, and to live with timber is to live with a material that yields at every step. The house is a living presence as a stone house never can be, responding to temperature in all its joists and floorboards, creaking, allowing you to follow every step sometimes, in every room. Imagine an old staircase and magnify its physical presence till it becomes a whole dwelling.

Children discover, among their first sensual experiences in the world of touch, the feel of tongue-and-groove boards, the soft places where they have rotted, the way paint flakes and the wood underneath will release sometimes, if you press it, a trickle of spicy reddish dust. ...

You learn in such houses to listen. You build up a map of the house in sound, that allows you to know exactly where everyone is and to predict approaches. You also learn what not to hear, what is not-to-be-heard, because it is a condition of such houses that everything can be heard. Strict conventions exist about what should be listened to and these soon become habits of not-listening, not-hearing. So too, habits grow up of not-seeing.

Wooden houses in Brisbane are open. That is, they often have no doors, and one of the conventions of the place (how it came about might be a study in itself) is that doors, for the most part, are not closed. Maybe it is a result of the weather. Maybe it has something to do with the insistence that life as it is lived up here has no secrets—or should have none. Though it does of course. ...

So there it is, this odd timber structure, often decorated with wooden fretwork and scrolls of great fantasy, raised on tree-stumps to leaf level and still having about it some quality of the tree—a kind of tree-house expanded. At the centre a nest of rooms, all opening on to a hallway that as often as not runs straight through from front to back, so that when you step up to the front door of the house you can see right through it to trees or sky. Around the nest of rooms, verandahs, mostly with crossed openwork below and lattice or rolled venetians above; an intermediary space between the house proper, which is itself only half closed in, and the world outside—garden, street, weather.

David Malouf, 'A First Place: The Mapping of a World', Herbert Blaiklock Memorial Lecture, 1984[1]

Queensland houses and their distinctive architecture have had a significant effect on the work of David Malouf. The writer, born in Brisbane in 1934, lived initially at 12 Edmondstone Street, South Brisbane, where his English-Lebanese family owned a shop on the corner of Edmondstone and Melbourne streets. Subsequently his parents moved to Arran Avenue in the north eastern suburb of Hamilton, where many of the town's middle class had moved to higher ground after the 1893 flood, and where some of Brisbane's finest houses were built. In his first novel *Johnno* (1975), in an apparently autobiographical passage, Malouf causes his narrator Dante to describe just such a childhood shift.

It wasn't the house of my childhood. We had moved there in 1947 when my father built the place, huge, ugly, show-offish, after his own design. I had never really cared for it. My memories were all of our old house in South Brisbane, with its wide latticed verandahs, its damp mysterious storerooms where sacks of potatoes and salt had been kept in the ever-dark, its washtubs and copper boiler under the porch, its vast garden that ran right through to the street behind, a wilderness that my grandfather before he died, had transformed into a suburban farmlet, with rows of spinach, tomatoes, lettuce, egg-plants, a shed where onions and garlic hung from rafters, and a wire coop full of fowls. The new house at Hamilton was stuffily and pretentiously over-furnished and depressingly modern. It represented an aspect of my father, of his earliest ambitions perhaps, that I had never understood, some vision of worldly success and splendour that I could find no model for. Victorian armchairs covered with French velvet, bevelled glass mirrors, brocade curtains, chandeliers. The only thing I could connect it with was a set of raw silk bed-covers that he had penpainted for my mother's glory chest. Blazoned all over with red and yellow poppies, the oil paint crusty, the oil seeping into the material with a brownish stain, these objects had always impressed me more with their gaudy opulence and seemed all the more extraordinary because my father had painted them at twenty when he was a member of Brisbane's toughest rugby push.

Now as I began to sort through his 'effects' it occurred to me how little I had really known him.

Dante (and Malouf?) disparagingly explores the accident of fate that leaves him living in Arran Avenue, Brisbane.

Brisbane is so sleepy, so slatternly, so sprawlingly unlovely! I have taken to wandering about after school looking for one simple object in it that might be romantic, or appalling even, but there is nothing. It is simply the most ordinary place in the world.

Arran Avenue, Hamilton, Brisbane, Queensland …

Queensland, of course, is a joke.

Malouf gives detailed descriptions of the Arran Avenue house, and by extension, many Queensland houses.

They were all enormous those houses. Huge one-storeyed weatherboard mansions that had been intended for more spacious days, and for larger families than we could manage, they were only half lived-in nowadays. Every house had its row of locked bedrooms on one side of the hall. You could look into them from long sash windows on the verandah, and believe (as I was told often enough) that people had died there—grandmothers, little brothers from scarlet fever or whooping-cough, bed-ridden uncles from injuries they had received in the First World War. The high beds had brass ends with superbly polished finials and little rows of porcelain balusters. Lace curtains, a lace coverlet and bolster, a washstand with doilies and a floral jug-and-basin. And often as not, as in my grandmother's house, a Sacred Heart of Jesus over the bed, and on shelves of the dressing table a whole series of extravagant saints among artificial flowers and candles. The kitchens were tiled, with walk-in pantries and an old wood range (for baking) beside a newer gas stove, perhaps an Early Kooka like ours, with its legs in tins of water to keep off ants. One huge room, always at the centre of the house, always darkly panelled and with a picture rail, was never opened except to visitors. Its curtains were kept drawn to preserve the carpets and the genoa velvet lounge chairs from the sun; there were chromium smokers' stands and brass jardinieres full of gladioli; on a heavy sideboard, cut-glass decanters of whisky, brandy, port; and a big central lamp-shade of silk brocade, with tassels, that gave a smoky gold light.[2]

In *12 Edmondstone Street* (1985), his later collection of autobiographical pieces, Malouf lovingly explores the earlier house in a long, finely observed passage which recalls Hal Porter's similar impeccable memory for childhood detail, and George Johnston's similar ability to reconstruct urban and family history from the features of a domestic interior. Here, Malouf documents the house as social indicator.

Like most people in those days, my father was ashamed of our house. He would have preferred a modern one made of brick. Weatherboard was too close to beginnings, to a dependence on what was merely local and near to hand rather than expensively imported. It was native, provincial, poverty-stricken—poor white. Real cities, as everyone knows, are made to last. They have foundations set firm in the earth. Weatherboard cities float above it on blocks or stumps. Weatherboard houses can be lifted if necessary, loaded on to the back of a lorry and set down again two suburbs or a thousand miles away. They have about them the improvised air of tree houses. Airy, open, often with no doors between the rooms, they are on such easy terms with breezes, with the thick foliage they break into at window level, with the lives of possums and flying-foxes, that living in them, barefoot for the most part, is like living in a reorganised forest. The creak of timber as the day's heat seeps away, the gradual adjustment in all its parts, like a faint instrument being tuned, of the house-frame on its stumps, is a condition of life that goes deep into consciousness. It makes the timberhouse-dweller, among the domesticated, a distinct species, somewhere between bushie and brick-and-mortar man.

As for verandahs. Well, their evocation of the raised tent flap gives the game away completely. They are a formal concession that you are just one step up from nomads.

The characteristic Queensland verandah forms a neutral ground between the interior and the outside world.

A verandah is not part of the house. Even a child knows this. It is what allows travelling salesmen, with one foot on the step to heave their cases over the threshold and show their wares with no embarrassment on either side, no sense of privacy violated. It has allowed my mother, with her strict notion of the forms, to bring a perfect stranger in off the street and settle her (for ever as it happens) in one of our squatter's chairs. Verandahs are no-man's-land, border zones that keep contact with the house and its activities on one face but are open on the other to the street, the night and all the vast, unknown areas beyond.

As in the southern capital cities, the pre-war suburban existence Malouf describes is determinedly Anglocentric, but in the sultry climate of Queensland this 'Englishness' becomes a test of moral courage.

The larger section is our dining-room. The smaller, beyond the arch, is Cassie's cooking and washing-up place. Here, on an Early Kooka gas-stove with a laughing jackass on the front, she prepares the enormous meals of those days, meals that defy latitude and the facts of climate and weather by reproducing the baked dinners, stews, hot-pots and boiled puddings of the Mother Country (our mother's country), which we continue to consume, after more than a century, as if a hundred degrees of humidity constituted a strictly moral challenge, and we had our real existence in a cold place on the other side of the globe. Physical bodies and the actual have nothing to do with it. In a properly British way we ignore them, as we ignore the view from the window on to a backyard that dazzles in sunlight, steams after rain, and is choked with tropical weeds out of which cannas burst in scarlet and golden flames.

What we are feeding when, at fixed hours—breakfast at seven, dinner at noon, tea at six-thirty—we assemble behind serviette-rings initialled with our names, are the spirits of the fathers. We are paying tribute to origins—even those of us whose origins are of another kind.[3]

Jessica Anderson, who was born in Brisbane in 1916 and lived in semi-rural Rocklea until her parents moved to 56 Villa Street in suburban Annerley in 1920, similarly sets one of her finest novels, *Tirra Lirra By the River* (1978), largely in Brisbane. Anderson's elderly protagonist Nora Roche returns to the Queensland house—'taken from life', according to the author's prefatory note—in which she spent her childhood. This house, says Anderson, is based on the turn-of-the-century 'Queenslander' houses built on the river flats at Yerongpilly near Annerley, which are approached by steeply angled wooden steps.[4]

The front stairs are just as I visualised them on the plane, fourteen planks spanning air, like a broad ladder propped against the verandah. ... As I follow the man across the verandah I hear my own footsteps, like a small calf on a quaking bridge ... the real house, a heavy wooden box stuck twelve feet in the air on posts.[5]

And again, in an autobiographical story called 'Under the House' from her later collection *Stories From the Warm Zone* (1987), Anderson recreates a World War I childhood in a house on Moolabin Creek (which she calls 'Mooloolabin') near Rocklea, in which she stresses the importance of 'under-the-house'.

I could never go alone into the under-the-house at Mooloolabin without an uneasiness, a dogged little depression. Unless it was raining, no lines of washing hung there, and nor did my father use that space for his workbench, as he would do in the suburban house to which we were soon to move, for at Mooloolabin all such needs were filled by the Old Barn, the first shelter my father's parents had put up on their arrival with their family from Ireland.

So, in the under-the-house at Mooloolabin, there was no extension of the busy house above except for the meat safe hanging from a rafter, the boxes of wood cut for the stove, and the tins of kerosene used for the lamps. These objects, dull and grey in themselves, left dominant to my eyes the sterile dust at my feet, the rows of tall sombre posts with blackened bases, and the dark vertical slats splintering the sunlight outside. Broken cobwebby flowerpots were piled in one corner. From a nail in a post hung the studded collar of the dog Sancho, who had had to be shot, and from another hung the leg irons dug up by my grandfather, relic of 'some poor fellow' from the days when Brisbane was a penal colony.[6]

*   *   *

David Malouf recreates wartime Brisbane both in *Johnno* and in his fifth novel *Harland's Half Acre* (1984), a family saga that encompasses, as well as the life of an artist, much of the historical development of Queensland. Brisbane was the American general Douglas MacArthur's headquarters in Australia, and because of its northerly position was felt to be in much greater danger of Japanese bombing than the southern capitals. It was also the location of the 'Brisbane Line', the point at which, in the event of invasion, the Australian Government felt that territory further north would have to be abandoned to the enemy. In *Johnno*:

The Japanese struck Pearl Harbour, and my father took me one brilliant morning to see the first American warships come grandly upriver and swing at anchor off Newstead Park. What I remember is the whiteness of the sailors' uniforms as they stood in dazzling rows on the deck, and the light of that moment floods all the years ahead. There was greater danger of course. But danger is open and easy to deal with. Better any day than dread. The war, now that it was with us, turned out to be quite an exciting affair. A bit frightening at times, but mostly comic and commonplace.

We were given air-raid kits that we took to school with us. They contained rubber mouthpieces that we were to bite on during the raids and rolls of bandage that got used almost as soon as they were given out for bloody knees. My father was made Senior A.R.P. Warden for South Brisbane. He wore a white helmet, a red felt armband, had a gasmask and rattle, and went out each evening after dark to inspect the blackout. … Brisbane was suddenly at the centre of things. Though we hardly knew it at the time, our city was having its moment of greatness, its encounter with History: General MacArthur had arrived and the whole Pacific campaign was being directed from his office in the A.M.P. building.

All night now the troop transports rumbled past our house, and in the early dusk, with mosquitoes beginning to dance under the bushes and flying foxes in the mango trees tearing into the pulpy fruit, I sat out on our lawn away from the sprinkler and counted them. They went on long after I had run out of numbers, and all through tea and the radio serials I listened to afterwards, and went droning on in my sleep. Neighbours began to evacuate to places like Coonabarabran, and the big houses along the park, where I used play in the afternoon, were boarded up with chains on the gate or turned over to the Yanks. When my father decided we should stay put our house was fortified with sandbags and workmen came to dig a trench in our tennis court.

… But our one-storeyed weatherboard wasn't the only one to be fortified. The whole city had taken on the aspect of an armed camp, and there were rumours that when the Japanese landed the whole country to the north would be scorched and abandoned in the Russian manner and a last stand made at Brisbane. We were suddenly in the front line. Concrete pill-boxes appeared in the streets and became places where people 'did things' after school, or where children who took sweets from strangers were discovered with their heads cut off—victims (now that all the swaggies and metho drinkers had been drafted) of the negroes who congregated round the Trocadero in Melbourne Street and the brothels along the south side of the bridge. Troop transports rumbled day and night, ferrying soldiers from the Interstate Station, where the New South Wales line ended, to Roma Street, where they would board one of the slow narrow-gauge lines to the north. Anti-aircraft guns were set up on the city's high places and the sky at night was crisscrossed with the shafts of giant searchlights, moving pale among the clouds, creating in the blackout a ghostly reflected light that you could actually read by if you half-opened the venetians. Our sleepy sub-tropical town, with its feathery palm trees and its miles of sprawling weatherboard, was on the news-reels. It was the gateway to that part of The War that was raging all over the islands now, just a thousand miles away. Brisbane had, for a time, the heady atmosphere of a last stopping place before the unknown, and there were service clubs, canteens, big dancehalls like Cloudland and the Troc where girls who might otherwise have been teaching Sunday school were encouraged by the movies they had seen, the hysteria of the times, the words of sentimental Tin Pan Alley tunes, and the mock moonbeams of many-faceted glass ball that revolved slowly in the ceilings of darkened ballrooms, to give the boys 'something to remember' before they were mustered (forever perhaps) into the dawn.[7]

In *Harland's Half Acre*, however, Brisbane, seen through the eyes of an artist, has a texture both more immediate and more adult.

Once, walking home late, he had slipped into a laneway here to take a leak and had come upon two figures fucking in the rain. A soldier in a long army greatcoat had a woman against the wall, with her feet in the greatcoat pockets and her bare thighs damp with moonlight. They moved slowly like figures in a dream, making a single creature with two locked and moaning heads, a mythological beast to which he couldn't have given a name, born out of the times, the war, as evidenced by the rough woollen material of the greatcoat with raindrops on its hairs.

It was a street of intense casual encounters and farewells. The high wall down one side, with its giant billboards, was the Interstate station—you could hear the crashing and clashing of carriages, and later, in the city stillness, the shunting of engines. The other side of the street, after the Trocadero dance-hall, was terraced houses approached by steep stone steps with railings: all brothels. Beyond, in a newer building on a corner, the junk shop.[8]

David Malouf, along with other Queensland writers such as Thea Astley, has emphasised the sensation of *difference* experienced by those growing up in Queensland. However, inherent in this observation is the implication that to become fully conscious of such a difference, it is necessary to experience life elsewhere to make a comparison. Malouf left Australia in 1959, aged twenty-four, to live in England and Europe. *Johnno* appeared seven years after his return to Australia in 1968. Anderson, except for a few years in London, spent her later life in Sydney, where she began her fiction-writing career. Like the distinctively Western Australian writers Elizabeth Jolley (who came in later life to Western Australia) and Randolph Stow (who as an adult lived in England), Anderson and Malouf wrote their 'distinctively Queensland' books after experiencing life elsewhere.

Tony Maniaty, born in Brisbane in 1949 to a Greek father and Irish mother, takes up Malouf's mantle in an autobiographical piece, 'All Over the Shop', extracted from his novel *Smyrna* (1990). This describes growing up in the late 1950s and 1960s in the 'peculiar sub-tropical world of my grandfather's house' and 'a succession of corner shops' including the Astoria cafe, where his father ran the kitchen— 'steak and chips, banana sundaes, nothing Greek at all …' —from which he was already planning to escape.

Brisbane, 1957. In those halcyon late fifties, we had still not discovered Bob Menzies and the Cold War and never would; for us, Vietnam was waiting. In the interregnum, we studied British coalfields at school and thought of Australia itself as curiously remote: as a distant place. We bought glass scientific apparatus with our pocket-money, and launched a rubber meteorological balloon with the tag: 'Please return to Maniaty's shop, at the ferry'. One day a man brought it back, strangely; without a word. Only a thin wall separated the public from our semi-private lives, which were forever being interrupted and consumed by the public. 'Shop!' my father would yell, or Mum; or both, and then I would serve too. You handled money all day and saw where it came from, and learned crudely about economics—the overdraft, the day's takings, the 'till' itself took on iconic qualities in my mind; and a mysterious man called Stelmak, who seemed to finance us, was mentioned rarely by my father with a swallowing in his throat, like an executioner. Is he still alive today, this man who lived off the loan interest of shopkeepers like us?

All this happened and grew in the fifties and sixties—their heyday, really—when Dad built the wooden cart for deliveries, then bought a car (a Holden, deep red and white) about which a dry old woman, after buying a packet of 'Craven A' every night, would hand over her shilling and say: 'That's for my hubcap …' If they needed us, they owned us too, twelve hours a day and seven days a week. Christmas and Good Fridays off.[9]

In *Tirra Lirra By the River*, Jessica Anderson's Nora Roche eventually leaves the city to escape the constrictions of small town life and a mean-spirited husband. In the post-World War II Brisbane of David Malouf's *Johnno*, Dante's friend Johnno, along with Dante, will also find it necessary to leave.

'What a place!' Johnno would snarl, exasperated by the dust and the packed heat of an afternoon when even the glossy black mynah birds, picking about between the roots of the Moreton Bay figs, were too dispirited to dart out of the way of his boot. 'This must be the bloody arsehole of the universe!'

And I had to admit then that it was difficult to see how anything could be made of Brisbane. … Nothing seemed permanent here. Brisbane was a huge shanty-town, set down in the middle of nowhere. I was reminded sometimes of ghost-towns in the north that had once had a population of twenty thousand souls and were now completely deserted—the houses one morning simply lifted down from their stumps, loaded on to the back of a lorry, and carted away to create another town a hundred miles off. In my childhood I had often seen houses being carried through the streets, creaking and swaying on the back of a truck. It wouldn't have surprised anyone, I think, to wake up one morning and find that Brisbane too had died overnight. Its corrugated iron would be sold off for scrap. The weatherboard houses would rot in the damp, be carted away, or fall victim to the voraciousness of white ants. Animals would nest in upturned water tanks.

And who, Johnno asked, would know the difference? Brisbane was nothing: a city that blew neither hot nor cold, a place where nothing happened, and where nothing ever would happen, because it had no soul. People suffered here without significance. It was too mediocre even to a province of hell. It would have defeated even Baudelaire! A place where poetry could never occur.

Perhaps.

Like Anderson's Nora Roche, Malouf's Dante later returns to a Brisbane (in this case of the 1960s) which is much changed.

If Johnno had intended us somehow to revive the exploits of our youth, Brisbane itself had taken measures to prevent us. … The brothels … were gone—closed by the new government as part of a campaign to destroy the city's reputation as a tropical backwater, sluggish, colonial, degenerate, and force it into the present. The menagerie in the Gardens had been removed at the same time, and the unfortunate animals, a few scrawny monkeys, a demented ape, some moth-eaten wallabies, and several cages of parakeets and lovebirds, had been carted off and exterminated. Their cages were burned and the gardens reorganized to make pretty walks, with lily ponds and a cascade. …

It was the same all over. The sprawling weatherboard city we had grown up in was being torn down at last to make way for something grander and more solid. Old pubs like the Treasury, with their wooden verandahs hung with ferns, were unrecognizable now behind glazed brick facades. Whole blocks in the inner city had been excavated to make carparks, and there would eventually be open concrete squares filled with potted palms, where people could sit about in Brisbane's blazing sun. Even Victoria Bridge was doomed. There were plans for a new bridge fifty yards upstream, and the old blue-grey metal structure was closed to heavy traffic, publicly unsafe. There would eventually be freeways along both banks of the river that would remove forever the sweetish stench of the mangroves that festered here, putting their roots down in the mud; the old boathouse where we had gone to dances was burnt-out and the pontoon where Johnno had swum that night during the flood had been dismantled and taken for scrap. Huge pieces of earth-moving equipment and cranes with iron-jawed buckets and hooks presided in the moonlight over dirt piles that seemed more extensive in some parts of the town than what was still standing.

It is a sobering thing, at just thirty, to have outlived the landmarks of your youth. And to have them go, not in some violent cataclysm, an act of God, or under the fury of bombardment, but in the quiet way of our generation: by council ordinance and by-law; through shady land deals; in the name of order, and progress, and in contempt (or is it small-town embarrassment?) of all that is untidy and shabbily individual. Brisbane was on the way to becoming a minor metropolis. In ten years it would look impressively like everywhere else.

When the narrator of *Johnno* returns in still later years, however, the natural textures of Queensland life seem to come freshly alive as he surveys them with the newly awakened sense of the returning native.

It was September, and the roughstone terraces with their thickets of tiny white daisies were aswarm with insects. The whole garden sizzled and hummed. Big slow-flying grasshoppers, so heavy they could barely stay airborne, barged across the lawn or lofted over a wall to the hibiscus. The air glittered, and bees were busy in the cups of creepers that were just bursting into flower, cascading over a trellis or choking a fence. Occasionally one of the local cats strolled through on its way to the waste ground next door and sniffed about for scraps; or a big waterbird floated in from the mangroves downriver and perched for a moment on a dahlia stake. Once I saw a good-sized goanna. Deserted for just a fortnight, my father's garden was already half wild. The darkness under the thickening boughs was alive with midges and heavy with the smell of rotting vegetation, jungle-damp and sickeningly sweet.

Upstairs, in the afternoon stillness, I worked through my mother's linen press and moved on to the spare room at the back of the house that had once been mine.[10]

By the mid-1980s, when Peter Corris's private detective Cliff Hardy visits Brisbane in pursuit of a missing person, he finds a city that seems to have settled into a characteristic warm amiability recognisable to anyone who has visited it—and there are still at least a few of the wooden houses on stilts.

I like Brisbane: I like the warm air and the houses up on stilts and the suburban gardens that are like small jungles. I hired a yellow Ford Laser at the airport, which was the least gaudy car they had. It had nothing on the clock but its springs were shot: it had a Brisbane street directory though, and at that time of night I was glad to be hiring that as well as the springs. I drove to the address I had for Chris Guthrie in the suburb of Paddington.

The same things have happened to Paddington, Brisbane, as have happened to Paddington, Sydney (and Paddington, London, for all I know): it's an inner-city suburb, once intensely working class, now saved or ruined by a middle class invasion, depending on your point of view. Unlike a lot of Brisbane, it is hilly and I noticed an encouraging number of pubs while I got myself lost in the dark, leafy streets. There had been some rain and the gardens gave off a moist, lush smell that would have gone better with the growling of tigers than the barking of dogs, which was what I got as I stumbled around looking for numbers on fence posts.

It was after midnight when I found the house: it was set high up on stilts with a lot of discarded furniture and machinery quietly mouldering and rusting beneath it. The garden was overgrown and fragrant with the wet, night smell. No dog. I pushed through the undergrowth and went up a set of rickety steps to a wide verandah.[11]

'A PATCH OF COASTLINE':
The Queensland Towns

Let me draw you a little map.

Take a patch of coastline and its hinterland, put it just north of twenty and one hundred and forty-six east, make it hot and wet and sprinkle it with people who feel they've been forgotten by the rest of the country—and don't really care. Where there aren't hills and unswimmable water, plant cane. There's this largish place called Reeftown on the coast and in the purple hills behind there are smaller towns that grow tobacco and maize and stories that ripen and wither and repeat themselves as cautions against being human. Human! Ah! There's the rub! It's not the dreaming that matters, as the poet man insisted. He couldn't have been more wrong. It's the reality that rubs. And rubs. And rubs.

Everything is very green here. Very blue and very green, and the depth of its coloration whacks out this response, not only from me but from the rest of us, who, having chosen, ripen and wither and repeat ourselves in stories. Which are re-lived by others. Over. Over. Maybe it's only a second-rate Eden with its rain-forest and waterfalls, its mountain-climbing burrower of a railway and sea-bitten rind of coast—a kind of limbo for those who've lost direction and have pitched a last-stand tent.

Take me.

Let me draw you a little map.

Thea Astley, *Hunting the Wild Pineapple*, 1979[1]

Outside Brisbane, Queensland provides a variety of terrain ranging from tropical rainforest to the anthill plains of the inland cattle country, to remote mining settlements and coral islands, which writers as diverse as Xavier Herbert, E.J. Banfield, Ernestine Hill and John Blight have made their territory. A string of coastal settlements—Rockhampton, Mackay, Townsville, Cairns, Cooktown—accessible only by ship until the railway was built in the 1860s and 1870s, dot the sugar cane-growing and small farming areas of the fertile coastal plain.

The vast pastoral lands to the northwest of Brisbane, newly opened up in the 1840s, as well as attracting settlers to Queensland became like a giant backyard to the squatters and entrepreneurs of New South Wales and Victoria, who often took up secondary pasturelands here. The Palmer River gold rush of the 1870s brought population to the far north of the state, and the frontier atmosphere of the remote Queensland towns, picturesquely called Roma and Birdsville and Julia Creek, and the exotically named rivers such as the Barcoo and Burdekin, Condamine and Maranoa, cast a shadow of romanticism down through the early literature of the state.

*   *   *

In 1885 Edmond Marin La Meslée visited Limestone Hill—or Ipswich—settled when limestone for building was first discovered there by Patrick Logan in 1827, and an important river port before the railway was built. After traversing eighty kilometres of the meandering Brisbane and Bremer rivers by paddle steamer to reach it, La Meslée encountered an early settler, a successful pioneer, who even then took delight in recounting tales of the 'bad old days' when Ipswich was a frontier town.

These old inhabitants of Ipswich had been in the country for thirty-five years. They had come out at a time when the district was peopled only by fierce, man-eating tribes of Aborigines who have since disappeared. The old man told us how one day the cannibals attacked the village, then inhabited by only about twenty settlers, who defended themselves energetically. After a struggle that raged throughout the whole day, the miscreants succeeded in taking prisoner a very fat settler. Many others perished at the hand of the savages during the fight.

On the evening of the battle, the blacks took their prisoner to the top of the hill where Ipswich Grammar School now stands. They killed, roasted and ate him, in full view of the horrified colonists who could do nothing to save him.

'We saw them dancing round their victim,' the old Irishman told us. 'We heard them chanting their savage yells, as they argued with each other over the tit-bits. But what could we do? There were four hundred of them, and only ten of us left.'

Recounting the story, with the enthusiastic corroboration of his wife, seemed to excite the old man; and he appeared to enjoy telling us about his early struggles, and painting a picture of the success which had crowned them.

'I came here a full thirty-five years ago, not worth a penny, and for the last twenty years I have been married to the good mate you see before you,' he added, pointing to the old lady who smiled at her happy memories. 'Well, I took my pick and shovel and went into the wilderness, for that is all it was in those days. I sowed, planted and ploughed. Later I was able to buy a plough and my crops grew heavier. I always had a good market at the Moreton Bay convict establishment. Over the years other settlers arrived: the land around me was cleared, and one day I saw the railway line pass by my front door. That was fifteen years ago now. Today all the land you see on both sides of the line is mine. My children are well established and, at the age of eighty-five, I can still walk three leagues in a day without becoming too tired. I have worked hard, and today I am much better off than I ever dreamed of being under my father's roof, when as a boy I used to mind our animals in the fields of old Ireland.'

Could anyone help admiring the energy of this hardy race, for whom twenty years had been almost enough to snatch from barbarism these countries destined to a limitlessly expanding future?[2]

In her novel *It's Raining In Mango* (1987) Thea Astley ranges across a century and a half of Queensland history from the 1860s onwards but, perhaps not unexpectedly from the often-acerbic Astley in a publication that almost coincided with celebrations of Australia's bicentennial year of European settlement, her family saga is one of heroic but often idealistic personal failures. Australia's national failures are also recorded, from the destruction of Queensland's rainforest to the early, almost casual killings of Aborigines.

In this passage describing Cooktown, the most northerly town on the Cape York Peninsula and in the 1860s a canvas slum of tents packed along the waterfront, the journalist and failed newspaper editor, Cornelius Laffey, and his wife Jessica Olive and their children prepare to move on to a newer and even smaller inland gold-mining settlement.

Midmorning of the second day, the packer pulled up to boil the billy. While Cornelius sat jotting notes in his diary, George wandered off from the track for kindling and, stopping to relieve himself behind a hump of boulders, found there a bonefield where the half-rotted bodies of a dozen black men lay in a fly-swarming putrescence. Above his head a hand stuck up from a crevice in the rocks, bits of drying skin flapping from the outthrust fingers. Stalking plump birds screeched into a high circle and George wet his pants in his bubbling terror as he fled back down the hillside, gobbling with incoherence and pointing, pointing.

The packer went on stirring the billy with a twig of green eucalypt while Cornelius went to inspect for himself. George vomited beside the track.

'That, George,' Cornelius said in a kind of footnote on his return, his face pulled awry with discovery, 'is the repository of the by-products of our Christian greed.'

George was only eight. He kept retching into the grass.

'They were men,' Cornelius went on, 'with prior rights. Here, wipe your face on this.'

The stench seemed to have followed them.

'For every white man killed, we kill a score of blacks, my boy. That's the government-approved ratio. There's an imbalance of justice for you.'

'Drink your tea, mate,' the packer ordered, thrusting a mug at Cornelius, 'and we'll be making a move, eh? You can smell the damn stink down here.'

As they neared another parody of a township, George whispered to his father, 'Are blacks worth less?' There was nothing left to vomit but questions.

'One would think so,' Cornelius replied. Momentarily he enjoyed the angry twist to the packer's mouth. 'We are trained to believe so. I hope you won't.'

They stayed a week in Byerstown, trapped in the turmoil of new arrivals pouring in from the Palmer. It was the most liberal of educations. Trotting behind his father, who moved through mobs of drunks and brawlers with journalistic detachment, George witnessed the bloodiest of fistfights and a lynching. At night he was kept awake by the screams of beaten women. There seemed to be no police. And through it all he also kept seeing the half-rotted bodies of the blacks and that pleading decayed hand, whose fingers formed a white bone barrier behind which protective grille these other horrors were minimised.[3]

Robert Drewe, in *The Savage Crows* (1976) has his narrator Stephen Crisp, a young ABC journalist, take his friend Anna on a holiday to far north Queensland. A hundred years later, the settlement— near the site where Captain Cook spent seven weeks repairing his ship the *Endeavour* after holing it on the Barrier Reef in June 1770—has not much improved in manners.

Cooktown squatted between the rain forest and the flat tropical sea. In the bar of their wide-veran-dahed hotel drovers, maintenance dodgers, unhung murderers and miscellaneous frontier drifters cracked stockwhips, drank, vomited and sang old favourites to the accompaniment of ukelele, lagerphone and bush bass. Wild whistles and wolfish leers greeted their arrival.

'It would be a good idea to put on a bra,' Crisp advised. 'Or I'll have to fight everyone.'

'Nonsense.'

'What room are you in sweetie?' This came from a menacing dark lounger in a black singlet. He extracted a cigarette stub from his mouth with two finger stumps, blew smoke into the air with movie-western-saloon braggadocio. 'I'll be up for a night cap.'

'Or a Dutch cap,' called a freckled youth. The boy was small and thin with lined eyes and ginger hair. Crisp gave him a level television current affairs stare.

The bar subsided, friendly enough, and by the time they returned downstairs and ordered drinks— Crisp anticipating perhaps a little trouble, Anna brassiered under her shirt—enough other women had arrived, stringy local wives and an off-duty barmaid as well as the passengers from the launch, to take the heat off Anna. They drank schooners of beer with nervous gusto, Crisp broadening his accent (as he did almost unconsciously in such circumstances), shedding dulcet Commission tones, even leavening them with the occasional obscenity.

To his relief the dark singleted man reeled off into the night with an armload of bottles. The ginger youth though stumbled into their conversation just before closing time.

'The name's Andy from the Isa,' he said. 'I want to get married.'

'Good for you,' Crisp said.

The boy clutched a glass of rum and performed a little dance on the spot. 'I wanta get married,' he sang, and broke into quiet tears. No one else paid any attention to his crying. Anna looked fazed.

'Steady on there,' Crisp said.

'I got my girl up the stick and she pissed me off. I thought, shit, I'm lucky escaping the shotgun. But I miss her. Never seen the kid. Rhonda's engaged to a fucking bank teller now. Fucking cuntuver bank teller! Drives a bloody Toyota and goes water skiing. Rhonda pissed me off. Can you beat that?'

Andy from the Isa turned creased and cloudy eyes on Anna, blinking pale freckled lids. 'Will you marry me? You've got nice big tits.' He stretched a hand out to touch her.

'Piss off mate,' Crisp instructed.

Cooktown is apparently still living off its colourful frontier legends, Crisp finds.

They tramped around the town, absorbing its gold rush history, avoiding the long grass (pythons: shy but fond of domestic tabbies), swamps (crocodiles: smallish but sharp-toothed) and coastal shallows (stone fish, box jellyfish and similar fatal stingers). The local Aborigines, they learned from an old fisherman pulling in abundant rock cod on an old gut line from the town jetty, used to prefer the Chinese miners in their cannibal days.

'Stands to reason,' he said, squinting at the tourists with a canny and faded eye. 'Sweeter tasting, full of rice and bamboo shoots. Not like your Aussie miners. Too stringy and full of booze and tobacco. Meat eaters. Their sweat stinks.'

'I didn't know they were cannibals. I've never heard,' Crisp said.

'Wild country up here mate. This isn't your Brisbane, you know. Funny things happen up here. Ask Norm over there.' He pointed to a thin blue-black man in a red T-shirt hunched over a line at the end of the jetty. 'Hey Norman. You used to eat Chows, didn't you?'

'What?'

'Chows. Chinks. Chinamen. You blokes ate them, you black bastards.'

'Too fuckin' right,' Norman muttered, without turning around.

'Good bloke, Norm,' the old man confided. 'Salt of the earth. Not like some of them.'

He sent his father a postcard. It was coloured photograph of stone memorial to a local heroine, a housewife who had fought off savage Aborigines while her husband was away from home. They had speared her Chinese servant and pursued her in canoes when she escaped with her baby into the tropical ocean, paddling a water tank. She and the baby evaded the blacks but her breast milk dried up and they died of thirst, the baby first. The tank floated on into the Coral Sea, collecting much rainwater. On the memorial was engraved:

> In Memoriam
> Mrs Watson
> The Heroine of Lizard Island
> Cooktown
> North Queensland
> A.D. 1881
> *Five fearful days beneath*
> *The scorching glare*
> *Her babe she nursed.*
> *God knows the pangs that*
> *Woman had to bear,*
> *Whose last sad entry showed*
> *A mother's care,*
> *Then— 'Near dead with thirst.'*
> JOHN DAVIS, *Mayor 1886*

It was the only postcard the store carried. On it he wrote, 'Dear Dad, Cooktown is fascinating. Almost a ghost town. How are things with you?'[4]

However, Astley's and Drewe's generally black picture of Queensland humanity is somewhat balanced by the gregarious English composer and broadcaster Thomas Wood, who in a volume of travel memoirs called *Cobbers* (1934), equated the delights of the mango with the friendliness he encountered in Townsville in 1930.

It is the mango which stands alone among tropical fruits; not because of its supreme quality, but because it sums up, in itself, without any outside aid, all that you have imagined beforehand a tropical fruit should taste like. There are people who say it is a sponge tied up with packthread and soaked in turpentine. Dull clods! They have never taken the trouble to find the right kind of mango, nor have they learnt the only way to eat it. You must slash the thick hard skin with your knife, criss-cross, right round: peel it off in quarters and take four bites, two on one side, two on the other and let the fruit melt into you, lusciously; cheering your sense with the flavour and aroma of orange and jasmine, mellowed into a poem by golden sunlight.

This is the mango. It reaches perfection in Townsville. Men grow it lovingly. They have cultivated it as a friend. I do not think it is too fanciful to say that in doing so they have caught its prodigality, its warmhearted generous spirit. Certainly, no place I know, anywhere, is so good to the stranger as Townsville.[5]

Between 1930 and 1932 Wood travelled throughout Australia, after which he wrote in his autobiography *True Thomas* (1936) of his experience in a north Queensland hotel just before his departure for the Barrier Reef.

We had got to the traditional last course of port-wine jelly which ends that meal called by Australian hotels, astonishingly, tea. It was seven o'clock. The brief tropical dusk had deepened into sultry and airless and humid night; even the plates were hot; and to and fro the fans swung rhythmically, each in its gilt wire cage, humming as they swung. There were four of us at one very small table: all strangers. Australia is like that. Australia and hospitality mean the same thing, but no-one yet ever had a table to himself in any of the smaller Australian hotels. Even if the rest of the room is empty he is ordered to sit among fellow-sufferers in a corner and forbidden to move out of it—a custom of the country that needs some getting used to. Two of the four were commercial travellers. I guessed that from their dress. They ate stolidly and in silence 'mutton broth, roast lamb or cold beef, two veg.'; and between one course and the next they absent-mindedly sucked their teeth.[6]

This same small-town atmosphere survived to be evoked by Thea Astley in her Blaiklock lecture 'Being a Queenslander' (1978): 'Queensland means living in townships called Dingo and Banana and Gunpowder. Means country pubs with twelve-foot ceilings and sagging floors, pubs which, while bending gently and sadly sideways, still keep up the starched white tablecloths, the heavy duty silver, the typed menu …'[7]

The depression of the 1930s, when thousands of out-of-work men ranged up and down the Queensland coast looking for jobs that did not exist, figures strongly in novels by both David Malouf and Thea Astley. In *Harland's Half Acre*, Malouf's Frank Harland, an artist, also takes to the road.

The villages and larger townships where they appeared were never happy to have bands of homeless men in the vicinity and they were often hunted. They slept in ruined homesteads and barns, or on farms deserted by men much like themselves, or in camps that sprang up overnight in clay-pans, along a creek-bed or in the showgrounds of country towns, and if they developed the threat of permanency would be broken up by citizens armed with the law.

He shared a fire or a good sleeping-pozzie and talk with many different sorts of men and heard their stories: drought stories, mortgage stories, wife stories; stories of the war that had been raging during his years in his aunt's house and which some men were still fighting, deep to the eyeballs in mud; stories of prickly-pear, of rust in wheat, of diphtheria and whooping-cough epidemics, of prison terms—the ordinary miseries of the poor. Though it wasn't always misery. Sometimes there were tunes on a mouth organ or on the button accordion some man had saved out of the ruins of his settled life, or old jokes made new with a turn of phrase or a different situation; even, on occasion, snatches of philosophy, since it wasn't only the unskilled who were driven to this nomadic existence, and he began to wonder if even education and the profession he wanted for the last of his brothers was a fence against disaster. The open air did him good when it didn't reduce him to bone-shaking fever, and he saw something of the land he had been born to: cane fields waving their plumes under the moon, and so sweet-smelling you could get drunk on them (that was rum country); greyish plains where ant hills taller than a man were stacked all the way to the horizon—buried cathedrals showing only the tops of their blood-red spires; sea inlets fringed with glossy-leafed mangroves, thunderous surf.

These scenes fed his senses. They were of a grandeur that caught all his blood up in a display of cloud and colour that could transfigure the most ordinary day. They were a drama that had never been expressed. Well he would find forms for it. Great eloquent evenings as solemn and still as the brows of women—that sort of grandeur. Whiplike dawns: a crack of sunlight from sinewy arms. It made up a little for the shame he felt at being ragged, and dependent always on the charity or pity or wary suspicion of women at whose back verandahs he sought the chance to clear up a bit of untidy yard or to clean out gutters or chop wood, and for the hostility he saw in the eyes of men who were still settled and safe but might not be.

He found a companionship in misery that he had never known when he was in work, or which in those days he had not needed. He discovered that he belonged. But with those who were outside.[8]

In Astley's *It's Raining in Mango*, Harry Laffey, impoverished himself, finds it impossible to turn away such men, forced by the 'susso laws' to keep moving from town to town to pick up their welfare cheques.

All month and the month before that the men on the road had been going through looking for work. Their humility was stunning, corrosive. They would do anything, for a few bob. He'd had to turn all of them down. After the first dozen or so, he got tired of explaining that the crop had failed that year with unseasonal rain. He got tired of telling them he was nearly through himself. He couldn't bring out the words to describe the week there'd been nothing to eat but pumpkin. Couldn't. He simply couldn't cry poor mouth when he had a roof over his head.

The men stood hopelessly before him their eyes with that dead look peculiar to long hungers, their clothes baggy around the skinny bodies.

He simply waved his arms in an emptying gesture that took in the flattened soggy fields.

There was always a sandwich up at the house he said. And a cuppa.

Most accepted. Some didn't. Those who followed him to the kitchen nibbled at Clytie's 'doorstep' as if the sight of food made them shy. 'Thanks, missus,' they mumbled, looking past her, because this beggary had reduced them below humanity. One of them had even wept.[9]

Thea Astley, born in Brisbane in 1925, taught until 1967 at a number of schools in northern New South Wales and Queensland, including Ipswich and Townsville, the latter of which figures in her early novel *Girl With a Monkey* (1958). Her mastery of the geography and social politics of the small community is also evident in her second novel *A Descant for Gossips* (1960), which reiterates the theme of the persecution of the outsider. Astley's 1974 novel *A Kindness Cup* draws on an incident of brutality and small-town power-mongering that occurred in Mackay in the second half of the nineteenth century, when a number of Aborigines were massacred and a white man injured by the same party of vigilantes. In her later collection of short stories, *Hunting the Wild Pineapple,* she turns a sharply ironic eye on the Steinbeckian drifters—this time the hippies, alternative lifers, dole-bludgers and misfits in the northern rainforest—who gather in small towns such as Kuranda in the hills above Cairns.

There's the newness of the language, too, in luxuriant subcultures filled with Balmain and South Yarra drop-outs who fester in rain-forest patches or on tableland acres. They're lean people and arrogantly young. They groove and they say 'don't heavy-scene me, man' and they despise bread. Not give us this daily variety. They're all for that. There's a great trade in wholemeal and wheat germ in these parts and every dole day you'll see them heading back to the scrub with their government-sponsored goodie boxes jam-packed with the products of a society they reject. I like that, too. I like their cheek. I think of

The key colour is green, and of a particular density …

DAVID MALOUF, 1984 (PAGE 93)

GLASSHOUSE MOUNTAINS, QUEENSLAND. PHOTOGRAPHER: OLIVER STREWE

As for verandahs. Well, their evocation of the raised tent flap gives the game away completely. They are a formal concession that you are just one step up from nomads.

DAVID MALOUF, 1985 (PAGE 102)

NEW FARM, BRISBANE, QUEENSLAND. PHOTOGRAPHER: CAROLYN JOHNS

one particular po-faced tick who manages his defence theories on life-style with a base vocabulary that leaves me gasping.

'Vegetables are a superior form of life to humans,' he decides … 'You can groove with trees. Man, can you groove!'[10]

In *An Item From the Late News* (1982), Astley once again documents an instance of the sort of malicious brutality that figures in *A Kindness Cup*, but this time the setting is an inland mining settlement, all but deserted, and ironically named Allbut. The story is narrated by Gabby, an unsuccessful artist, who views the action from the sidelines.

The town. Just to think of it. Hardly a town.

A cluster of thirty buildings, a blackened fringe of aboriginal humpies as far out as we can persuade them and the carcasses of seven dead cars. Metal junk rusts easy under the scrub beside the bitumen and has become part of the landscape this town cannot bear to part with.

Under the bounce of December heat.

There's a school, two-teacher, closed for summer, a shire office soon to be closed down, a police station, a store that sells canned food, frozen meat and melting ice-cream with a petrol pump by the kerb. Once a month a tanker comes here to fill it. There's a hardware store garnished with chain-saws cans of paint snake boots bolts of material hurricane lamps tanned with dust sundowners' hats twill britches and plastic garbage bins. They have never shifted those garbage bins. There's a small haberdasher's and one-box room that doubles as post-office and bank. We all double here. There's a leaning two-storeyed pub called The Wowser. Jesus! I ask you. I know now why mother wept, wept under the slow bombardment of sun and cicada, white-ant and dry-rot.

'Don't slam that door,' the locals warn each other, heavily jokey. 'The whole place will fall down.'

I am going to slam that door.

There's the shell of a school of arts, a bit of architectural lah-di-dah with its verandahs and porches and carved architraves, a left-over from the days when the town was prosperous with tin mines back in the hills and travelling shows came through. Red yellow white the streamers dangle still above the stage piano where a three piece band used to whine through the numbers chalked up on a blackboard. They're still there. I see them there, saw them for years, unchanged through my childhood: Your Cheatin Heart, says the copper-plate, Spanish Eyes, South of the Border!

I'd cry, but for heat-dazzle.

I blink time back and read again the notice that stayed for years chalked up under the dance list: The most beautiful workmanship displayed on the figures of the dancers was the costumes. Thank you Mrs Amherst. The Allbut Concert Company Ballet.

Why has no one rubbed it out? Perhaps I am not the only one fighting clocks.

In the middle of the bitumen strip on a fenced patch of weeds and Noogoora burr is our war memorial. I know the names—there are only six—like a mnemonic for dying: Wieland, Amherst, Swan, Nehmer, Jerrold, Rider.

Across the bottom of the stone someone has scrawled in red paint: 'This is what we fought for' with an arrow pointing downtown. The words and the arrow have been there a long time and no one has bothered to remove them.

All in a bracelet of stumpies.[11]

There is an echo of this sense of dilapidation and impermanence in the townships in Queensland-born writer Vance Palmer's novel *The Passage* (1930). Based loosely on the fishing port of Caloundra. *The Passage* describes life in a small coastal fishing settlement which receives the temporary attention of a developer, but which is then left behind as the neighbouring township becomes popular.

'T'at Osborne,' said Kunkel, 'he know when to get out at the right time. Places like Lavinia, they go ahead and then they don't. It's always the same in this country. A thing looks like it's growing, and then it wither up. Not enough dept' in the soil! T'at Osborne always knows ahead when the change come, and he gets out rightside up.'

He repeated his ideas monotonously to the men who came in to hear the gramophone in the evenings. There was a blight on new countries that made things wither up before they were properly rooted. Look at the deserted orange-orchards on the road to the township, with broken fences and trees running to wood! Look at the tumble-down shanties you saw in places that were just opened up! People hadn't the spunk to stick in one place and leave their mark on it; they were blown about by the wind. Not like in old countries, where there was a rich soil nourished by the bones of millions and men didn't always have the itch to go off and scratch at fresh spots like a fowl!'[12]

And the novelist and journalist Charmian Clift, travelling in Queensland in the 1960s to gather material for her newspaper essays in the *Sydney Morning Herald,* finds the landscape on the western side of Cape York Peninsula even more daunting.

Karumba is a name on the map of Australia, a dot on the mouth of the Norman River where it flows into the Gulf of Carpentaria and the hawks hang over the mangroves and the sandflies are murder. Thousands of jellyfish pulse on the tide, and the opaque, oily-looking water harbours huge coarse repellent fish, mud-crabs as big as large platters, sharks by the hundred, and still, they tell me, a few old and sagacious crocodiles.

It is sinister country. Evil to me since I react violently to landscape and am repelled utterly by this one that seems to be saturated with a sort of thick grey heat. The river here is broad, sluggish, yellow-grey in colour. The mangroves are green-grey, dense and heavy. And the blue-grey sky seems to have a tangible skin of heat on it. From the river the baking salt-pans and the sparse grey scrub recede into for ever. It is not, one would think, a desirable residential area. Still, I suppose frontier towns never did spring up because of residential qualifications. A frontier is pushed forward because of the commercial possibilities of the terrain—because of gold or minerals or gems or hides or seals or ships or apes or ebony or peacocks.[13]

PERTH

FROM EARLY SETTLEMENT TO THE ARCADIAN AGE

I walked down the track
to where the camp place used to be
and voices, laughing, singing
came surging back to me.

It was situated on the Swan
not far from the old homestead.
That's gone too.
Kindly old man Hammersley,
they can stay there as long as they like,
he said.
Now he too is dead.
Billy Kimberley used to corroboree
there weekends
for a tin of Lucky Hit,
then share it with his friends.

Now we who were there
who were young,
are now old and live in suburbia,
and my longing is an echo
a re-occurring dream,
coming back along the track
from where the campfires used to gleam.

Jack Davis, from *The Dreamers*, 1982[1]

There is a strong sense of the regional in Western Australian writing; insofar as there are contemporary writers who continue to live in that state, produce work about Western Australia, and are identified with it in the minds of other Australians. Yet, on analysis, this impression of regionalism rests on the work of a few. In the past, Western Australian-born writers Randolph Stow and Katharine Susannah Prichard have made a significant impression on the national psyche, and more recently Elizabeth Jolley and Tim Winton, whose style and subject matter are particularly distinctive, have been prominent on the national literary scene. The name Peter Cowan—less well known nationally—would have to be added to any such list of authors whose work is uniquely connected to the Western Australian landscape.

Western Australian regional writing has traditionally concentrated on the sparsely populated seaboard and inland regions. Although its population, like that of the other states, is largely urban, Western Australia, unlike for example Queensland, has produced very little writing about its capital city, Perth. Once again, the reasons for this may lie in geography. Western Australia occupies about one-third of the land mass of the Australian continent, and most of this is waterless desert. It has the longest coastline of all the states. Until the aeroplane came into common use for passenger transport after World War II, vast belts of almost uninhabited land in Western Australia, along with the Northern Territory and far north Queensland, remained—in European terms—the most physically isolated regions in the country. To reach even the capital city from the eastern states it was necessary to undertake a lengthy journey by sea, or overland by unsealed road or railway (opened in 1917) across thousand of miles of desert and salt lake country. A sensation of isolation from the rest of Australia seems to have engendered in Western Australian writers a powerful awareness of the hinterland and the sea. Western Australian writers such as Katharine Susannah Prichard, Randolph Stow and Peter Cowan have produced works of fiction largely moulded by the contemplation of just that arid interior landscape, and Western Australian writers including Tim Winton, Robert Drewe and Nicholas Hasluck have taken as their subjects the coastline and its associated early history of shipping and whaling.

From the earliest times the landscape of Western Australia has been presented in terms of romance, mystery and contradictions. The southern coast, mapped by Dutch explorers and traders such as Peter Nuyts in 1627, figured in Jonathan Swift's *Gulliver's Travels* (1726): the island of the Houyhnhynms was located south of 'Nuyt's Land', and Gulliver later disembarks on the 'south west point of New Holland', where he feeds on shellfish and is chased by bow-and-arrow-wielding Aborigines. In another odd literary detour, the isolated regions of northwestern Australia were used as a jumping-off point for the fantasies of Henri Louis Grin, a Swiss who was butler to the governor of Western Australia for a short period in 1875. Grin's *The True Tales of Louis de Rougemont, As Told By Himself* (1899) recounts how he allegedly spent some thirty years as a castaway on the shores of King Sound more than 2500 kilometres northeast of Perth, where he rescued two shipwrecked girls who had been living with Aborigines in the Derby area. Grin, who had been pearling in the north and later turned up at Cooktown in Queensland, was familiar with the region, but 'de Rougemont' and his 'true adventures' were revealed as a hoax.

In *Moondyne Joe: A Story of the Underworld* (1879) (which was, like Marcus Clarke's *For the Term of His Natural Life*, a melodramatic convict tale originally written as a newspaper serial), John Boyle O'Reilly describes Western Australia a 'vast and unknown country, almost mysterious in its solitude and unlikeness to any other part of the earth. It is the greatest of the Australias in extent, and in many features the richest and loveliest ...'

O'Reilly also coined the term 'the Cinderella State' for Western Australia because of her (then) lack of gold.

The biologist and naturalist E. L. Grant Watson, writing in 1948 of his time in Western Australia in 1910–11, recalls the account given to him then by descendants of John Bussell and his brothers about the early days of settlement at Busselton on the shores of Geographe Bay, 230 kilometres south of Perth, in 1834.

> When the first English colonists came to Australia, many of them had but the vaguest idea of the sort of conditions they were going to find. Many brought with them the accustomed conveniences of civilisation, things which would be of little use to settlers in a virgin forest. Such things as family coaches were shipped, their owners fondly imagining that they would be able to drive from place to place in the new continent ... Many of the ships were run ashore on the beaches, and the cargoes landed as best they could be. Men, women and children, horses and cattle, coaches, carts and household goods, were littered along the seaboard. The settlers slept in the coaches and rigged up tablecloths, sheets and blankets to protect them from the sun's heat. The cattle and horses strayed; some were lost; some found good pasture ... and when the owners followed and saw the good grass, the little river and the pleasant prospect, they had built their farm and called it 'Cattle Chosen'.[2]

Critic Dorothy Green disputes this account, noting that the settlers arrived before the cattle;[3] but the banks of the deep and slow-moving Vasse river (named for a sailor from Nicolas Baudin's expedition of 1801 in the ships *Geographe* and *Naturaliste*, who was drowned in its bay) were so pleasant that

John Garret Bussell, inspecting the area in 1830–32, wrote that '… here was a spot that the creative fantasy of a Greek would have people with Dryad and Naiad and all the beautiful phantoms and wild imagery of his sylvan mythology. Wide waving lawns were sloping down to the water's edge. Trees thick and entangled were stooping over the banks …'[4]

D.H. Lawrence, who in 1922 stayed with his wife Frieda in Mollie Skinner's guesthouse, Leithdale, in Lukin Avenue, Darlington, a small town in the vineyard and orchard area of the Darling Ranges, was more mixed in his initial reaction to Western Australia. On 15 May, after eleven days in the country, he wrote to E.H. Brewster, an American painter with whom the couple had stayed in Ceylon:

> We are here about 16 miles out of Perth—bush all around—strange, vast empty country—hoary unending 'bush' with a primeval ghost in it—apples ripe and good, also pears. And we could have a nice little bungalow—but … I just don't want to stay, that's all. It is all so democratic, it feels to me infra dig. In so free a land, it is humiliating to keep a house and cook still another mutton chop …[5]

In *Kangaroo* (1923), however, written later in the year while the Lawrences were staying at Thirroul on the south coast of New South Wales, Lawrence gave a rather different response to his character Richard Somers, whose progress in Australia paralleled Lawrence's own. Europe, Somers has decided, is 'done for, played out, finished', and he must go to a new country.

> The newest country: young Australia! Now he had tried Western Australia, and had looked at Adelaide and Melbourne. And the vast, uninhabited land frightened him. It seemed so hoary and lost, so unapproachable. The sky was pure, crystal pure and blue, of a lovely pale blue colour: the air was wonderful, new and unbreathed: and there were great distances. But the bush, the grey, charred bush. It scared him. … It was so phantom-like, so ghostly, with its tall pale trees and many dead trees, like corpses, partly charred by Bush fires: and then the foliage so dark, like grey-green iron. And then it was so deathly still. Even the few birds seemed to be swamped in silence. Waiting, waiting—the bush seemed to be hoarily waiting. And he could not penetrate into its secret. He couldn't get at it. Nobody could get at it. What was it waiting for?[6]

In this case Lawrence might have had in mind the Avon Valley, a hundred kilometres east of Perth on the east of the Darling Ranges—sheep and wheat and mixed farming country—the setting of *The Boy in the Bush* (1924), his and Mollie Skinner's account of the pioneering period of Western Australia.

Since vast tracts of previously unexploited territory were opened up to European occupation after World War II, the language, imagery and literary symbolism of the Western Australian inland have begun to show signs of change. Huge agricultural and mining operations now scar the face of the geologically ancient land mass, and the establishment on the North West Cape of satellite surveillance stations has had global political implications. Phenomena such as these have begun to appear in works by writers including Robert Drewe and Tom Flood — both of whom were born in eastern Australia but grew up in the West.

*   *   *

The capital city of Western Australia, Perth, was founded in 1829 on the banks of the Swan River sixteen kilometres inland at the base of Mount Eliza, a site chosen by Captain James Stirling, Western Australia's first governor, who had made an exploratory visit to the area two years before. This part of the coast had been previously touched upon by the Dutch and French, and Willem de Vlamingh of the Dutch East India Company had in 1697 named the Swan River after the black swans he discovered there. The Dutch navigator found the region—with its long sandy beaches backed by low hills and similarly low, sandy islands offshore—too 'arid, barren and wild' to be of interest.

However, in 1827, Captain Stirling wrote of sailing upstream from the mouth of the Swan River to find that '… the richness of the soil, the bright foliage of the shrubs, the majesty of the surrounding trees, the abrupt and red-closured banks of the river occasionally seen, and the view of the blue summits of the mountains, from which we were not far distant, make the scenery around the spot as beautiful as anything of the kind I have witnessed'.

Subsequently, Perth itself caused less extremes of literary effusion. The optimistic descriptions by Stirling of the landscape immediately around the site proved a little misleading as to the availability of agricultural land, and the settlement progressed slowly. Perth in 1830, according to a visitor, resembled a 'straggling tented field', and to another in 1843 'a large straggling village'. A Lieutenant Bunbury, serving in the British army in the mid 1830s, found the settlement 'dismal', and 'duller than anything you can imagine', but reflected that as he had come of his own accord he had 'no right to grumble'. Convict labour was introduced in the mid 1850s, which stimulated development and a significant program of public building, and the settlement was proclaimed a city in 1856. In 1872, by which time the population had grown to 6000, Trollope pronounced it 'a pretty town', but disparaged its mosquitoes. From then until the new century, the embryonic city seems to have made little impression on visiting or local writers: most remarked politely on the view from Mount Eliza and left it at that.

The discovery of gold in the Kimberleys in 1872 and, more importantly, at Coolgardie in 1892 meant that Perth was once again marked with tent cities, but this time they were overcrowded ones. There was an influx of population for the whole state, rising from about 29,000 in 1890, when the colony achieved self-government, to 110,000 a decade later. This development also gave a boost to the state's literary beginnings: some sixty newspapers flourished on the goldfields between 1892 and 1909, although only nine still existed after 1920. These published populist verse, satire and humour as well as current events and the occasional radical view, causing 'Grant Hervey' (George Henry Cochrane), blacksmith turned journalist, poet and soon-to-be-notorious forger, to note in the *Kalgoorlie Sun* of 1904: 'considering the strict limitedness of the population, journalism flourishes like measles, and you can hardly throw a defunct marine out of the window without abrasing a scribe'. Katharine Susannah Prichard documented this era in her trilogy *The Roaring Nineties* (1946), *Golden Miles* (1948) and *Winged Seeds* (1950)—the last of which coincided with Gavin Casey's *City of Men* (1950), which deals with the same theme. The prosperity of the state, initially based on gold, wheat and wool, was subsequently founded on fortunes made, often in one generation, by the discovery of other mineral wealth in the vast hinterland.

The influx of population that came with the gold rushes, forty years after those of New South Wales and Victoria, also turned the Western Australian capital into a city of suburbs. Perth itself, like most other Australian cities, was marked by a kind of provincial Englishness in its architecture and its cultural aspirations. John Keith Ewer, a schoolteacher in Perth from 1924 to 1947, in his novel *Money Street* (1933), describes a suburb with '… houses … closely packed, many of them semi-detached and crowded with ornate frontal decorations that were relics of the late Victorian age of cottage architecture. Here a wreath of flowers, species unknown, set in the masonry. There a pillared balustrade hid a receding gable. Quaint houses they were, each breathing a definite personality'.

Thomas Arthur Guy Hungerford, born in Perth in 1915 and growing up in its semi-rural southern suburbia in the depression years of the late 1920s and 1930s, affectionately recreates the period in his collection of semi-autobiographical short pieces *Stories From Suburban Road* (1983). Along with collecting birds' eggs, catching goannas and stealing vegetables from the Chinese market gardens, Hungerford exercised racehorses for a local stable.

> Rudolph and I used to leave the stables in the starlit freezing darkness just before the sun came up over the Darling Ranges. It was easy and comfortable going for us and the horses on the deep-red loam washed off the gravel surface of Suburban Road by more winters than I could remember. The thud of hooves was softened by the foliage of the trees above us—gums and kurrajongs and peppercorns, lillipillis and Moreton Bay figs made that section of our ride almost like a tunnel.
>
> On the river side of the road there were an orchard or two with a house here and there, but mainly paddocks that went right down to the Chinamen's gardens. They were covered with dandelions in the spring, so golden-thick that you had to be careful walking through them or you'd tread on a bee and get stung. As we rode past the houses I often wondered if the people sleeping inside them could hear—maybe through their skin—the soft clip-clop and jingle-jangle-jingle of Rudolph and me ghosting by on the other side of their fences.
>
> We never changed the way we went—right down through South Perth to Mill Point, out onto the beach near the old mill, off for a gallop on the hard sand and then into the water for a splash. As we trotted back along the beach to the point the first sunlight would be trickling down the sides of Mount Eliza like water. The brewery's big horse-drawn lorries would be creaking out onto Mount's Bay Road, heading for pubs all over Perth—every one was pulled by two enormous Clydesdale horses that always took a part in processions through Perth, dressed up with shiny horse-brasses and red-white-and-blue ribbons plaited into their manes and tails. The first trams would be clanking and clanging around the bays under the side of the Mount, and every now and then a motorcar would chug along like someone with a bad cold. It was so still as we sat on the horses, giving them a breather before we started home, we could hear the soft hum of the city from across the water, like the sound of a travelling swarm of bees.

T. A. G. Hungerford remembers an early childhood redolent with stories from the Pilbara gold-digging days of the 1890s, when his family, like so many others, had come to the state.

> I would sit behind the sofa in our living-room when they had visitors on Sunday afternoons, old friends from the Goldfields, all blue serge and lace fronts and gold all over them, big lumps of it on the men's tiepins and pretty funny-shaped nuggets fastened to the ladies' shoulders on gold chains so fine you could hardly see them. My mother had a gold nugget almost the perfect shape of Australia: it was given to her by a man in a place called the Nullagine, but she sold it once when, as she said, we were on the way to Queer Street. They'd say: *The Yilgarn, Arthur! When we were in the Yilgarn, Min!* and it made my hair stand on end the way it did when I looked at the racehorse goanna.
>
> He and the place they called the Yilgarn seemed to belong to some world I could never know about, hard and scaly and secret, older than Australia and deeper than the mines at Kalgoorlie, and hotter than any summer I'd ever known.

There are also stories of the city's convict history.

> Hierrison Island was where the Causeway crossed that part of the river, from the carbarn in Perth to where the Albany Road and Fremantle Road met on the South Perth side. For years we'd called it Harrison Island because a boy at school told us it was called after his grandfather, who had been an early settler. When he was skiting about it down at our place one day my mother said 'Oh? Did he help to build the Town Hall too?' and my father said: 'Now, Min!'
>
> He told us afterward the Town Hall had been built by convicts, and while the boss wasn't looking they'd shaped some of the windows like the broad arrows they wore on their clothes, just so you'd know it had been built by convicts. Next time I went into town I remembered to have a good look at the Town Hall and the little windows in the tower were shaped like broad arrows. So my father hadn't been pulling our legs, as I'd thought.

As in other coastal Australian cities, going to the beach is an important ritual.

> After my father sold his carrying business and went into the shop we always went to Como by tram. It seemed that nearly everyone in Perth did—tram-load after tram-load of them, often with two trams hitched together to carry the mobs, until when you got there you couldn't find a shed to sit in or a bush to sit under: unless you got in early, so that on holidays such as Boxing Day and even on ordinary Sundays, sometimes, when it looked like being a scorcher, I was sent down first thing with a rug and a few lunch baskets to spread around and grab a good spot. I didn't mind—I'd take a Tarzan book and have a good read while I was waiting for my mother and my sisters to come down. My brother Mickie never came with us, but we often brought Mrs Moodie and her girl Jean, who was my first sweetheart.
>
> There were swings and a merry-go-round and sometimes hoop-la and a coconut-shy, and a man with little donkeys for penny rides on the beach, and fish-and-chips and other shops for everything you'd need. There was even one with a long line of coppers, always on the boil for people buying water to make their tea. Every Sunday was like People's Day at the Show, with crowds running around and skylarking and just sitting on the grass or the sand, in bathers and pretty summer colours that always reminded me of my mother's cinerarias.
>
> If we went early, we'd swim until lunchtime and then we'd wolf into hot tea and sandwiches. Mrs Moodie called them 'sangwidges'. She'd say to my mother: 'What sort of sangwidges have you brought, Mrs Hungerford? I've got tomato-and-onion.'
>
> 'Well, Mrs Moodie—I've got mutton-and-mustard—and I put a few tomatoes in whole. Shall we share and share alike?'
>
> They said that every week, and they shared every week, but they never did it without discussing it first …
>
> When it began to get a bit cool someone would say: *Last one down the jetty's a black-gin!* Someone else would always say: *Five more dives!* and after we'd had them we'd scramble out and race down that long stretch of rough boards with the cold wind at our backs and the great red sun touching the tops of the trees on the other side of Melville Water. When we'd got dressed we'd line up again for more sandwiches and hot, sweet, milky tea, and every time I buried my face in my enamel mug I told myself there couldn't be anything better in the whole world.[7]

Like Hungerford, the poet and novelist Kenneth Seaforth Mackenzie shares with many Western Australian writers a memory of a peaceful Arcadian childhood in pre-war Perth and its environs. Mackenzie was born in Perth in 1913 and grew up in the timber, cattle and dairy-farming countryside around Pinjarra, a town on the Murray River 85 kilometres south of the city, where he moved with his mother at the age of nine and lived until he went to school in the city at thirteen. To this landscape he ascribes an urgent sensuality. In *The Young Desire It* (1937), Mackenzie's extraordinary novel of adolescent love, his character Charles Fox awaits the time when he will go to boarding school in Perth.

> During those days of examination the first heat of summer came westward like the waves of a tide rising over the hills. Passing a long morning alone at the far end of the parade ground, waiting for dinner and an afternoon sitting, Charles was surprised to see the grass already dry where it stood up golden from the hardened earth. October had worked with sweet secret poison, and a week of heat would reveal the final dryness at the heart of all free, untended growth. In the misty gold of the tall grass, wild oats drooping ripe against the sky, sorrel turning purple and brown and hard on the red earth, cicadas had, as it were in one day, reached their swooning full chorus, and rushed the trembling light with a sound like blood heard in the ears. The whole earth and all nature sank into a still swoon beneath the eternal ravishment of the sun, and the ceaseless, passionate susurrus of the insects gave sound to the heat, as already mirage was giving it a shaking visibility, clear and refractory like water.
>
> Lying on the hard ground, he smelt the earth and the warped dead leaves. Tough hardy grasses, slippery and matted like hair now, had ripened their transparent pods of seeds hurriedly; they were the

fruit of those pink and white threepenny stars, the only reminder and proof that they had ever shone in the balmy October sunlight. Looking carefully into the tiny maze of shadows beneath the silken strands, he could still find seeds not yet baked by the sun. They were sweet and bulging with juice, so full that between his teeth they cracked and burst like minute grapes in March. Above him the sky arched, enamelled dry and bright. No cloud troubled its mighty concentration. It had not yet the bleached pallor of full summer, but winter's depth and the hazy velvet of spring were already gone from it, and its cerulean sweep was glossy and hard.[8]

Mackenzie lived in Perth until 1934 when he moved to Sydney. *The Young Desire It*, published three years later, was regarded as controversial for its open treatment of both heterosexual and homosexual love and its thinly disguised portrait of Guildford Grammar School, which Mackenzie calls 'Chatterton' in the novel.

Kenneth Seaforth Mackenzie's Arcadian sensuality is echoed in the work of Randolph Stow, born in the northern coastal town of Geraldton in 1935, where he lived in Gregory Street until he was evacuated during World War II to Sandsprings, a family sheep station forty kilometres inland in the Murchison region. Stow was educated at state schools, until, like Mackenzie before him, he left for Guildford Grammar School in Perth and then the University of Western Australia. He drew on these experiences for his partly autobiographical novel *The Merry-Go-Round in the Sea* (1965).

Stow describes Perth as it appears during the years of Word War II to a young boy, six or seven years old, from a small town further up the coast.

Rob stopped dead in the middle of London Court. 'Look!' he called out. 'Nan, look,' pointing at Dick Whittington and his cat.

Nan stopped and stared, open-mouthed. People were jostling to get past them.

'You are a pair of country bumpkins,' Margaret Coram said, embarrassed. 'Don't just stand gawping.'

'I want to see St George kill the dragon,' Nan said.

'Darling, we can't just stand and watch the clock for a quarter of an hour.'

'Isn't London Court *romantic*,' Rob said, dragging against his mother's arm, which was trying to tear him away from Mr de Bernales's Elizabethan folly.

'Really,' Margaret Coram said, 'people will think you've never been anywhere.'

'We *have* never been anywhere,' Nan pointed out.

'Do come along,' begged their mother.

The children trailed after her, down the cobbled street of the Court, of which visitors were wont to say that there was nothing like that in London. In the Terrace they met a damp wind off the brown and choppy river.

'Perth's cold,' Rob said, shivering. He had never been really cold in his life before.

'People wear coats,' Nan said, studying the passing natives, 'even when it isn't raining.'

They were fascinated by the customs of this new country. …

'It's really only about twenty years older than Geraldton,' his mother said. But that meant nothing, emotively. Perth was ancient, an ancient civilization. Soot-darkened buildings proved it, and the existence of a Museum, and the fact that it was a seat of government. And it was a very special city, cut off from other cities by sea and desert, so that there was not another city for two thousand miles. Among all Australian cities it had proved itself the most special, by a romantic act called the Secession, which the other cities had stuffily ignored.

Cinderella State, he thought, feeling indignant. That was the reason for the Secession. Because they had ignored his poor Cinderella State, all one million square miles of it.

Maybe after this war there'd be another war. Western Australia against the world, the Black Swan flying.

'We shouldn't have gone to Parliament House,' his mother had remarked, 'it seems to have made you political.'

'I'm a Rebel,' he had replied. 'I've got Rebel blood.'

Walking down the Terrace he dreamed of Eureka, the torn banner of the Republic of Victoria.

But not a republic, not in Western Australia. That would hurt the feelings of Princess Elizabeth and Princess Margaret Rose.

'When will Western Australia be free?' he wondered.

'I don't know,' said his mother. 'Perhaps when Bonnie Prince Charlie comes over.'

'Aww.' He grew disgusted at her flippancy.

The buildings in the Terrace were tall and clean and pale, and the small trees that grew there were so neat that they could never have suffered the south wind. At a corner a boy was shouting: 'Py*per*-her! Py*per*-her!' He was not much older than Rob, and Rob stopped to stare at him, admiring this loud-voiced insouciance. Then the newspaper boy noticed, and made a face like a Jap, and Rob moved away.

Encapsulating both the lyrical hinterland and coastal images that permeate much of his writing about the state, particularly in this novel, Stow writes of a childhood in which his *alter ego* Rob seems little aware of the existence of imperfection in his world.

The sunflowers followed the road a long way out of town. When the road ran by the white sandhills they were still there, they stood up tall and yellow against the dunes, which were dazzling, like Scottish snow. Sometimes they were framed against a broad triangle of bright sea. But as the coast fell away and the road cut across the river flats the sunflowers thinned out and vanished. Now the land was pale and flat, littered with empty farmhouses, tobacco-bush leaning in through glassless windows. Here and there clumps of gumtrees and palms sheltered houses where people still lived. But there were many, many empty houses, big houses too, like the tall, staring shell of the Old Brewery, a ruined castle.

'Look, Rob,' his mother said, taking her hand off the steering wheel and pointing. 'What are they?'

He followed her finger, and saw the familiar gumtrees, crippled and stooped by the southerly, bowing northward and trailing their leaves on the ground.

'Oh,' he said, with a sort of embarrassment. 'The ladies washing their hair.'

That was something he had said when he was very small, and he supposed that they were teasing when they reminded him. But it was still true, and the bent trees always would look like Grandma and Aunt Kay washing their hair; like Grandma and Aunt Kay trailing their long weeping leaves in the basin.

But later, the possibility of evil intrudes.

The spring country flowered in stunning profusion, flowered like chintz with flowers whose multitudinous names one could never hope to learn, flowers whose names were 'that mauve one with three petals' and 'that pink one like the ones Little Red Riding Hood was carrying in the picture-book'. As a grown man Rob would discover that the pink ones were trigger-plants, but he would never know the names of more than a fraction of the flowers of his country, which would continue to be called by different names by different individuals within different families.

In the spring pasture and among the maturing wheat red and blue wild geraniums flowered. Unflowering wild strawberry leaves draped dead wood and lichened rock, and everlastings rustled in one-coloured drifts of pink and white and yellow.

By rock pools and creeks the delicate mauve-petalled wild hibiscus opened, and the gold-dust of the wattles floated on water. Wild duck were about, and in trees and in fox-holes by water he looked for the nests, staring in at the grey-white eggs, but touching nothing. Climbing a York gum, he was startled when a grey broken-off stump of branch suddenly opened golden eyes at him. He gazed into the angry day-dazzled eyes of the nesting frogmouth and felt that he had witnessed a metamorphosis.

Under the wattles, between the flowering shapes, he plunged in the cold rock pools. But his cousin Peter, when he was there, would not take his clothes off. Peter said it was a mortal sin or something. He was boarding now at the convent, and learning boxing and football from someone called Sister Catherine, who was in the habit of beating kids around the head with her empty beer bottles, according to Peter.

On the small, rocky hills, among the keening flowering sheoaks, Rob walked old Bob and drank in the country above Bob's ears. For the first time in his life he knew that he was young, and knew, with agreeable sadness, that he would not be young for long.

Time and death could stain the bright day, and the leaf-brown foxes that traced green paths in the dew could die poisoned and in agony among the flowers. He stood by the body of a young fox, and watched the capeweed and horseradish flowers bend in the wind against it, pollen clinging to the stippled hide. Furry-silvery fingered leaves of lupins dipped and swayed, and the new blue flowerheads nodded. Out of the tender blue sea of the lupin paddock a windmill rose, sandy-tawny with rust, spinning against the lupin-blue sky. Lupins withered and foxes rotted, and the windmill whirled and whirled against all seasons of the sky, drinking from the filled dark caves below the earth.[9]

Stow later travelled extensively, and from 1966 lived permanently in England. This removal from the landscape and the perspective it may bring possibly contributes to the highly symbolic nature of his fiction; the Australian landscape, particularly that of Western Australia, is poetically evoked and used as a canvas against which to explore large themes. While his descriptions of his childhood and later environment are acute and accurate, both the landscapes in their intensity and the characters who move within them are often larger than life.

In a significant verse from a poem in his first collection, *Act One: Poems* (1957), Stow writes

My childhood was seashells and sandalwood, windmills
And yachts in the southerly, ploughshares and keels,
Fostered by hills and by waves on the breakwater,
Sunflowers and ant-orchids, surfboards and wheels,
Gulls and green parakeets, sandhills and haystacks, and
Brief subtle things that a child does not realize,
Horses and porpoises, aloes and clematis—
Do I idealize?
Then—I idealize.[10]

For Randolph Stow's youthful character Rob, the return of a favourite cousin from the war into his life brings both a loss of innocence and a disturbing encroachment into his sheltered life of knowledge of the world outside. For T.A.G. Hungerford, born a generation before Stow, it is the uninterrupted sensuality of long summer days by the sea a few dozen kilometres outside Perth with a group of like-minded companions which represents the golden age.

We spent a lot of time up at Triggs every summer. It was a mile or so north of Scarborough, with nothing but bush either side of it or behind it. Hardly anyone but us ever went up there, apart from the occasional rod-fisherman trying out the Blue Hole, and now and then knowledgeable folk after mutton-fish—spend half-an-hour wading around with a tyre-lever on the reef on the north side of the island, and if the surf wasn't running too high you'd fill a chaff-bag without getting your knees wet.

There was a convent retreat house in a deep fold of the dunes just behind the island, a long, low, unpainted weatherboard barn of a place with a faded red iron roof and heavy lattice work all around the verandah—just like Josie's Bungalow down in Roe Street, only Josie did at least keep her joint painted. You never saw much of the nuns. Only once, one weekday when I was on holiday, I'd gone for a walk up to Triggs by myself and surprised a flock of them having a swim. They probably thought nobody would be about, midweek. You never saw such a sight—I'd say the only article of their clothing they'd taken off to go in was that great heavy leather belt they wear to hang their crucifixes and their wooden rosaries on. They were flapping around like penguins at the very edge of the Blue Hole, and as I watched them from the sandhills I wondered if they had any idea of just how dangerous it was. Maybe they just put their faith in God—but then, so had a lot of people: it seemed that every few weeks in summer the papers carried stories about someone in trouble at the Blue Hole, either dragged out of it or drowned in it. …

During the summer we prowled around the reef, under-water, for hours, picking up shells and bits of coral, watching the fish—but not that Sunday. Although the day was so hot the water was freezing—I can only describe it as like swimming in champagne. We didn't loiter, though. When we began to turn blue we clambered out and burrowed into the hot sand again. Mack, who really didn't care about much apart from the pub and the SP shop, said that if ever he got to Heaven and didn't see something like Triggs just inside the Pearly Gates, he'd turn around and come straight back.

This Arcadian period, and the possibility of an innocent sea-born romance, is lost forever with the approach of war.

It's curious how safe and normal things seemed when I climbed the wooden steps to the front door of the flat. The old place had been built God-only-knows-when, a couple of hundred yards off the end of the promenade, in the sandhills. Not another house around, which suited us fine. More lately, though, it had been divided into three flats at thirty shillings a time, and very nice money for whoever owned it. The dividing walls were only single sheets of asbestos, and we used to reckon you could hear the lady next door putting on her cold cream. It didn't worry us, generally, but one summer a couple of years earlier it did worry the lady next door. She was from Kalgoorlie, and a real Tartar. She wrote to one of the papers saying how disgusting it was that the girls stayed in our flat with us over the weekends, and what was the younger generation coming to, and did our mothers know, and much else beside. Actually the girls all slept in the inner room, while we were ranged around the verandah wherever we could fit in: but you'd have had a hard time making that stick with the lady from Kal. Dirt is like beauty—most times it's in the eye of the beholder.

About mid-morning Mary and I went for a ride. There was a pretty rough sort of 'riding academy' in the sandhills a couple of miles in from the beach, with a flock of very savvy horses that knew much more about being ridden than most of the customers knew about riding. Apart from the few which we always reserved, they were barely-broken brumbies from further up the coast which had been caught and belted into submission. When they could be got to move at all it was a slow, dogged walk—until you turned them toward home: then Darby Munro himself couldn't have held them in for a thousand quid. It cost us five-bob a throw, but it was worth it. Outside the ball season it was usually our only big outlay for the week.

There was a lovely stretch of country behind the dunes between Scarborough and North Beach, and we used to take off into it—gum forests and deep gullies, and sand-hills so high you could sit your horse on top of one and see Fremantle in one direction and almost up to Whitford's in the other. There wasn't much in the way of buildings in either direction to stop the view, in any case.

We usually rode for about an hour-and-a-half out and the same back. On the way home we were walking the horses along a sandy bush track, almost stoned on the scent of gumtrees and wildflowers—perhaps partly just with being together and alive on that unexpectedly glorious spring day.

'Are you going, Hungie?' Mary said, without looking at me.[11]

Hungerford served in World War II as a commando in New Guinea, New Britain and Bougainville and, like the Victorian writer Hal Porter, travelled to Japan with the occupying forces after the war. He returned to Sydney to work as a freelance journalist and writer and travelled extensively before eventually moving back to Perth and working for the Western Australian government.

*   ·*   *

In the postwar years, as for other Australian cities and writers, the mood changes. Randolph Stow in *The Merry-Go-Round in the Sea* explores the postwar malaise or discontent that sets some to chafing against a suburban conformity that did not seem so onerous in the previous idyllic age.

In *The Young Desire It*, set in the 1920s, Guildford Grammar School, called Chatterton by Seaforth Mackenzie, was described as it appeared when the thirteen-year-old Charles Fox arrived there for the first term.

> In the late afternoon of a day in February, that hottest of Australian summer months, when a brutal sun stood bronze above the river flats which you may see from the dormitory windows of Chatterton, Charles came to the School with his mother, walking from the railway station to the gates by a private path across a burnt, untidy field, overhung with Cape lilacs that still drooped, dusty and melancholy, in the late heat of afternoon.
> It was February, with three more months of summer yet to come.[12]

Mackenzie himself had run away from Guildford at sixteen and never returned. Now, in *The Merry-Go-Round in the Sea,* Stow writes of the 'tall, narrow and pale' chapel rising behind Cape lilacs which in spring 'would pile in drifts in certain corridors, drowning the school smell with sweetness, floating even in inkwells'. Rob's cousin Rick discusses the previous publication with a friend.

> 'Did you ever read *The Young Desire It*?' he asked idly.
> 'Quiet,' Tom said. 'You don't mention that round here.'
> 'It's good,' Rick said. 'It feels young. He was young. So I suppose they can't forgive him.'

In the same novel, a character called Hugh Mackay, a wartime comrade of Rick's, rents a house near the river 'in the brick-and-tile and conscientiously gardened suburb' of Nedlands near the University of Western Australia (from which Stow graduated in 1956 and worked as a temporary lecturer in the English department in the early 1960s). The imagery of banality is reminiscent of George Johnston's reaction against confining suburbia in *My Brother Jack*, as is Rob's hero-worship of his older cousin and his friend reminiscent of the young David Meredith's admiration of Jack. 'There was something brilliant and buccaneering about Hughie, that ought not be shut up in a shop, that ought to be out in the Territory hunting camels among the legendary gallant bones, or shooting crocodiles in the wet north', thinks Rob.

But Hughie, of working-class origin, is pragmatically prepared to settle down and make the best of it. Rick, however, who like Hughie has nearly lost his life in a Japanese prison camp in Thailand, rages against his boredom and 'wants to be young before he is old' in a passage which reflects Johnno's similar rage against Australia in David Malouf's *Johnno* (1975).

> 'I can't stand this,' Rick said, 'this—ah, this arrogant mediocrity. The shoddiness and the wowserism and the smug wild-boyos in the bars. And the unspeakable bloody boredom of belonging to a country that keeps up a sort of chorus: Relax, mate, relax, don't make the pace too hot. Relax, you bastard, before you get clobbered.'[13]

At the end of the novel, Rick leaves for London. Like Johnston, like Malouf, Stow also left, and after a brief return lived permanently in England.

A RISING TIDE OF HEAT:
The Western Australian Landscape

To live in Western Australia is to be strongly aware of a physical landscape—one behind the urban facade, and even though the population is overwhelmingly urban. This sense of another environment comes through to the sprawling suburbs, it comes in the smoke of the forest department's endless burning fires, the lack of water, the heat, the distances …

And out beyond metropolitan Perth there are stretches of quite pitiless but utterly attractive landscape. Even here we put down instant towns and suburbs that are replicas of Perth—or other such Australian models. The new iron ore towns of the north, for instance. This does make for sameness, yes. Yet a few miles outside their airconditioning and supermarts one can die in a couple of days, left alone.

Peter Cowan, 'Regionalism in Contemporary Australia', *Westerly*,1978[1]

When Western Australian writers move in their maturity into the regions beyond the immediate vicinity of Perth for their subject matter, the picture changes dramatically. Peter Cowan's stories of rootless, affectless rural workers and the inhabitants of the mundane landscapes of small Western Australian towns stand in sharp contrast to the lyrically nostalgic work of his contemporary Seaforth Mackenzie.

Peter Cowan was born in Western Australian in 1914 and, after leaving a job as an insurance clerk, worked in the depression years of the 1930s as an itinerant farm labourer before enrolling at the University of Western Australia in 1938, where he later taught. He now lives in Mount Claremont, Perth. His fictional landscapes, ranging from the jarrah forests of the southwest to the central and eastern wheat belt and small farming areas, the goldfields and Perth itself, are based on those areas which he directly knows. Like Stow in his novel *Tourmaline* and Katharine Susannah Prichard in *Coonardoo*, Cowan takes as his territory the 'quite pitiless but utterly attractive landscape' beyond Perth where the wilderness itself has a major effect on man's designs. Where a Western Australian regional writer such as Elizabeth Jolley or a Queensland regional writer such as Thea Astley might use the patterns of weather and season and physical landscape to reflect and reiterate the state of mind of her characters for satiric or dramatic effect, Cowan—and to a lesser extent Stow and Prichard—lets the landscape dominate their actions.

'The Collector', from Cowan's 1979 collection *Mobiles*, is set near Cue, an old mining town not far from Mount Magnet, 645 kilometres northeast of Perth on the Yilgarn plateau, a geologically ancient region of spectacular granite landforms. A man and a woman, emotionally warring, have strayed from their car while collecting antique bottles, and are lost.

The ground was bare, only the broken boughs of the mulga, near them the few clumps of dry cassia. As they came closer to the breakaway, quartz, white, clear in the light, scattered across the surface. Bright, she thought. Like water, Snow. Snow out here. By the base of the outcrop long bands of white clay, stained where the brown ochre ran from the upper surface. Broken to make a white floor in the clefts and caves that were shadowed.

Shade, she said. At least it's cool.

She heard his boots scrape on the rock as he began to climb one of the clefts to the top, and then in the silence she looked out over the flat, hot ground they had walked over, the spread of the mulga bushes, more distant now as the light lessened. All the same, she said, hearing her voice against the soft walls of the cave, the silence. Everywhere. The smell of the soil beneath her feet. Sharp. Brown, dry pellets.

We can't stay here. She looked at him standing in the opening. It's filthy. Look at it.

Kangaroos, he said. We'll have to stay here.

No.

Well, you can sit outside.

You didn't see the track?

No. It's higher ground over towards the west. Tomorrow we can go back. We'll come on the machine.

Machine, she said. Her hand beat at the flies. Why do you always call it that?

Do I? I don't know. I didn't know I did.

And come on it is the only way we'll ever find it now.

We'll have to stay here. In the dark we'll be completely lost.

I'm glad, she said. Glad we're not completely lost.

You don't make it any better. And I did suggest you stay in the machine. Sorry. Whatever you call it.

By myself. Waiting for you. What would I have done.

The shade from the rock moved out in a long angle. In the sun, near the dry trunk of a mulga, a lizard stretched, without movement. Like the stones. He said: It's cooler, now. We could walk.

If there's any point. She moved slowly. I was going to work tomorrow. But I don't remember what day it is.

It was almost dark when they reached the round mass of broken red boulders, and the mulgas seemed thicker, branches scattered on the ground. The sharp green of poverty bush.

It's the same place, she said.

Not again. Please.[2]

In being stripped to the most elemental physical states, Cowan's characters are also reduced to the most elemental of emotions, and in this story the woman's realisation that the man may have deliberately stranded them to cause her death leaves her strangely indifferent, just as the bare, austere and silent landscape is indifferent. The landscape is unforgiving and distant—so are the man and woman, who are nevertheless at its mercy.

Like Peter Cowan in his fiction of small towns and suburbia, Randolph Stow also seems ultimately to find civilisation an ambiguous blessing. In *The Merry-Go-Round in the Sea*, the European towns and pastoral settlements carved out of the Western Australian wilderness are represented as a haven of love and almost idyllic family life. However, in his earlier works, the loss of innocence which is found at the end of that book is firmly implanted from the beginning. Nowhere here does he attempt to evoke such childhood certainties and such faith in human nature. In other novels where he draws on the landscape of Western Australia he frequently places his characters uneasily between the small settlement and the wilderness, as if to recognise the precariousness of European occupation. This precariousness is represented also in the spiritual and emotional state of his European characters.

At the beginning of his fourth and fifth novels *To the Islands* (1958) and *Tourmaline* (1963), Stow sketches the natural landscape as an isolated and desolate environment against which his protagonists try to work out their individual destinies in the eyes of their individual God. In both cases (as in *The Merry-Go-Round in the Sea*) the arrival of a stranger, or the return of someone long gone, causes a crisis in the lives of those who have remained. In *To the Islands*, set on an Aboriginal mission station in the northwestern wilderness, the aged missionary Heriot wakens to a scene where the landscape and its inhabitants conjoin to reflect his emotions.

A child dragged a stick along the corrugated-iron wall of a hut, and Heriot woke and found the morning standing at his bed like a valet, holding out his daylight self to be put on again, his name, his age, his

The spring country flowered in stunning profusion, flowered like chintz with flowers whose multitudinous names one could never hope to learn, flowers whose names were 'that mauve one with three petals' and 'that pink one like the ones Little Red Riding Hood was carrying in the picture-book'. … he would never know the names of more than a fraction of the flowers of his country.

RANDOLPH STOW, 1965 (PAGE 123)

WILDFLOWERS, WESTERN AUSTRALIA. PHOTOGRAPHER: OLIVER STREWE

Scattered between … are a few lonely farms tucked and folded as if sewn neatly into the landscape for many years. In places, where the road rises, the dark seams of these farms can be seen in the distance. It is as if they are embroidered with rich green wool or silk on a golden background. …

ELIZABETH JOLLEY, 1985 (PAGE 134)

WHEAT CROP. PHOTOGRAPHER: PHILIP QUIRK

vague and wearying occupation. His eyes, not yet broken to the light, rested on the mud-brick wall beside his bed, drifted slowly upwards to the grass-thatched roof. From a rafter an organ-grinder lizard peered sidelong over its pulsing throat.

Collecting himself from sleep, returning to his life, he said to the lizard: 'I am Heriot. This is the sixty-seventh year of my age. *Rien n'égale en longueur les boiteuses journées -*'

Outside the crows had begun their restless crying over the settlement, tearing at his nerves. The women were coming up to the kitchen. He could hear their laughing, their rich beautiful voices. Already the heat was pressing down on him, the sheet under him clung to the skin of his back, and it not yet six o'clock and a long day.

'When shall I be cool again?' demanded Heriot of the lizard. 'Soon the weather must change, the Wet is over, an old man can begin to live again.' He tore aside further his sagging mosquito net, and the lizard took fright, dropped down, scurried to the doorway and froze there, waving a frantic paw.

… Deep in fading grass the country stretched away from the hut, between the rocky ridge and the far blue ranges, dotted with white gums, yellow flowering green-trees, baobabs yet in full possession of their foliage. And from the grass, which harboured also goats, creepers and all rustling reptiles, rose the Mission, the ramshackle hamlet of huts and houses, iron and mud-brick and thatch, quiet below the quiet sky.

So still, so still in the early heat. Standing at the door of the shower, pulling on his shirt, he watched Mabel walking through the grass, Djimbulangari slowly following. They moved like he did, loosely and tiredly, two old women with their hair tied in kerchiefs, their dresses hanging straight on their thin bodies. Looking at Mabel he thought that he had never seen her in any clothes but these, the dirty coloured skirt sewn to a flour bag bodice on which the mill brand was still bright green and legible. Picking their way like cranes through the grass talking occasionally, not looking at one another. Old, dried-out women, useless and unwanted.[3]

Randolph Stow worked at the Anglican mission at Forrest River in the Kimberleys for six months in 1957, among the Umbalgari people. *To the Islands* deals with the last days in the life of the director of such a mission, who believes he has killed one of the Aborigines, a newly-returned outcast, with a stone thrown in anger. He sets out to find his own fate in the 'islands' of death, which the local Aborigines believe await them. He is accompanied by Justin, another Aboriginal, who only agrees to leave him when he knows that Heriot has found a form of peace and is therefore ready to die.

Tourmaline, completed five years later, also begins with a description of a small settlement, the value of which seems equally ambiguous in the eyes of the narrator, the local policeman.

I say we have a bitter heritage, but that is not to run it down. Tourmaline is the estate, and if I call it heritage I do not mean that we are free in it. More truly we are tenants; tenants of shanties rented from the wind, tenants of the sunstruck miles. Nevertheless I do not scorn Tourmaline. Even here there is something to be learned; even groping through the red wind, after the blinds of dust have clattered down, we discover the taste of perfunctory acts of brotherhood: warm, acidic, undemanding, fitting a derelict independence. Furthermore, I am not young.

There is no stretch of land on earth more ancient than this. And so it is blunt and red and barren, littered with the fragments of broken mountains, flat, waterless. Spinifex grows here, but sere and yellow, and trees are rare, hardly to be called trees, some kind of myall with leaves starved to needles that fans out from the root and gives no shade.

At times, in the early morning, you would call this a gentle country. The new light softens it, tones flow a little, away from the stark forms. It is at dawn that the sons of Tourmaline feel for their heritage. Grey of dead wood, grey-green of leaves, set off a soil bright and tender, the tint of blood in water. Those are the colours of Tourmaline. There is a fourth, to the far west, the deep blue of hills barely climbing the horizon. But that is the colour of distance, and no part of Tourmaline, belonging more to the sky.

It is not the same country at five in the afternoon. That is the hardest time, when all the heat of the day rises, and every pebble glares, wounding the eyes, shortening the breath; the time when the practice of living is hardest to defend, and nothing seems easier than to cease, to become a stone, hot and still. At five in the afternoon there is one colour only, and that is brick-red, burning. After sunset, the blue dusk, and later the stars. The sky is the garden of Tourmaline.

The fictional town of Tourmaline is reputedly based on the old gold mining settlement of Sandstone, about 665 kilometres north of Perth on the Great Northern Highway. The present town of Sandstone has a population of about sixty, in comparison with some six thousand in its heyday at the beginning of the century.

> To describe the town, I must begin with the sun. The sun is close here. If you look at Tourmaline, shade your eyes. It is a town of corrugated iron, and in the heat the corrugations shimmer and twine, strangely immaterial. This is hard to watch, and the glare of the stony ground is cruel.
> The road ends here. There is a broken fence to show it, its posts leaning, its barbed wire trailing to the ground. Facing this, the Tourmaline war memorial, a modest obelisk, convenient for dogs and the weary. Some sons of Tourmaline, it seems, patronized the empire in the days of the Boer War, but not much is remembered. To the right is Tom Spring's store, the white paint flaking from its iron and the purple paint from its ancient advertisement for Bushell's tea. In the window, shaded by a rough veranda, tinned food, soap, cutlery and boots cradle the immemorial cat of T. & M. Spring.
> On the left is Kestrel's Tourmaline Hotel, of stone and rough plaster, once whitewashed, but now reddened with dust. The roofing iron is also red, and advertises a brand of beer no longer brewed. A veranda shades the bare dirt on three sides. In this hot metallic shade Kestrel's dog wakes and yawns, and sleeps again. The windows are closed, and painted inside. It is dim in there.
> Following the raw red streak of the road are the houses of Tourmaline: uniform, dilapidated, stained with the red dust. There are not many.[4]

The lives of the few remaining inhabitants of Tourmaline are altered by the arrival of a newcomer, almost dead from exposure, who comes to be regarded by some as a saviour through his supposed ability to divine water. There are strong parallels here between *Tourmaline* and Thea Astley's novel *An Item From the Late News*, published in 1982, in which a similar isolated mining town reacts to the same circumstances—the coming of a man perceived to be a saviour—with unmitigated evil. Both novels end in the defeat or the destruction of the man-made saviour. 'Tourmaline' is a real Western Australian town in its detail, but Stow uses it to allegorical purpose or as a means of contrasting the ideas of Taoism and Christianity, according to one critical interpretation of his sequence of poems 'From the Testament of Tourmaline: Variations on the Themes of Tao Teh Ching' (1966).

In both *To the Islands* and *Tourmaline* the Western Australian landscape is delicately and poetically evoked, even when it is at its most harsh, and it is a terrain of the mind as much as the body. For Stow, like Patrick White in *Voss* (published in 1957, but of which Stow was unaware at the time of writing *To the Islands*), the wilderness outside the fragile settlements is a proving ground where man confronts or seeks out death, or learns to reconcile himself with the rest of humanity. 'Man' here is the appropriate word: Stow's women may be catalysts, or may represent sources of an abstract kind of love, but they are never an active principal in his parables of spiritual seeking.

It is interesting to compare Stow's work in this aspect with the writings of E. L. Grant Watson. In 1910, in company with the anthropologist Alfred Radcliffe Brown and Mrs Daisy Bates, Watson, then 25 and just out of Trinity College, Cambridge, set off with a Swedish cook, two Aboriginal guides and a horse and cart for Geraldton, then to the goldmining town of Sandstone. Outside Sandstone they set up camp in the bush between two large Aboriginal groups numbering several hundred, who were not on particularly friendly terms. Watson describes this Australian experience, which he later said changed his life, in his autobiographical accounts *But to What Purpose* (1946) and *Journey under the Southern Stars* (1968).

Like that of Randolph Stow, E. L. Grant Watson's thought was influenced by Tao philosophy. After his return to England, he mixed in the cultural and intellectual circles of T. S. Eliot and Joseph Conrad, and corresponded with Jung and Havelock Ellis. Among more than 40 works of fiction, poetry, scientific observation, and psychoanalytic and spiritual searching, Watson wrote six novels with Australian settings, several of which were concerned with themes of the spiritual purity of the desert landscape, and the effect of its solitude on the European mind. In recent years, wrote critic Dorothy Green, Watson's biological work has become interesting to scientists who do not find the purely mechanistic explanation of the phenomena of evolution wholly convincing.

Of a period occurring a decade before the appearance of D. H. Lawrence's Australian work and nearly half a century before both Patrick White's *Voss* and Stow's early novels, Watson writes:

> Those first weeks I spent in the bush were rich, not so much in outer as in inner experience … In psychological interpretation, it is, I suppose, that the mild, innocent and aloof quality of that virgin territory appears as a symbol of the unconscious, as a symbol of all that civilisation has chosen to disregard. It is a vast interrogation mark, questioning itself, and more than consciousness can know of itself, or indeed of life. It says to man: thou insignificant spark, where art thou? How is it with thee in thy soul? Canst thou sustain my vast and indifferent regard? Or wilt thou shrivel into nothingness, rather than listen to my silences?[5]

In *To the Islands*, Stow uses his secondary characters to explore a variety of attitudes to the landscape, and the power of the country to make its inhabitants peculiarly its own. One such character, Dixon, finds that his isolation on the Aboriginal settlement has made him too much part of the landscape, and like the Aborigines, he feels alienated in the coastal towns.

> Dixon wandered away and turned at the end of the street towards the foreshore where the brown sea lapped at the brown mud. The sea and the hills hemmed in the town, it could never be more than one street wide. The boat rode at anchor on the water, and he could see one of the boys asleep on the deck, but there was no sign of the others. But he guessed where they would be, and strolled on towards the citadel of empty petrol drums on the shore, and through its passages walled higher than his head, until a whispered: *'Djuari brambun!'* stopped him. White man, devil-devil coming.
>
> They were sitting in a kind of room inside their labyrinth, and seeing a few scattered cards he knew that they had been playing their own peculiar form of poker with a few natives from town. The cowboy hat that he remembered seeing on Arthur's head, during the trip in, now sat on a stranger, but Matthew had acquired a girl's scarf and had it knotted round his throat. Around them lay broken and empty beer bottles, relics of ancient parties. But they had not been drinking, they swore it with their defiantly innocent eyes as they watched him.
>
> 'When we go, brother?' Matthew asked eventually.
>
> 'Tonight's tide. That's what I came to tell you. Should be about ten past eleven if I've worked it out right.'
>
> Arthur gave an exaggerated sigh. 'I happy now, brother. This country make me sick.'
>
> Wish it made you sicker, Dixon thought. Wish it made you sick enough to stay on the boat and leave these town blokes alone. Lines from a song that Gunn sang with the children round the piano came back to him.
>
> *Tell Bill, when he leaves home,*
> *To let them down-town coons alone.*
> *This morning. This evening. So soon.*

'This isn't our country,' he said. 'The mission's our country.'

'I feeling homesick,' Matthew complained, 'away from my country.'

Dixon grinned. He liked them, he would have preferred to stay with them, but he was more than ever foreign to them, and unwanted, here. He was foreign everywhere, and disliked it, being a friendly man and anxious to be in no way different from the rest of the human race. At such times he recognized, without congratulating or pitying himself, the extent of his sacrifice. But there was no help for it, he could only go and fix up about a few stores, then drift into the hotel and sit about somewhere waiting for the tide. It's a dog's life, he thought. I feel homesick for Matthew's country, too.[6]

In vivid contrast is Randolph Stow's treatment of the Western Australian fishing port of Geraldton in *The Merry-Go-Round in the Sea*. Seen through the lens of an idyllic childhood in the 1940s, the small northern town in wartime seems old and mysterious, and to possess an almost mythic quality. It is from this illusion of age that Stow's six-year-old main character Rob obtains his sense of history.

His mother was in the Library, getting books. He could see her now, coming out on to the veranda. The Library was a big place with an upstairs. It used to be the railway station in the Old Days, which made it very old indeed. … Across the street the convict-built courthouse crumbled away, sunflowers sprouting from the cracked steps. The great stone barn at the next corner was Wainwright's store, where the early ships had landed supplies. That, too, was crumbling, like the jetty and the courthouse and the bougainvillaea-torn shed, like the upturned boat on the foreshore with sunflowers blossoming through its ribs.

The boy was not aware of living in a young country. He knew that he lived in a very old town, full of empty shops with dirty windows and houses with falling fences. He knew that he lived in an old, haunted land, where big stone flourmills and small stone farmhouses stood windowless and staring among twisted trees. The land had been young once, like the Sleeping Beauty, but it had been stricken, like the Sleeping Beauty, with a curse, called sometimes the Depression and sometimes the Duration, which would never end, which he would never wish to end, because what was was what should be, and safe.[7]

Geraldton itself, a fishing port on the Indian Ocean 421 kilometres north of Perth, was established in 1849 on the recommendation of the explorer George Grey. Xavier Herbert, who was born in Walkaway, a township a dozen miles southeast, in 1901, has described the town in that era somewhat scathingly in his autobiographical work *Disturbing Element* (1963):

… it consisted of a long jetty jutting out into the waters of the Indian Ocean, a struggling main street that crookedly followed the shore line, a little railway depot, and two or three cross streets that ended in a sandy scrubby waste in which there was a fringe settlement of Afghan camel drivers and the dispossessed aboriginal blacks.[8]

Nene Gare also touches on the Aboriginal camp at Geraldton in her novel *The Fringe Dwellers* (1961). There are

[a] half dozen humpies … clustered on the outside edge of a half-circle of bare brown earth and beside them, shielding them from the wind, grew bushy stunted wattles. The humpies were of wood and rusty iron reinforced with rotting grey canvas, and none of them looked large enough to contain more than two small rooms. From each humpy projected the inevitable rough bough shelter which also served as a kitchen.[9]

However, these images are missing from Stow's childhood memories. In his idyllic, dreamlike existence there are only the scents and sensations of an upper-middle-class semi-rural life in which to rejoice.

The tennis balls were going *pong! pong!* against the racquets, and galahs were screeching over the orchard, wheeling in the late sun, eyeing the nectarines. Aunt Mary was sitting on a rug under the gumtree, surrounded by children. The gumtree reached right over the tennis court fence and dropped small nuts on the court just in front of the bench where the boy was sitting. He was sitting forward on the bench, bending over, watching the bull-ants swarm among the gumnuts.

... It seemed to the boy that he had spent a lot of time watching people play tennis: on the tennis court at home, when his mother gave the ladies the yellow drink with the passionfruit in it, and at Susan's, and at Sandalwood. In summer they played tennis, and in winter they played golf. Everybody's house was full of silver cups and silver spoons for playing tennis and golf: and Andarra, which was Gordon's house, was full of even bigger cups for playing cricket. Rick and Gordon had so many cups between them that Aunt Mary had told them not to win any more, she was sick of cleaning them. Gordon was famous for playing cricket. Even now, when he was at the war, they remembered that.

However, there is still a yearning for something intangible beyond the town, a certain restlessness for the alternative life of the inland sheep stations—or, perhaps, the terrain of the mysterious 'Australia' that lies beyond the world Rob knows.

The town had its scents and seasons like the country. In the dry, burning easterly of March, leaves wilted, grass dried. The southerly was damp and salt off the sea. And winter smelled of crushed wintergrass, crushed clover, crushed, sour and bitter weeds, fat-hen and mallows.

And spring smelled of capeweed and wattle, the pollen of capeweed and wattle, the green-white flowers of peppertrees, the flowering sandhill scrub. Spring smelled of wild jonquils in his grandmother's lawns, of freezias tended by Aunt Kay, of that elusive scent from the black-columned mauve-petalled sprays of Cape lilac. And autumn smelled of the first rain, bitter and clean; and bitterly, cleanly, of chrysanthemums in Aunt Kay's garden.

In every season the boy exulted in his senses, in his body. He exulted in the heavy sweetness of jonquils and in the frail scent of tomato leaves; in the harsh rasp of leaves on his skin as he climbed a figtree, and in the waxy dusty smoothness of the minute datepalm flowers; in the cold sea of early morning, and in the warm sea under the rain. He loved the rough taste of gumleaves and the sweetness in tecoma flowers; the red jewels in pomegranates, and the shells of rainbow beetles in the grey tuart bark. The boy then was little more than a body, a set of sense organs. To himself he had little identity, and to his friends none at all, as they had none to him. They knew each other by sight and hearing, by certain mannerisms. In absence, they ceased to exist for one another.

They exulted in their bodies, in their senses; in skills of movement that they were learning, in appetites that were new. Their bodies, their senses did not fail them. Only with the first sunburn of summer could a body be a liability; and when that was over it became a new fascination, as absorbedly they peeled the skin from each other's backs.

The boy exulted in his senses. The town had its sounds and its scents and its seasons. But at times he raged against the town, feeling dispossessed, feeling exiled from the country where he knew his body belonged.

But like Malouf's Brisbane, and Robert Drewe's Perth, even Geraldton in the 1940s undergoes its rites of passage, and passes from the golden age of Rob's childhood to a new and brash modernity.

Mrs Maplestead's palms had chambers and passages, reptile-haunted brown caves where secret societies flourished briefly and then died of boredom. At the risk of snake-bite or impaling an eye, it seemed that boys might extend the labyrinth forever.

Then the bulldozers came, and the palms went down in a roar and a fume of sand. Flames leaped, and the palm-trunks lay under sun and rain in the devastated paddock like black-scaled basking dinosaurs.

Mrs Maplestead had sold her palms, and people were going to build houses. The town was growing in a sudden spasm. It was going to be a *real* town, everyone said.

Fury reddened the boy's face, and choked his speech. 'Oh, you conservative old thing,' said his mother, surprised.

She could not understand, and he could not explain to her, that the bulldozers were worse than the Japs.

The sandy town shook with explosions as the piles of the jetty, the first of the two rotting jetties to go, were blasted from their bed. There was more sea without the jetty, but still the boy was resentful. Huge grey timbers washed up on the surf beaches, and the boy stood on them, thinking that he had probably stood on them before.

He supposed that some day, in South Africa or somewhere, other boys would stand on the timbers of the jetty and wonder where they had come from, as he always wondered about driftwood and bottles.

He walked in the town and watched the town change: the empty, dirty-windowed shops restored, the poky, shabby shops growing Yankee-flash, the swinging doors coming off the pubs, the verandas and wrought-iron balconies over the street torn down by order of the council. In time, the whole run-down haunted town would be reborn, remade, according to standards of beauty and elegance proper in a nation which had done its pioneering in hovels. ...

So he walked the streets, and then rode his bicycle through the streets, and at last drove his car through the streets, asking himself how a country town on the sea had become a provincial seaport, how a world so congruent, so close-knit by history and blood and old acquaintance, had become fragmented into a mere municipality. But he knew the answer, by that time.[10]

In her novel *Foxybaby* (1985), Elizabeth Jolley makes use of the small towns of the southern wheat belt to provide a sensation of isolation similar to that made familiar by Cowan and Stow. This is the world, removed from ordinary urban existence, into which Jolley catapults her protagonist Alma Porch, a novelist teaching at an isolated summer school.

There is only one road going east from the township of Cheathem West and this road after approximately two hours of sedate driving (one hour for the reckless) becomes the main high street of Cheathem East.

There are scarcely any houses in Cheathem East as very few people live there. There is no hotel and no shop.

Scattered between the two Cheathems are a few lonely farms tucked and folded as if sewn neatly into the landscape for many years. In places, where the road rises, the dark seams of these farms can be seen in the distance. It is as if they are embroidered with rich green wool or silk on a golden background. In the design of the embroidery are a few trees and some silent houses and sheds. Narrow places, fenced off and watered sparingly, produce a little more of the dark-green effect in the picture.

At intervals, sometimes as if they do not belong to anyone in particular, there are unsupervised windmills turning and clicking with a kind of solemn and honest obedience.

On reaching Cheathem East the high street, after passing directly alongside the surprisingly unexpected windows and front door of a very large old house, very soon becomes a gravel track which, with many twists and turns through endless paddocks of wheat, turns finally and like a river without any water reaches the sea. This place where land and water meet is scattered with enormous rocks as if some enormous children, the sons and daughters of a pair of happily married giants, suddenly tiring of their playthings, have hurled them into the sea. The dark waves running and rising wash endlessly over the rocks from both sides giving an impression of two oceans meeting in conflict. Some of the rocks dropped by the little giants rest where they have fallen along the wide sandy bay.

But where Cowan's landscape is spare and ungiving, Jolley, with her distinctive European vision, turns the prosaic region of the southwest wheatfields into an eerie Gothic dreamscape, so that Alma Porch might explore the intangible boundaries between fact and fiction like an elderly, idiosyncratic Alice.

There are strange things about driving alone on long lonely roads through the wheat. Old, grey, bent men and women wait indefinitely on green misleading corners, becoming part of the bushy roadside undergrowth as soon as the helpful traveller stops to investigate. Comfortable inviting tracks in the twilight, lined with soft sand and leaves, appear to lead off easily to the right as the main road curves to the left. And, as dusk advances, more gnarled old men march in formation, keeping up a remarkable speed, alongside, in the shadowy fringes of the saltbush. Occasionally a solitary driver pulls off on to the shoulder of the road to allow a ship to cross in front of him from one moonlit paddock to another.[11]

The younger Western Australian writer Tim Winton, in his second novel *Shallows* (1985), also turns a fictional eye to the same region, this time to the whaling town of Albany, 408 kilometres from Perth on the state's southernmost tip. *Shallows* is the story of several generations in a small coastal town, here called Angelus, and of the conflict between conservationists fighting to save the remaining breeding whales that frequent the area and the townspeople whose livelihood has traditionally depended on their slaughter. In a prologue, Winton provides a vignette of life in the town on the day of an annual football game.

And now it is the year 1978 in Angelus, Western Australia.

The town's station wagons form a metal and glass perimeter around Angelus Oval. Two teams of men slog about in the turfy mud, upending one another, punching and kicking the soggy leather kernel from one end of the swampy ground to the other. The townspeople cheer and jeer from the waxed fenders of their Kingswoods.

Behind the circle of cars, boys play in the muddy gravel, damming up the ochre water, and girls spatter their floral frocks in games of hopscotch, watched by the old woman with the shopping bags who leans on one leg—bun slipping from the side of her head like a cherry from the tip of a melting cupcake—a cathedral of noise in her ears.

After the game, half the town is ushered into the belly of the rickety grandstand for celebrations. Des Pustling, patron of both teams, spits another tooth and sucks his bloodless gums as he signals for the spearing of the kegs. For an hour, this afternoon, black footballers and their families drink with the whites.

In the evening, the pubs are full of talk. Old drinkers in the Royal Albert fling stories and froth and wisdom about; grey-suited men in the London talk through the crooks of their arms and scatter void betting slips. Businessmen in the Amity and the World shout with their heads low to the bar, and at the Black & White words and darts fly thudding into the walls. And down by the waterfront, as he waits for the whalemen to arrive, Hassa Staats, the Aryan publican of the Bright Star, pours watery beers for old men whose fathers drank here, and whose sons will drink here when they tire of youth. His anticipation is boyish, incongruous to his sixteen stone. He almost hops from foot to foot. On the floor, head resting in the sputum and ash of the foot-tray, Earnie Easton, Staat's oldest customer, unable to remount his bucking stool, is dreaming of 1915...

In the main street an old woman sleeps against the window display of the new Woolworths store.

Off the main street, an old man writes his sermon for the morning, eyes stung to tears by the fumes from a kerosene heater.

Those Aborigines strong and sober enough wander down from the Reserve, caught in the chiaroscuro of lightning, to sit outside the pubs and beg bottles of Brandovino from the white men who fall and fight on the footpaths. On a deserted golf links across town, their sons fight a rival group from out of town with crowbars and reticulation spikes.[12]

However, it is left to Robert Drewe, who grew up in Perth, to bring the remoter settlements of the Western Australian coastline into the modern age in *Fortune* (1986), his adventurously constructed novel of marine treasure-seeking. Drewe's most amusing character here is Colonel Tripwire, American ex-Vietnam veteran and nuclear refugee, who has settled in the early 1980s in what he considers the safest place in the world. Along with a number of other refugees of various types, Tripwire—so called because of his habit of guarding his perimeter with alarm systems of fishing line and tin cans— lives at a place Drewe calls Thirsty Point on the south coast of Western Australia, where he hoses his turkeys to keep them alive in the summer heat and awaits 'the Doctor', a cooling afternoon wind.

> He could cope with landscape; he had beaten landscape. He must remember he'd chosen this particular landscape for its advantages. In this landscape he could escape and evade. Now he must learn to cope with weather. Unremitting extremes in weather were, after all, to expected in the long, or short, term. Therefore, squirting his turkeys that morning, the colonel had concentrated on weather.
>
> The present weather could be broken down into two factors: wind and heat. The Easterly and the sun. Even before sunrise the Easterly was quickly drying the night's sweat on his scalp and chest. He faced into it. Over the great red tableland and the scrubby escarpment it brought him (personally, he thought) subtle whiffs of gravel dust, eucalypts and dead grasses. When the sun was up the wind delivered him traces of bushfire smoke, hot granite surfaces and something else—a scaly, furry, musty odour that lodged high in his sinuses and reminded him of panicky animals kept in a confined space. Then it scorched and battered his farm, seared his vegetables and sent his turkeys' temperatures soaring.
>
> It was odd remembering how benignly he had regarded the Easterly until the Doctor ceased to arrive. Only the summer before the land and sea breezes had complemented each other. You could set your clock on the Doctor calling at two, bringing relief from the midday heat, and on a calmer, balmy Easterly returning at seven.
>
> His wind chimes always changed tune. When the wind turned around, the brass chimes from Thailand or Kampuchea or wherever changed sound from *tinkle* to *ping*. After their Wednesday tennis game he and Leon would sit on the terrace watching the crayfish boats returning home, laughing at the idea of Flora and Fauna, the crayfish inspectors, spreadeagled in the dunes, spying on the fishermen through their binoculars.
>
> The boats ploughed through the choppy waves. The chimes changed from *tinkle* to *ping*. The smell of the backyard barbecues wafted up from Treasure Point. Behind their deck chairs stretched a continent of ancient desert. In front of them the sun sank into the horizon. The glass of beer was cold in the hand; its condensation dampened the palm. How delightful was that first swallow, the prickling in the throat, the hops tangy on the tongue! At these moments there was peace in the soft warm wind and the idea of the land taking over from the sea.
>
> The colonel missed these serene evenings. This summer the heat had stopped their tennis, called a halt to most socialising, especially among the Americans.[13]

The unspoken irony is, of course, that at North West Cape a major joint Australian-American satellite tracking station is located, making Western Australia, in some opinions, a major communications target in the event of nuclear war.

THE FRAGRANCE OF DUST:
Modern Perth

Every morning the sprinklers make water snakes in the dust. The fragrance of this water on the dust is sharp like an anaesthetic. There is too the smell of petrol and of dogs' dirt and of empty champagne bottles, of scented groins and burning toast. Sometimes there is the sweetness of cut grass drying in little brown ridges. Then there is the aromatic scent of the yellow broom and of roses and of a lemon, bruising slightly, as it falls. The spindles of rosemary straggling by the gate post, brushing the legs of people as they pass, add their fragrance to the Chinese privet and the datura—those long white bells whose perfume can lift the passer-by into a temporary forgetfulness. …

In an area between the railway line and the sea there are certain places, a bend or a gentle rising of the road, from where it is possible to see the sea. Serene blue surprises in glimpses between the trees and the houses. This smooth blue sea, beyond immediate reach and yet visible from time to time, seems to meet the sky in a quiet gentleness only possible in dreams. And dreams are necessary in the suburbs …

Elizabeth Jolley, 'A Sort of Gift: Images of Perth' 1988[1]

In Peter Cowan's story 'Seminar' from his later collection *Mobiles* (1979), the aridity of the farming country which is this writer's usual literary territory is replaced by the suburban aridity of provincial university life. Here, an unnamed academic who has come to deliver a paper on Faulkner meets his former lover at a literary seminar in Adelaide. He finds that, despite a resurgence in sensual pleasure, their inability to touch each other's essential being is now set in aspic. In an exchange typical of Cowan's laconic dialogue, Lena, the former lover, setting out to return to husband and domesticity while the academic himself prepares to leave for the self-professed nullity of his unnamed home city, asks:

And you are on your way back?
To my far western frontier. My lawn green suburban land.
You like it there? You must, I suppose, staying there. As you do.
Yes. I have time to think.
What do you think about?
Nothing really.
I see.
But I have time for it.[2]

In modern Perth it would seem that there is little to sustain the contemporary novelist or short fiction writer. For most other Australians today, Perth conjures up an image of blue skies, constant sunshine, the white sails of racing yachts, and less-than-reputable business entrepreneurs and politicians. Even the senior Western Australian writer T.A.G. Hungerford, in his evocations of prewar Perth childhood in *Stories from Suburban Road* (1989), tends to concentrate on the Arcadian rural edges of the city and rarely penetrates the centre. Perth, on the literary evidence available, appears a centreless place. There are few inner-city Perth writers, it seems, except perhaps contradictorily, Aboriginal ones. And with the exception of Elizabeth Jolley, even Perth suburbia has few modern day champions.

However, as the Western Australian critic and academic Bruce Bennett has pointed out, this recent distaste for suburbia may be in the eye of the beholder. When well-known Perth radio broadcaster Walter Murdoch was writing his attack on what he called the 'suburban spirit' in his preface to his collection of journalistic essays *Speaking Personally* (1930), he described his previous home city of Melbourne in terms of '… the awful sameness of … suburban streets, with their red-tiled houses, neat lawns, gravel paths, Pittosporum hedges, reflecting a uniformity of spirit, a complacency, a positive fear of originality or difference'.[3]

Bennett notes that at the time Murdoch was living in a pleasant two-storeyed house named Blithedale (after a novel by Walter Scott) in the same Suburban Road, South Perth, that figures in the same period in T.A.G. Hungerford's lyrical autobiographical memoir.

This is where, ironically, the poetically inclined young Tom Hungerford, as he recounts in *Suburban Road*, visits the famous Murdoch and is advised—to his displeasure—to write about things he knows: 'A poem must be a statement of fact'. As Bennett comments, 'One man's Arcadia is another's imagined hell.' While Hungerford went on to learn to use his eyes and his other senses on the immediate world, the older academic, living in the same street, suffered the mental tortures of perceived cultural deprivation in the provinces. (Bennett omits to emphasise, however, that Hungerford's idyllic visions of Suburban Road were recreated in nostalgic old age, and that the romantic and escapist literary efforts he showed to Murdoch in the 1930s might have indicated as much contemporary dissatisfaction with Perth as Murdoch's own.)[4]

Nevertheless, this suburban malaise continues to contaminate the views of most of Perth's contemporary writers. Western Australian poet William Grono conjures up a quietly desperate, if comfortable, monotony in his 'Postcard from Perth' (1980).

> Between the long white shore
> and the pillaged hills
> the haze of roses
> in the aching suburbs
>
>   Your mother and I
> are keeping well, touch wood.
> The garden's looking very nice. It seems
> to be getting through the summer
> all right. Old Harry has finally
> passed away, which was a blessed release
> all things considered.
>   Please let's hear
> from you soon …[5]

While Peter Cowan begins 'The Island' (1965), a short story about suburban psychological imprisonment, with similar images: 'The square of lawn before the house, and the small rose bed, were neat and clean, but as he looked at them he thought that the roses needed pruning …'[6]

Elizabeth Jolley, who lives in Claremont, a riverside suburb southwest of the city, has immortalised the area in *The Newspaper of Claremont Street* (1981) for its older, rambling houses and big old trees—'Norfolk Island Pines, Moreton Bay fig trees and the gigantic mulberries in the old gardens'—which, she continues, will 'be bulldozed and burnt and cleared away' to build 'new houses in Spanish or Mediterranean style … together with the two-storey town houses with white walls and red tiles, built in squares around car parks …'. Nevertheless, Jolley shows signs of an ultimate optimism. In an essay for the *Bulletin* in 1988 she writes:

> For the lonely or the heavy hearted the neat streets with well kept lawns, brick and tile houses with

closed doors, blank venetians and drawn curtains, as in other parts of the world, seem to be unpeopled and without exuberance of any sort. In other words they seem to be the most sad and depressing places to be in, especially on a Sunday. And, now that the fashion for high brick walls is coming in, escape is essential. Perhaps the imagination can come to the rescue and a mulberry tree be inhabited by people having a mulberry fight, arms, legs, faces, hair and clothes purple-red with mulberry laughter. Or, at a given time every afternoon, at a certain time of the year, a shepherd complete with lamb and crook appears on the roof of the house next door. A trick of light and shade and chimney? but a Blessing all the same.

The imagination is the saviour, implies Jolley, even in suburbia. In a gentler reiteration of the theme of the bush invading the city, which has surfaced in the work of many of Australia's urban writers— or perhaps, as Peter Cowan has put it, of the 'landscape behind the urban facade'—Jolley writes movingly of the effect of the Australian city on her sensibility as a writer.

> The mood of the river can change very quickly. Is it possible to hear an image? Something unforgettable is the screaming and complaining of a flock of black cockatoos as they fly low over waters changed by gale and heavy rain. One of the questions I am asked from time to time is, has it made a difference to my writing coming to live in Western Australia? And what would my writing be like if I had stayed in Britain. There is no answer to the second question, I am unable to answer it. To the first, of course, there is a difference. Until I came to Western Australia I had never seen or heard a flock of cockatoos. These marauding birds, heralding a mysteriousness unfathomable to us, fly low, almost breasting the choppy waves of the river swollen with rainstorm and purple brown with top soil washed down from the vineyards in the Swan valley. As the cockatoos disappear the rain bird calls, little phrases of bird notes climbing up in among the flame tree flowers brilliant against the dark clouds. Drops of water quiver on the fencing wire and the thin narrow leaves of the eucalypts tremble. To come to this country is to come to foreign land.
>
> How can I be the same person after the flight of the cockatoos?

A little later she notes:

> There was a time when writers, some writers, felt they had to deny their regions. But it is in the very places where you live and walk and carry out the small things of living that the imagination, from some small half-seen or half-remembered awareness, springs to life and goes on living.[7]

Hal Colebatch, writing in the mid-1970s in a poem entitled 'Sestina on Taking a Bus into Perth past the Narrows Bridge', enjoys the same quiet optimism.

> Now here at the end of the West the concrete
> thing is slipping away. But in this dust of waiting afternoon
> there is no history between the City and the glass
> of the river. Little girls hold kittens. The wind
> chases buttefly yachts. Like comfort, small red and blue
> wrapped presents are put on Christmas trees. Such banal love
>
> is still something concrete. In this wind
> is more than afternoon. Sharp into the blue
> air hangs the city of glass. We can come to
> terms with love.[8]

However, a more cynical portrait of suburban Perth is contained in Robert Drewe's novel *The Savage Crows* (1976). Here, the focus is once more on the perils of development, new-found wealth, and undigested modernisation. When the journalist Stephen Crisp returns to the city of his childhood to visit his brother, he finds that he, as well as his childhood suburb, has experienced a new prosperity.

> He had it made, Geoff: the Dalkeith house and pool (on which his brother presently glided on a squashy Li-Lo reversible mattress—green one side, blue the other—absorbing the ultra-violets and cogitating, a fallen gum twig between the teeth, on such matters as Geoff's genial bigotry, moderate wealth and local social cachet), clubby and undemanding business interests, a pretty ('best in the State!') and doting wife, the affection and respect of his peers. Such as they were.
> 'How do you do it?' Stephen had asked him early on. 'How does everyone do it? Why are you all so bloody rich?'
> 'Rich! This is W.A. my boy. Personally I haven't got much dough but this place is rich in every mineral under the sun. This is the biggest quarry in the southern hemisphere. We support the rest of the country.'
> 'Oh, come on.'
> 'We could be the biggest at everything in the southern hemisphere if the East didn't bleed us dry. No bullshit.'

Geoff has married a rich wife, a savvy former beauty quest winner and charity queen whose father, conveniently dead, owned an earth-moving business in three states.

> And Geoff still in his brother's mind's eye a small kid whingeing along behind him on his tricycle, had married this long-legged El Dorado and never looked back. Denise's reputation had launched the Claudia fashion boutique into a sound proposition; buying Spiro's, the wine bar next door, had been his idea but its close proximity to the University made it a sound one. He spent only two mornings a week in his city office, leaving the land speculation (the main money spinner) to his manager. He might play a round or two of golf at Lake Karrinyup or sit around drinking beer with Bernie Caravousonos, who trained his two geldings, in Bernie's kitchen at Belmont. Or he might drive over to the Claudia-Spiro complex, run an eye over the accounts and take the salesgirls from the boutique next door for *quiche lorraine*, sending them reeling back to work full of the house hock. Geoff might stay drinking with a few regulars until the bar filled with students at four o'clock and then take Lady and Cheyenne for a surf and a run along the beach. At night, if there wasn't a lodge meeting, he and Denise usually ate out, either at a favourite steakhouse at Fremantle or at a seafood restaurant built over the river bank where the oysters were flown in daily in ice-packed crates from the East. (The creatures refused to grow locally.) Every Saturday there were the races—and regular mid-week country meetings. There were parties and balls and friends popped in with convivial armloads of bottles. There was never a spare moment.[9]

Drewe spent much of his childhood at Leon Road and then at Circe Circle in riverside Dalkeith, near Claremont, a suburb of quarter-acre blocks developed in the 1920s and 1930s. Born in 1943 in Melbourne, but educated in Western Australia, Drewe worked briefly for the *West Australian* newspaper, but left to move to Sydney in 1961. In his novels and short stories, he draws frequently on these early experiences, particularly in the early memories of his character Stephen Crisp in *The Savage Crows*.

Drewe bases two of his stories in *The Bodysurfers* (1983) at Cottesloe Beach: 'The Manageress and the Mirage', where the young protagonist has a family Christmas dinner at the Seaview Hotel, which is apparently based on the Ocean Beach Hotel, and 'The Silver Medallist', which describes the decline of an ex-Olympic swimming champion who has a beach equipment hire and suntan oil business there.

*   *   *

A small but significant body of Aboriginal writing in English has also emerged in recent years in

Western Australia, much of it originating in Perth. Beginning with the novel *Wildcat Falling* (1965) by Colin Johnson (later known as Mudrooroo Narogin), which opens as its protagonist is released from Fremantle gaol, it continues with Archie Weller's *Day of the Dog* (1981), which describes the world of street fights, stolen cars, booze and police brutality that frequently makes up urban Aboriginal existence. Both novels describe a violent rejection of the white suburban lifestyle, demonstrating that younger Aborigines have moved on from the more resigned imagery presented by Western Australian poet and playwright Jack Davis in his poem 'Whither', published in 1970.

> The Park
> Became our home in the after-dark.
> We sat with our arms around our knees;
> Some of us curled up and died around the boles of
> transplanted trees:
> We crept under bridges looking for vacant ground.
> Here we were safe from the sound of the city street
> And the rat-van on its nightly beat,
> And the empty bottles shrieked a sign
> Of a stupefied brain and red, red wine ...[10]

Long Live Sandawara (1979), Colin Johnson's second novel, continues the imagery of *Wildcat Falling* by juxtaposing the story of the Aboriginal warrior Sandawara, or Pidgin, with the efforts of a part-Aboriginal youth to establish a resistance group in Perth. Once again, urban life is portrayed as a largely negative experience.

However, the relative loyalty of Perth's writers, in that most continue to live there, demonstrates that there are grounds for renewed hope. Western Australian academic Veronica Brady, surveying contemporary Perth Literature and mores in a 1987 essay entitled 'A Postmodernist City', draws a parallel between Perth today and at the time of arrival of the earliest settlers 150 years ago. Then, she writes, arriving settlers had been misled by the glowing reports of Captain Stirling, who had explored the lower reaches of the Swan River before hastening back to England to found a joint stock company to exploit the new territory. Instead of pastures of verdant green, the colonists discovered expanses of brilliant white sand stretching as far as the eye could see, which their puzzled children mistook for snow.

> Today, your first impressions are dazzling, somehow at odds with expectation. There is a sense of light, of glittering display, and somehow also of fragility. The city's glass towers of office blocks and luxury hotels rising beside the glitter of the river seem somehow strangely transient beneath the large sky. Nearby, in springtime, King's Park, several hundred acres of bushland, blazes with wild flowers, gold, blue, crimson and ghostly white, and further off sprinklers play on suburban lawns, flower beds of roses, hydrangeas, chrysanthemums, English daisies, flowers from the other side of the world blooming on the edge of the desert. Everything seems prosperous, an earthly paradise. The pace is easy, people move slowly, dress casually and seem friendly, glad to meet you and ready to help, delighted now to be famous, 'home of the America's Cup'. Signs by the side of the Great Eastern Highway winding into the city from the airport and, if you've driven the 2000 or so miles, from the rest of Australia, welcome you: 'G'day from WA'. The houses by the river are substantial, surrounded by lawns and flowering gardens. Nearly everyone has a car, many families two or three, and in parts of the city, like fashionable Dalkeith or Peppermint Grove, there seems to be a Mercedes to a house. Yachts at anchor in the river, power boats on the verge and swimming pools at the back, glass towers on the city skyline and cranes of new luxury hotels rising to meet them, all suggest that after all the sands were fertile.

Many people in Perth, Brady argues, like the first settlers prefer illusion to reality. Nevertheless, Brady shares some of the quiet optimism of Elizabeth Jolley about the saving qualities of the imagination, and posits that in such a young and developing society, there remains the possibility of alternative voices being heard.

> … Perth is still a frontier, there are possibilities still here, and not just possibilities of making money, but possibilities of somehow filling the absences, of making the spaces fruitful, in a human sense, of tapping the real story of the city, removing the mask, and letting the other voices speak.

In a relatively isolated city, she concludes, there is still the possibility of clarity of vision.

> Here in the West so many pretences which protect other Australians are falling away and it becomes clear who and where we are, a people a long way from home. The spaces open out, our sense of isolation and vulnerability expands. Space, like peace, Les Murray has said, is one of the great, poorly explored spiritual resources of Australia.[11]

MELBOURNE

'A Flat Place, Divided up into a Grid':
Nineteenth-Century Melbourne

Melbourne, in case you did not know, has its charms: botanical gardens, splendid churches, a high-domed public library where an old man can read the newspapers and stay cool on a hot day, etc. But there is no use denying that it is a flat place, divided up into a grid of streets by a draughtsman with a ruler and set square. The names of streets are just as orderly. King precedes William, neatly, exactly parallel. Queen lies straight in bed beside Elizabeth and meets Bourke (the explorer) and Latrobe (the governor) briefly on corners whose angles measure precisely 90 degrees.

Melbourne has a railway station famous for showing fifteen clocks on its front door, like a Victorian matron with a passion for punctuality, all bustle, crinolines and dirty underwear. It has Collins Street which is famous, in Melbourne at least, for resembling Paris, by which it is meant that the street has trees and exclusive shops where women in black with violently red lips and too much powder on their ageing cheeks are able to intimidate women like Molly McGrath by calling them 'modom'.

Oh, it's a good enough town, but it can take a while to realise it.

There is a passion in Melbourne you might not easily notice on a casual visit and I must not make it sound a dull thing, or sneer at it, for it is a passion I share—Melbourne has a passion for owning land and building houses. There is nothing the people of Melbourne care for as much as their red-tiled roofs, their lemon tree in the back garden, their hens, their Sunday dinners. You will not learn much about the city strolling around the deserted streets on a Sunday, no more than you will learn about an ants' nest by walking over it. Thus, when I seek something peaceful to think of, some quiet corner to escape into, I do not think of sandy beaches or rivers or green paddocks, I imagine myself in a suburban street in Melbourne on a chilly autumn afternoon, the postman blowing his whistle, a dog crossing the road to pee on those three-feet-wide strips of grass beside the road that are known as 'nature strips'.

The people of Melbourne understand the value of a piece of land. They do not leave it around for thistles to grow on, or cars to be dumped on. And this makes it a very difficult place for a man with no money to take possession of his necessary acre.

Peter Carey, Illywhacker, 1985[1]

In this deft summation, Peter Carey purports to recreate Melbourne as it might have appeared in the 1920s, but perhaps it is in the nature of the city that the description is still apt seventy years later. Melbourne is notable for the fact that most of its best-known writers—Martin Boyd, Hal Porter, George Johnston, Frank Hardy and playwrights David Williamson, Alex Buzo and Barry Oakley among them—have left it. Carey, born in Bacchus Marsh in Victoria in 1943 and educated at Geelong Grammar and Monash University, typifies a Melbourne writer in that he also departed early in his career.

> Melbourne, this mastodon of bleeding stone
> Sprawled on earth's dugs, repeats its type again,
> Boasting as archaeologists have shown,
> A giant's body and a louse's brain[2]

wrote poet Kenneth Slessor in 1925, having moved south to work for Melbourne *Punch* for a year before beating a hasty retreat back to Sydney.

This tradition of serious and semi-serious abuse by its own and visiting writers is continued by Peter Corris in his detective novel *Deal Me Out* (1986).

Today, your first impressions are dazzling, somehow at odds with
expectation. There is a sense of light, of glittering display, and somehow
also of fragility. The city's glass towers of office blocks and
luxury hotels rising beside the glitter of the river seem
somehow strangely transient beneath the large sky.

VERONICA BRADY, 1987 (PAGE 141)

CITY FROM MATILDA BAY, PERTH, WESTERN AUSTRALIA. PHOTOGRAPHER: PHILIP QUIRK

*Melbourne is the phenomenal city of Australia, and its
people have in it a pride which is a passion.*

FRANCIS ADAMS, 1893 (PAGE 154)

BOURKE STREET, MELBOURNE, VICTORIA. PHOTOGRAPHER: CAROLYN JOHNS

The Tullamarine freeway must be one of the most boring stretches of road on the planet; either they picked a boring landscape to run it through or they made it that way in the process. Anyway, there was nothing on the run to occupy my thoughts or delight my eye until we reached the city, which looked pretty good in the afternoon sun, if you like broad, tree-lined streets and a flat landscape ... It was after three when I reached Brewers Road. Kids were straggling home from school, battling a wind that whipped at the tails of their raincoats and shook the trees and shrubs of their well-tended gardens. Bentleigh was one of those flat Melbourne suburbs, with the odd suggestion of rise and fall in the landscape, which made it just possible to imagine it as a pleasant place before 1835.[3]

Corris was born in Stawell, 230 kilometres from Melbourne, in 1942, but like Peter Carey he subsequently moved to Sydney.

The topographical basis for these criticisms from both home and abroad is probably best expressed by Melbourne academic and poet Chris Wallace-Crabbe in an essay entitled 'Melbourne in 1963', published in *Melbourne or the Bush* (1974):

One cannot focus on any place, any situation, and say, 'Here indeed is the true centre of Melbourne', for the city is not merely large, but to an extraordinary degree, sprawling and centreless. The mile-square grid of the city proper is somehow far less real, less permanent, than the hundreds of square miles of suburbia into which the population flees in the evening, draws down the puritan blinds and settles itself before the blue shimmer of the television set.[4]

Wallace-Crabbe, born in the suburb of Richmond in 1934, reiterates some of the same ideas in his poem 'Melbourne', also dating from the early 1960s.

Ideas are grown in other gardens while
This chocolate soil throws up its harvest of
Imported and deciduous platitudes,
None of them flowering boldly or for long;
And we, the gardeners, securely smile
Humming a bar or two of rusty song.

... Highway by highway, the remorseless cars
Strangle the city, put it out of pain,
Its limbs still kicking feebly on the hills.
Nobody cares. The artists sail at dawn
For brisker ports, or rot in public bars.
Though much has died here, nothing has been born.[5]

Wallace-Crabbe is mourning a lack of cultural autonomy in his native city, both in relation to larger and older Sydney and to the rest of the Western world. Australian cities, he believed, were still to some extent 'spiritual suburbs' of London, Paris and New York. This latter thesis has also been frequently taken out for exercise in literary Sydney (and in fact the rest of Australia), but as the playwright David Williamson remarked not entirely seriously in 1980, Melburnians are more vociferous in their sense of aggrievement:

Melbourne is a much more belligerent city. Its dinner parties are more violent. The trouble with Melbourne is that it's made up of Scots stockbrokers and Irish publicans.[6]

'One of the remarkable things about Melbourne is that until recently it had virtually no definable literature of its own at all,' wrote Melbourne academic Laurie Clancy—taking his life into his hands—

a year later in 1981. He made exceptions of the work of Henry Handel Richardson, Henry Lawson in the 'Arvie Aspinall' stories, and more recently that of Alan Marshall, Judah Waten and Frank Hardy—noting, however, that the Melbourne they wrote about would be largely unrecognisable today. Clancy went on to make the point that 'a flood of imaginative literature' about Melbourne had emerged since Chris Wallace-Crabbe made his criticisms in the 1960s—partly, he claimed, because Melbourne had become 'a slightly more interesting' place in which to live.[7]

However, the general survival of this self-flagellating attitude—still evident in Clancy's disparaging comment that Melbourne was now only 'slightly' more interesting a dwelling place—may explain why (with notable exceptions in the work of Hal Porter, John Morrison and Helen Garner) it is not easy to find unashamedly affectionate accounts of the city in its home-grown literature.

But if sympathetic contemporary chroniclers of Melbourne life seem relatively scarce, another type of evocation has emerged to fill the gap. As is made evident by the lack of connection between, say, the South Yarra or Toorak of Martin Boyd and the Carlton of Judah Waten and Barry Oakley, Melburnians are perceived, by outsiders at least, to cling more closely and exclusively than other Australians to their social and political certainties. The more socially stratified an environment, the more it invites satire, as English comedy so successfully demonstrates. As a result, the more subversive elements of comedy and satire are probably richer in Melbourne today than in any other Australian city.

Satiric writers such as Barry Oakley—who in his novels pillories suburban man, forever mowing his front lawn—and Barry Dickins, a more slapstick comic, can caricature city life in Melbourne where it may be argued that in other, more brash and easy-going Australian cities there is less need and less object. In some aspects, Barry Humphries' ancient *alter ego* Sandy Stone, of suburban Glen Iris, and the earlier incarnations of Edna Everage of Moonee Ponds, could only be creations of Melbourne.

*   *   *

Historically Melbourne, founded forty-seven years after the penal settlement of Sydney, had the disadvantage of prosaic origins. Neither the cradle of the Australian colonial experiment, nor the oldest and largest settlement, Melbourne had its beginnings in financial entrepreneurship. Then known as Port Phillip, the city was founded in 1835 when two businessmen (one of whom was John Batman, who allegedly pronounced, 'Here is the place for a village') made a negotiation of dubious legality with the Yarra Aborigines for 600,000 acres in return for blankets, flour and knives. The settlement at Port Phillip was to service the rich pastoral hinterland, called *Australia Felix* by the explorer Major Thomas Mitchell in 1836, to which livestock would be overlanded from the north. Many of the settlers were well off and well connected: some of these English gentlemen farmers established homes in Melbourne from which to oversee their investments. In 1851 this southern portion of New South Wales was proclaimed a separate colony, called Victoria, with Melbourne as its capital. The smallest mainland state, it is second only to New South Wales in size of population.

In *The Imagined City: Melbourne in the Minds of its Writers*, John Arnold accounts for the earlier scarcity of distinctively Melburnian works by explaining that the colonial novels of the period tended to be disguised factual accounts written for an English market eager for information on the Australian colonies, especially the pastoral and mining Eldorado of Melbourne's hinterland. 'The stories concentrate on life on the goldfields or squatting or bushranging activities,' Arnold writes. 'Melbourne is usually mentioned in passing, or in a glowing chapter on the progress made in such a short time.'[8]

One such writer was the English novelist Henry Kingsley, who arrived in Melbourne in 1853 and spent some time on the goldfields before returning to England in 1857, where he published *The Rec-ollections of Geoffry Hamlyn* (1859). A little in the manner of a tourist guide, he causes one of his characters, Major Buckley, to describe to a newcomer the area as it appeared in 1836. Buckley is an impoverished member of the English gentry who has become a successful pastoralist in Australia.

'Port Phillip again,' said Frank; 'I have heard of nothing else throughout my journey. I am getting bored with it. Will you tell me what you know about it for certain?'

'Well,' said the Major, 'it lies about 250 miles south of this, though we cannot get at it without cross-ing the mountains, in consequence of some terribly dense scrub on some low ranges close to it, which they call, I believe, the Dandenong. It appears, however, when you are there, that there is a great har-bour, about forty miles long, surrounded with splendid pastures, which stretch west further than any man has been yet. Take it all in all, I should say it was the best watered and most available piece of country yet discovered in New Holland.'

'Any good rivers?' asked the Dean.

'Plenty of small ones, only one of any size, apparently, which seems to rise somewhere in this direc-tion, and goes in at the head of the bay. They tried years ago to form a settlement on this bay, but Collins, the man entrusted with it, could find no fresh water, which seems strange, as there is according to all accounts, a fine full-flowing river running by the town.'

'They have formed a town there, then?' said the Dean.

'There are a few wooden houses gone up by the riverside. I believe they are going to make a town there, and call it Melbourne; we may live to see it a thriving place.'[9]

By the 1840s, according to Rolf Boldrewood (Thomas Alexander Browne) 'sufficient time had elapsed to erect many weatherboard, and a few brick houses'—in one of which he and his family awaited the construction of a two-storeyed mansion in Flinders Street, not far from where Prince's Bridge would be built. Boldrewood had arrived in Australia in 1831 aged five, and his father overlanded cattle to Port Phillip in 1838. In *Old Melbourne Memories* (1884) he gives an account of arriving at the little outlying port of Williamstown with his family as prosperous settlers—complete with their own small steamboat to start a ferry service—in 1840:

The moderate-sized schooner which carried us safely hither in a few hours under a week, had been chartered by Paterfamilias, so that we were unrestricted as to many matters not usually left to the discre-tion of passengers. It was a floating home. Colonists of ten years' standing, we had many things to bear with us, which under other circumstances of transit must have been left behind. There were carriage horses and cows, the boys' ponies, the children's canaries, poultry, and pigeons, dogs and cats, babies and nurses, furniture, flower-pots, workmen, house servants—all the component portions of a large household shifted bodily from a suburban home, and ready to be transferred to the first suitable dwell-ing in the new settlement. One can easily imagine to what a state of misery and confusion such a freight would have been reduced had bad weather come on. But the winds and the waves were kind, and on Saturday afternoon the harbour-master of Williamstown partook of some slight alcoholic refreshment on board, and welcomed us to Port Phillip. Well is remembered even now the richly-green appearance of the under-stocked grassy flat upon which the particularly small village of Williamstown stood. A few cottages, more huts—with certain public-houses of course—made up the township. More distinctly marked even was the succulence and juiciness of the first Port Phillip mutton-chops upon which was regaled our keenly hungry party. We had just quitted the enfeebled meat markets of Sydney, scarce recovered from that terrible drought which wasted the years of 1837, 1838 and 1839. We had reached a land of Goshen evidently—a land of milk and butter, if not of honey—a land of chops and steaks, of sirloins and 'under-cuts'—of all youthful luxuries well nigh forgotten—of late unattainable in New South Wales as strawberry ice in a canebrake.[10]

Another witness to the period was George Henry Haydon, who arrived in Australia at the age of eighteen and spent the years from 1840 to 1845 working in Victoria as an architect, storeman, bookseller and drawing master. In 1859 Haydon published *The Australian Emigrant*, a humorous novel subtitled 'a rambling story, containing as much fact as fiction', in which he describes the reactions of new arrivals watching from the deck as the boat rounds a bend of the Yarra and the town is revealed before them.

'The next turn of the river, gentlemen,' said the skipper, 'and you'll see Melbourne.'

Then it was that the passengers became acutely sensible of their sluggish progress. On reaching that part of the Yarra indicated, several low huts were seen, on either bank of the river, standing close to the water's edge. On a beautiful green hill (Batman's Hill), which rose on their left they could distinguish a building of a better class; further up the stream, and on a parallel line with it, were several edifices built of brick; but the greater part of the best houses were of weatherboard. There were also some very doubtful-looking erections, unlike dwellings, but too good for piggeries. In reply to an inquiry addressed to the captain, he informed his passengers that they were merchants' stores. Amongst the buildings were large stumps, with the parent stems laid low by their sides, cumbering the ground. Gigantic trees dotted the undulating country in the distance, and with tents pitched here and there made the back-ground of the picture. Huge heaps of heavy timber, piled up high above some of the humbler huts, were burning furiously, and dense columns of smoke were so numerous, that one might easily have imagined the town was on fire. Thus does civilization mark her first inroads in a new country.

The skipper offers to point out the notable features, remarking that 'The shade of the largest trees left standing are our churches for the present' and at the same time delivering a vignette of the manners and mores of Melbourne society.

'Here's the wharf,' he continued, as the steamer bumped against the river's bank. 'Government house is not to be seen for trees; and up above there,' he said, pointing to the top of a shingled roof which appeared above the water, 'is the police office and lock-up, just under the falls d'ye see?'

'Under the falls!' said Hugh.

'Aye—that's the roof of it you see yonder with a hole in the top. Some of our jolly squatters—rough men, I tell you—being determined on a spree, thought the safest way to begin it would be to swamp the lock-up; and so being a pretty strong and united party, d'ye see, they defied the ten constables, stormed the police office, took it, and putting it on rolling logs of timber, they started it down the hill into the Yarra, and there 'tis now:—it nearly cost one or two of 'em their lives tho', for several of 'em would remain inside, and only saved themselves by tearing an opening in the roof.—Wild dogs!—wild dogs!' said the skipper, with a shake of his grey head;—'why that night they capsized half the wooden houses in the settlement.'[11]

By the 1850s, with the discovery of gold, Melbourne had developed rapidly and grown larger than Sydney. For a brief period it became a wild and energetic frontier town. Catherine Helen Spence's finely observed novel, *Clara Morison: A Tale of South Australia During the Gold Fever* (1854), incorporates a description of the frenetic atmosphere of the crowded city at the time. Clara Morison's friend Mrs Bantam recounts what happens when they leave the ship that has brought them from Adelaide.

We were directed first to one house, then to another, and another; but every place was full to overflowing. Then we tried the inns, and though I offered to wait upon myself, and to give no trouble if they would only give us house-room, it was of no use. The streets were crowded with noisy men and women driving about furiously in gigs; I counted five diggers' weddings while I was going about; the women were prodigiously smart, but the bridegrooms had only those horrid tartan things, like short smock-frocks— they call them jumpers here—which all the returned diggers seem to wear, and which looked very shabby beside white satin and lace veils. I was footsore and miserable, but still was determined not to go back to the ship if I could help it, when at the last public-house Mr Bantam inquired at, we were directed to go up a narrow lane, where there were apartments to let.

For the first time that day, we got an answer of 'Yes' to our inquiries. I was so glad that I sat down, but was vexed to hear that we could not have a room to ourselves. I could have a bed in a room where there were three ladies already, and Mr Bantam was to have a sofa in the parlour. However, there was no help for it, and Mr Bantam went back to see about getting our most necessary luggage brought to our lodgings, and left me staring at the six sofas, that were ranged all round the room, with an idea that each of them was slept on at night. I begged to be shown into my room that I might arrange my dress before dinner; and the landlady, who was a red-faced, vulgar woman, made a sort of apology for Mrs Tomkins, who was in bed, as she felt poorly. The bedroom was wretchedly dirty, and there was such a smell of spirits and tobacco in the house, that I felt quite sick. I had not got myself fairly tidy when I was summoned to dinner; but oh! what a scene met my eyes. Dirty, unshaved men, who were swearing at everything and nothing, and women who seemed unsexed altogether. It was a ship-load of convicts from Van Diemen's Land, with or without a ticket-of-leave; none of them deserved to be let loose on society, I was sure; the people who kept the house were old convicts; Mrs Tomkins was intoxicated; and there was I without my husband, exposed to every kind of insolence.

I thought Mr Bantam would never return; when he did, I begged him to take me back to the ship. Of course we forfeited the three pounds ten shillings we had advanced, and got a great deal of abuse besides from the people in the house. I was obliged to go into the bedroom for my bonnet and shawl, and found that Mrs Tomkins had got well enough to help herself to my handsome cameo, which I had left on the dressing-table. She used such dreadful language, that I did not dare to complain, but went off with Mr Bantam as fast as I could.

In a later chapter the Scottish-born Spence, who later worked as a journalist, causes the young South Australian clerk William Bell, who has gone to Bendigo to try his luck on the goldfields, to write back a carefully detailed description of the city as he found it in the same period.

'It is true that the best society in Melbourne has always been considered by Scotch people superior to its counterpart in Adelaide; but how was a stranger and a clerk, with such very slender social talents as you know I have, to get into it? Where should I find a place in the universal overturn of society which is taking place in Victoria? The aristocratic members of the community are retreating when they can to England, to keep out of the crowd and discomfort; the mercantile are turning over money with unexampled rapidity, large profits and quick returns being the order of the day; and there is the same keen money-making look about them, which you used to observe in the frequenters of your Exchange, but with more feverish anxiety about the Melbourne men.

'The town is densely crowded; places built in narrow lanes for stables are filled by human occupants, who live in dirt and discomfort, injuring the general health of the town. Owing to the stringent Building Act there have been many good streets built, because every man in buying his piece of land got the plan of the house to be erected towards the front; but as there was nothing to prevent the back being divided into lanes, the profit of the speculation has induced many to do it. It is shameful that with an unlimited extent of country, and in such a new town, people should be living in rows of houses only ten feet apart. You know a few such places in Adelaide; you know them to be nests of fever and sickness; and when I tell you that there are no fewer than two hundred and ninety of these private alleys in Melbourne not subject to the street regulations, you will not believe it can be a healthy city. Nor will you think so the more when you consider that a great proportion of the people are the sweepings of British jails, who have just made their way to a place where almost every description of crime may be committed with impunity. A feeble government, which is now led by a clique of squatters, a wretched police, and incompetent courts of law, is a great obstruction to the course of justice. I heard a gentleman say it was no bad thing for the colony that Melbourne was not a desirable place of residence; for that in a new state comfortable and luxurious cities impede the spread of the people and the subjugation of the soil. And there is some truth in that, but the only subjugation people think of now, is getting the gold out of the land; and every other description of industry is for the time paralyzed. I did not see much gambling in my peregrinations, at least not nearly so much as I expected, from our knowledge of its extent in California … But of drinking and swearing I saw more than enough. I thought Adelaide was not particularly moral, but it is infinitely better than this. Even gentlemen make a boast of swearing in Victoria, while few, except bullock-drivers, do so in South Australia.'[12]

According to Melbourne-born Henry Handel Richardson, however, the approach to gold-rush Melbourne by land provided obstacles of a more physical nature.

The two men ceased their trifling, and nudged by the fall of day began to ride at a more business-like pace, pushing forward through the deep basin of Bacchus's Marsh, and on for miles over wide, treeless plains, to where the road was joined by the main highway from the north, coming down from Mount Alexander and the Bendigo. Another hour, and from a gentle eminence the buildings of Melbourne were visible, the mastheads of the many vessels riding at anchor in Hobson's Bay. Here, too, the briny scent of the sea, carrying up over grassy flats, met their nostrils, and set Mahony hungrily sniffing. The brief twilight came and went, and it was already night when they urged their weary beasts over the Moonee ponds, a winding chain of brackish waterholes. The horses shambled along the broad, hilly tracks of North Melbourne; warily picked their steps through the city itself. Dingy oil-lamps, set here and there at the corners of roads so broad that you could hardly see across them, shed but a meagre light, and the further the riders advanced, the more difficult became their passage: the streets, in process of laying, were heaped with stones and intersected by trenches. Finally, dismounting, they thrust their arms through their bridles, and laboriously covered the last half-mile of the journey on foot. Having lodged the horses at a livery-stable, they repaired to a hotel in Little Collins Street. Here Purdy knew the proprietor, and they were fortunate enough to secure a small room for the use of themselves alone.

In *Australia Felix* (1917), the first volume of her trilogy *The Fortunes of Richard Mahony*, Richardson researched contemporary sources to recreate Melbourne during the early 1850s, when her own father worked on the Ballarat goldfields. The city she presents appears prosperous and fast-developing.

… they turned and descended Great Collins Street—a spacious thoroughfare that dipped into the hollow and rose again, and was so long that on its western height pedestrians looked no bigger than ants. In the heart of the city men were everywhere at work, laying gas and drain-pipes, macademising, paving, kerbing: no longer would the old wives' tale be credited of the infant drowned in the deeps of Swanston Street, or of the bullock which sank, inch by inch, before its owner's eyes in the Elizabeth Street bog. Massive erections of freestone were going up alongside here a primitive, canvas-fronted dwelling, there one formed wholly of galvanised iron. Fashionable shops, two storeys high, stood next tiny, dilapidated weatherboards. In the roadway, handsome chaises, landaus, four-in-hands made room for bullock-teams, eight and ten strong; for tumbrils carrying water or refuse—or worse; for droves of cattle, mobs of wild colts bound for auction, flocks of sheep on their way to be boiled down for tallow. Stock-riders and bull-punchers rubbed shoulders with elegants in skirted coats and shepherd's plaid trousers, who adroitly skipped heaps of stones and mortar, or crept along the narrow edging of kerb.

The visitors from up-country paused to listen to a brass band that played outside a horse-auction mart; to watch the shooting in a rifle-gallery. The many decently attired females they met also called for notice. Not a year ago, and no reputable woman walked abroad oftener than she could help: now, even at this hour, the streets were starred with them. Purdy, open-mouthed, his eyes a-dance, turned his head this way and that, pointed and exclaimed.[13]

Later in *Geoffry Hamlyn*, the novelist Henry Kingsley describes the city as it appeared at the time he left for England in 1857. Here he enthusiastically relates the rapid progress in the two decades since 'the Yarra rolled its clear waters to the sea through the unbroken solitude of a primeval forest'.

> Now there stands there a noble city, with crowded wharves, containing with its suburbs not less than 120,000 inhabitants. 1,000 vessels have lain at one time side by side, off the mouth of that little river; and through the low sandy heads that close the great port towards the sea, *thirteen millions sterling* of exports is carried away each year by the finest ships in the world. Here, too, are waterworks constructed at fabulous expense, a service of steamships, between this and the other great cities of Australia, vieing in speed and accommodation with the coasting steamers of Great Britain; noble churches, handsome theatres. In short, a great city, which, in its amazing rapidity of growth, utterly surpasses all human experience.
>
> … I never stood in Venice contemplating the decay of the grand palaces of her old merchant princes, whose time has gone by for ever. I never watched the slow downfall of a great commercial city; but I have seen what to him who thinks aright is an equally grand subject of contemplation—the rapid rise of one. I have seen what but a small moiety of the world, even in these days, has seen, and what, save in this generation, has never been seen before, and will, I think, never be seen again. I have seen Melbourne. Five years in succession did I visit that city, and watch each year how it spread and grew until it was beyond recognition. Every year the press became denser, and the roar of the congregated thousands grew louder, till at last the scream of the flying engine rose above the hubbub of the streets, and two thousand miles of electric wire began to move the clicking needles with ceaseless intelligence.
>
> Unromantic enough, but beyond all conception wonderful.[14]

The period of affluence and expansion following the gold rushes allowed many fortunes newly gained through pastoral and mining investment to be consolidated in real estate. The population of Melbourne increased by five times in the decade beginning in the year 1850, and the suburbs spread ever outwards. Melbourne, the business city, would always provide a ready mirror of the colony's financial fortunes, the booms and busts of the years that followed. Perhaps because of the generally self-aggrandising nature of business entrepreneurship and the aura of finance, and because of local pride in its non-convict origins—and perhaps also because of a perceived local necessity to promote its importance relative to its older sister to the north—Melbourne seems to have been the only Australian city to attract such excessive adjectival flourishes as 'noble' and 'great' from its chroniclers. At any rate, the city now developed an image of solidity and worthiness that flighty Sydney never achieved.

In the next decades the rapidly developing city gained the sobriquets 'Marvellous Melbourne' and 'Queen of the South' before subsiding again when the land and mining boom burst in the late 1880s. This is the city of which Patrick Moloney, or 'Australis', author of *Sonnets and Innuptum* (1879), felt inspired to write:

> O Sweet Queen-city of the Golden South,
> Piercing the evening with thy starlit spires,
> Thou wert a witness when I kissed the mouth
> Of her whose eyes outblazed the skiey fires.
> I saw the parallels of thy long streets
> With lamps like angels shining all a-row,
> While overhead the empyrean seats
> Of gods were steeped in paradisic glow …[15]

This is also the prosperous and expanding city of Jessie Couvreur's *Uncle Piper of Piper's Hill* (1889). Published under the pseudonym Tasma, and serialised as 'The Pipers of Piper's Hill' in the *Australasian* in 1888 before being released in book form a year later, the novel is set in Melbourne in the 1860s and follows the fortunes of the English and—by adoption—Australian members of the same family.

Tom Piper's substantial house, Piper's Hill, is built in South Yarra, a desirable area, and the plot traces the shift of power from old to new money as the prosperous ex-butcher Uncle Piper's poor but rather overly well-bred relatives arrive from England to live with him, and must swallow their pride and accept his benevolence.

> Mr Piper made his coachman drive the new-comers through the whole length of Collins Street on their way from Sandridge to South Yarra.
>
> 'I've seen it go up stone by stone,' he told them; 'I've seen it grow from the time that it was nothing but canvas, like a big fair. A few shanties and stores—that was the beginning. The best day's work I ever did was to buy that bit o' land at the corner. You wouldn't ha' given me a thank you for it then, and just you take and look at it now.'
>
> It was about eleven o'clock in the forenoon, and the street was very full. The part Mr Piper pointed out was the most crowded of all. It was a space on the broad pavement under a verandah in front of a row of offices and hotels—a kind of open-air Exchange in which the new-comers were assured that 'big fortunes were built up and pulled down daily.'

The impoverished Cavendish family are prepared to be condescending, but as they approach Piper's Hill the vain and beautiful Sara finds herself struck by the thought that their own appearances, so shabby-genteel, do not show to advantage in this setting.

> It came upon her with fresh force as the carriage turned suddenly off the main road through two wide-open gates of wrought iron, and rolled swiftly and smoothly up a broad and perfectly-kept avenue. To the right lay a lawn as soft as velvet pile, dotted with flower-beds. The first spring roses were already opening their pink and lemon-coloured buds. The orange shrubs, clustered round a fountain in the centre of the lawn, filled the air with wandering scents. The Moreton Bay figs and the Murray pines, which Mr Piper would fain have urged into speedier growth, looked to the unaccustomed eyes of the new-comers like rare tropic trees of rich beauty …
>
> The verandah seemed of great and marvellous breadth. notwithstanding the intrusion of deep bay windows on either side of the door. To the Cavendish family it seemed large enough as they approached to have held a whole row of London terrace-houses of the cramped kind to which they had been accustomed. Yet here, as everywhere else, flowers, shrubs caught the eye in every corner, and delicious lounges that looked fit place for the weaving of such fancies as 'youthful poets dream,' encompassed by 'Sabean odours,' stood next to the balustrade or against the wall. As the carriage stopped, being brought up with a kind of mathematical precision that spoke of long practice, at the very centre of the bottom step of the verandah flight, Mrs Cavendish looked at her husband for the second time. Surely there would be some little trace of surprise or amazement at such unlooked-for magnificence!
>
> … Then, as they stood for an instant in a great marble-floored hall, from which a grand circular staircase rose at the lower end, and took in vaguely the impression of a soft stained light, streaming through a painted window over the landing—of statues in niches holding candelabras in their extended hands, of gilded baskets filled with flowers, and porcelain vases filled with rose-leaves—Sara was aware that a door to the right was suddenly opened, and a figure that looked as though it might have descended from the very brightest of the vases, and grown into breathing flesh and blood, advanced towards them. Sara looked at her with prompt and critical curiosity, after the manner of her sex. Women are often more impartial judges of beauty than men allow them to be. Sara decided instantly in her own mind that Laura was extremely pretty. Query No. 1—'I wonder whether she is made up?' Query No. 2—'I wonder whether she gets her things from Paris?' Query No. 3—'I wonder whether she's engaged to our cousin George?'[16]

Couvreur is also documenting a period of social change and exchange. Uncle Piper, common, but secure in his wealth and business ability in relation to the feckless but well-bred Cavendishes, wishes for a socially improving marriage for his son George. The remainder of the novel traces, with some ironic humour, the progress of the young people's romantic intrigues.

Martin Boyd, in his later novel *The Cardboard Crown* (1952), the first volume of the Langton tetralogy, which deals with the same period in relation to his own family, further enlarges on the development of the then-prevailing social system.

It may be as well to explain the significance of Melbourne suburbs. Most of the early colonists of the better sort, the judges, army officers, and gentlepeople who somehow found themselves in Victoria, lived in East St Kilda, which remained an 'exclusive' neighbourhood up to some time in the 1920s. But long before that it had been overshadowed by Toorak, where the first large Government House was built, and where, as the wool-growers became richer, they raised their Italianate mansions. Toorak with its neighbour South Yarra became the Mayfair of Melbourne, and the descendants of the St Kilda gentry, as they felt their social importance in danger, fled to any cottage in the shadow of the mansions of squatters, many of whom their grandparents would have refused to know, in the same way that people would rather live in two rooms above a delicatessen shop behind Berkeley Square than in a house with a ballroom in Putney. It was also possible to live in Kew, Brighton, and other suburbs, but this was only done by people who preferred the river and the sea to 'society'. It is necessary that the reader should understand the character of these different suburbs, particularly the potent meaning of the word Toorak. If Alice had decided to build in Toorak instead of buying the house in East St Kilda, this story would probably have been very different. In fact I might not have been born, as it was that decision which brought about our close association with the Bynghams.[17]

However, if Melburnians were well aware of their own and their city's social importance, such was not necessarily the case overseas, where the city's image was evidently much more colourful. The romantic novelist Ada Cambridge, who came to Melbourne with her clergyman husband in the early 1870s from Norfolk, in her autobiography *Thirty Years in Australia* (1903) commented a little acerbically on the misrepresentations current in England at the time of her arrival.

No description that we had read or heard of, even from our fellow-passengers whose homes were there, had prepared us for the wonder that Melbourne was to us. As I remember our metropolis then, and see it now, I am not conscious of any striking general change, although of course, the changes in detail are innumerable. It was a greater city for its age thirty years ago than it is today, great as it is today. I lately read in some English magazine the statement that tree-stumps—likewise, if I mistake not, kangaroos—were features of Collins Street 'twenty-five years ago'. I can answer for it that in 1870 it was excellently paved and macadamised, thronged with its wagonette-cabs, omnibuses, and private carriages—a perfectly good and proper street, except for its open drainage gutters. The nearest kangaroo hopped in the Zoological Gardens at Royal Park. In 1870, also—although the theatrical proceedings of the Kelly gang took place later—bushranging was virtually a thing of the past. So was the Bret Harte mining-camp. We are credited still, I believe, with those romantic institutions, and our local storywriters love to pander to the delusion of some folks that Australia is made up of them; I can only say—and I ought to know—that in Victoria, at any rate, they have not existed in my time. Had they existed in the other colonies, I must have heard of it. The last real bushranger came to his inevitable bad end shortly before we arrived. The cowardly Kellys, murderers, and brigands as they were, and costlier than all their predecessors to hunt down, always seemed to me but imitation bushrangers. Mining has been a sober pursuit, weighted with expensive machinery. Indeed, we have been quite steady and respectable, so far as I know. In the way of public rowdyism I can recall nothing worth mentioning—unless it be the great strike of 1890.[18]

By now, Melbourne was settling inexorably into its self-created social classifications. It is arguable that the English class system survived longer in Melbourne than in any other Australian city, and some of its manifestations are still more apparent today. There were Australians, it seemed, and there were the inhabitants of Melbourne. As early as 1886 commentators were attempting to sum up the characteristics of the typical Melburnian. Francis Adams, who came to Australia for reasons of health in 1884 at the age of twenty-two and stayed for six years, in 1886 found Melbourne '… a city into whose hands wealth and power is suddenly phenomenally cast: a general sense of movement, of progress, of conscious power. This, I say, is Melbourne—Melbourne with its fine public buildings and tendency towards anarchy. And Melbourne is, after all, the Melbournians. Alas, then, how will this city and its civilisation stand the test of a really fine city and fine civilisation?'

Adams answered his own question by giving Melburnians a good mark for 'conduct', but finding 'enormous room for improvement' in 'intellect and knowledge'. However—allowing for thirty or forty years' higher education of the middle class—he felt the Australian version would 'leave the English middle class far behind'.

In a subsequent publication, *The Australians: A Social Sketch* (1893), written after his return to England, Adams strongly contrasted the people of the cities and those of the bush. In general, he noted, he found Australians alternatively energetic—sometimes overly so—and languorous; materialist, independent and friendly; with young women who were restless, frank, unprudish and 'well able to look after themselves'. Of Melbourne Adams wrote:

> Melbourne is the phenomenal city of Australia, and its people have in it a pride which is a passion.
> The old Anglo-Australian generation which founded its prosperity is quietly but swiftly passing away.
> The native Australians who follow on them have too often the self-sufficiency that is begotten on self-confidence by ignorance.
> … But in Melbourne, where much that is typically Australian is to be found, much also is a mere replica at second-hand of the older civilisation.
> The closeness to England is the obvious cause of this.
> The stream of European civilisation finds here its terminus, and threatens the city with a crude cosmopolitanism.
> Melbourne is in reality pagan, but a sort of worldly Presbyterianism has inflicted itself upon its official presentment as the social counterpart of the political stagnation.[19]

It is probably the novelist Martin Boyd who best represents the often uneasy literary transition between the visiting writer, usually English-born, who surveys the local scene and returns home, and the writer who, even if born elsewhere, places his primary loyalty in the new country. Boyd, inextricably caught between the two, straddles the cultural worlds of both.

Martin Boyd's upper-middle class Melbourne—as conveyed in his first major novel, *The Montforts* (1928), which fictionally traces the history of his mother's family from the 1850s until the First World War—was, according to the author, culturally inferior to English society. At the same time one suspects he would have resented, though understood, the visiting Francis Adams' rather high-handed generalisations.

At the beginning of *The Montforts* where, in a situation similar to that at the beginning of *Uncle Piper of Piper's Hill*, Henry and Letitia Montfort arrive in Melbourne after a troubled voyage to join Henry's brother Simon and his wife Sophie, Boyd is able to sympathise with the feelings of both parties. The family lands at Williamstown in the afternoon and takes a steamer up the Yarra to Melbourne, where on high ground ahead of them there is little else to see but a few substantial houses and some mean cottages surrounded by piggeries and cow pens. Simon and Sophie, who have come to meet them, point out the features of the town, including the newly developing St Kilda and, in a churchyard, the stump of a tree beneath which Simon pitched his tent when he first came to Port Phillip.

> 'This is Collins Street, our main thoroughfare,' said Sophie. The road was ill-made and not yet free of the stumps of trees. Amy, conscious of parental agreement with her criticism, said pertly:
> 'Is not Melbourne an odious place, mama?'
> An electric current of hostility passed round the carriage. Sophie's grey eyes flashed blue.
> 'You should never show a fool or a child anything unfinished,' she said.
> Letitia was annoyed at Sophie's brusqueness to Amy, and its justice increased her annoyance.
> 'You should not chatter so, Amy,' she said, and herself relapsed into silence. At length they came to the house, and the excitement of arrival dispelled some of their awkwardness.
> Sophie took Letitia to her room, which was large and light. The furniture was new and of heavy mahogany. Spiral pillars ran up the front of the wardrobe and supported the marble slab of the wash-stand. The wall-paper had a floral design, and on the chintz covers and curtains was a pattern of large yellow flowers of no botanical origin. On the carpet was yet another floral pattern.
> Letitia found it elegant and far more fashionable than she had expected. The band round her heart loosened.
> 'You cannot imagine how delightful it is to be in a house again,' she exclaimed to Sophie, with a genuine attempt at friendliness.

But for Letitia, the strangeness is too much.

> She pictured the house in Bedford Square where, a year ago, she had been so content with all her well-arranged possessions. Now they were scattered and the house was swept bare and reinhabited by strangers. With that scattering of her possessions it seemed that something of herself had been blown to the winds. Robbed of the background which she had so carefully built, she felt as if in some way there were less of herself as a definite human entity. This thought hardly was shaped in words in her mind, but it was the sensation she experienced.
> She reawoke to her surroundings. At her feet were twigs, fallen from the trees, and among them crawled ants, larger than any she had seen before. The air was full of hot dry scent of the eucalyptus trees. Beyond the house, silhouetted black against it, the setting sun had painted the sky blood-red.
> Suddenly, from a tree near by, came a sound like a prolonged grotesque laugh. Startled, she looked for its cause and found that the noise was made by a bird like a large grey kingfisher.
> Henry, breathing excitement and pleasure, came across the lawn to her.
> 'Well, Letty, my dear,' he said affectionately, 'how do you like our new country?'
> She tried to answer, but her lip quivered, and she burst into tears.[20]

Martin Boyd, who was born in Switzerland in 1893 while his parents were making a European tour, never resolved, either in his life or his fiction, the conflict of the Anglo-Australian family with one foot in Europe and one foot in Victoria, and constantly made uneasy transitions back and forth between the two hemispheres. When in England or Europe his characters characteristically long for the free-dom of Australia, when in Australia they resent their cultural exile. Always the cultural elitist, Boyd's loyalty to Australia lay largely within a small coterie of Melbourne society that jealously guarded its refinement and its distinguished ancestry. This was despite the fact that both his own and his fictional family—like Uncle Piper's—were relatively impoverished in England, and in Australia relied for their prosperity on sheep stations on 'remote burning plains, blasted by January sun … built of corrugated iron and (smelling) of sheep', about which some family members continued to be disparaging.

However, neither is Boyd blind to the follies—particularly the architectural ones—of that stratum of society he might seem to favour. In one passage he acerbically describes Melbourne's new cathedral as 'early English, relieved with Moorish stripes, Italian mosaic and bathroom tiles ...'

Boyd in his fiction consistently points to England and Europe as the arbiters of cultural standards, but he is also aware of the dangers of attempts to transplant 'civilisation'. In *Lucinda Brayford* (1946), another family saga straddling two hemispheres, Lucinda Brayford is introduced by Canon Chapman to old Mrs Talbot, who seems unaware that the England she left so long ago is no longer replicated in her crumbling house on the outskirts of Melbourne.

> The car bumped on for another mile, still through the pale brown honeycomb. They turned round a low hill and came in sight of Cape Furze House. Set amidst those parched, distorted paddocks on the edge of an arid coast, protected only by a few pines, almond trees and eucalyptus, it had no resemblance to an English country house. Its white stone gables and twin gothic towers had shed all association with the north. They had been bleached of its influence by the salt and the sun, and brought into affinity with the twisted white stems of the gum trees. They shone chalkily above the sombre pines. The only bright green was the harsh splash of some sea shrub down the cliff to the right of the house which blazed against the expanse of the sea itself, today wine-dark and vivid.
>
> When they were admitted to the house through a massive but blistered door, which the elderly parlourmaid shut quickly behind them to keep out the noontide heat, they found it difficult at first to see in an interior which was as cool and dark as the outside was dazzling. Two frightening suits of Japanese armour loomed in the darkness. These flanked the drawing-room door, which suddenly opened between them, framing in an oblong of light the tiny hobbling figure of Mrs Talbot, who in Australian fashion had come out to meet her guests instead of waiting for them to be shown in to her.
>
> Mrs Talbot led them back into the drawing-room, where nothing had been changed since the last century, and where the chintz covers, though freshly glazed, were fifty years old. Mrs Talbot had pale blue eyes which danced with excitement. Her frail old body seemed to be bubbling and shaking with quiet silvery laughter. She used to love parties but now she never left her home, and any visitor was an excitement at Cape Furze.
>
> ... The luncheon table in the bogus baronial dining-room was like an illustration from an early edition of Mrs Beeton. It was crowded with ruby glass and a great many objects of ornate silver filigree. An epergne was filled with yellow roses and surmounted by a pineapple. There were quantities of very good food. The delicate garfish had been caught by the gardener that morning in the sea below the house. There were three roast ducklings between the five of them, and, surprisingly in this outpost of civilisation, an iced pudding with strawberries in it. The claret was some of the last had been laid down by Mr Talbot.

Later:

> Mrs Talbot led Tony and Lucinda down a long stone passage and let them out by a door in which there were panes of red glass, which made the outside world, already sufficiently hot, appear as if it were on fire. Although the terraces were crumbling, and half the garden abandoned to gorse and brambles, she still behaved as if Cape Furze were a show place, and directed them to what had been its finest pleasures.[21]

Boyd, after an attempt to return permanently to Australia in 1948, left again for England in 1951, finally settling in Rome in 1958, where he lived until his death in 1972.

In the well-mannered novels of Martin Boyd and Jessie Couvreur, the darker side of a rapidly expanded—perhaps too-rapidly-expanded—city of the period is never glimpsed. Marcus Clarke, who worked as newspaper columnist in Melbourne after his arrival in 1863, noted in the *Argus* of 28 February 1868 that 'for its size, Melbourne is as vicious a city as any in the Southern Hemisphere ...' although he went on to add that 'the artificial impetus given to crime by the outbreak of the gold mania is subsiding, the permanent settlement of a large number of industrious persons having in a great measure absorbed the floating criminal population ...'

Nevertheless, it appeared that enough of the low life associated with the gold rushes had survived into the following decades for Fergus Hume convincingly to base his popular thriller *The Mystery of a Hansom Cab* (1886), in Little Bourke Street, Melbourne. Hume, born in England in 1859, trained in law in New Zealand and arrived in Melbourne as a solicitor's clerk in the mid-1880s. Drawing on his detailed knowledge of the street life of the city during that period for background, he succeeded in producing a forerunner of the classic whodunit. The picture presented, however, is of a Melbourne in which Sara Cavendish or Lucinda Brayford would never have set foot.

Bourke Street is always more crowded than Collins Street, especially at night. The theatres are there, and of course there is invariably a large crowd collected under the electric lights. Fashion does not come out after dark to walk about the streets, but prefers to roll along in her carriage; therefore the block in Bourke Street at night is slightly different from that of Collins Street in the day. The restless crowd which jostles and pushes along the pavements is grimy in the main, but the griminess is lightened in many places by the presence of the ladies of the demi-monde, who flaunt about in gorgeous robes of the brightest colours. These gay-plumaged birds of ill-omen collect at the corners of the streets, and converse loudly with their male acquaintances till desired by some white-helmeted policeman to move on, which they do, after a good deal of unnecessary chatter. Round the doors of the hotels, a number of ragged and shabby-looking individuals collect, who lean against the walls criticising the crowd, and waiting till some of their friends ask them to have a glass, a request they obey with suspicious alacrity. Further on, a crowd of horsy-looking men are standing under the Opera House verandah, and one hears nothing but sporting talk about the Cup, and odds being given and taken on the cracks of the day. Then here and there are ragged street Arabs, selling matches and newspapers; and against the verandah post, in the full blaze of the electric light, leans a weary, draggled-looking woman, one arm clasping a baby to her breast, and the other holding a pile of newspapers, while she drones out in a hoarse voice, "*Erald*, third 'dition, one penny!' until the ear wearies of the constant repetition. Cabs rattle incessantly along the street; here, a fast-looking hansom, with rakish horse, bearing some gilded youth to his Club—there, a dingy-looking vehicle, drawn by a lank quadruped, which straggles blindly down the street. Alternating with these, carriages dash along with their well-groomed horses, and within, the vision of bright eyes, white dresses, and the sparkle of diamonds. Then, further up, just on the verge of the pavement, a band, consisting of three violins and a harp, is stationed, which is playing a German waltz to an admiring crowd of attentive spectators.

... Now and then a mild-looking string of Chinamen stole along, clad in their dull-hued blue blouses, either chattering shrilly, like a lot of parrots, or moving silently down the alley with a stolid Oriental apathy on their yellow faces. Here and there came a stream of warm light through an open door, and within, the Mongolians were gathered round the gambling-tables, playing fan-tan, or leaving the seductions of their favourite pastime, to glide soft-footed to the many cook-shops, where enticing-looking fowls and turkeys already cooked were awaiting purchasers. Kilsip turning to the left, led the barrister down another and still narrower lane, the darkness and gloom of which made the lawyer shudder, as he wondered how human beings could live in such murky places.[22]

The Mystery of a Hansom Cab, a murder story set entirely in Melbourne, went on to sell more than half a million copies, despite the fact that Hume was initially forced to publish it himself, because it was a book local publishers 'refused even to look at ... on the ground that no colonial could write anything worth reading ...'—and, presumably, because conventional wisdom held that an urban novel could not be of interest unless set in one of the more important European cities.

THE NATIVE SONS

'Cheap to-day, lady; cheap to-day!'
Jostling water-melons roll
From fountains of Earth's mothering soul.
Tumbling from box and tray
Rosy, cascading apples play
Each with a glowing aureole
Caught from a split sun-ray.
'Cheap to-day, lady, cheap to-day.'
Hook the carcases from the dray!
(Where the dun bees hunt in droves
Apples ripen in the groves.)

An old horse broods in a Chinaman's cart
While from the throbbing mart
Go cheese and celery, pears and jam
In barrow, basket, bag or pram
To the last dram the purse affords—
Food, food for the hordes.

Shuffling in the driven crush
The souls and the bodies cry,
Rich and poor, skimped and flush,
'Spend or perish, buy or die!'

Food, food for the hordes!
Turksheads tumble on the boards.

Furnley Maurice, from 'The Victoria Markets Recollected in Tranquillity' (1934)[1]

After the land boom of the 1880s came the crash. All of Australia was thrown into financial depression. The perceived conservatism of Melbourne has sometimes been ascribed to the economic downturn after the wild days of the gold rushes. However, with Federation in 1901, Melbourne again experienced a brief moment of glory as the temporary capital of Australia, with national parliament sitting there until it moved to Canberra in 1927. Subsequently the city subsided once more into, according to one unsympathetic commentator, 'stolid suburban Scottish presbyterianism'. Then came the depression of the late 1920s and 1930s.

If Martin Boyd, with his conflict of loyalties between Europe and Australia, marked a transition of dominance from a literature culturally centred on England to the homegrown work of the native-born writers who were now beginning to emerge, he also marked, in Melbourne at least, the beginnings of a literary transition across class as well as national culture. Now literature in Melbourne was no longer entirely a middle-class preoccupation: the working-class suburbs were also finding their chroniclers. Boyd and his literary predecessors such as Ada Cambridge and Jessie Couvreur, with their concentration on upper-middle-class life and mores, gave ground to a group of writers, almost entirely male, from the other side of the political spectrum. Their concern was social realism rather than romance, and they gave voice to life as it was lived by the ordinary man in Melbourne during a downturn in economic conditions. Some of these writers, including Alan Marshall, John Morrison, Judah Waten, David Martin and Frank Hardy—all born in the first two decades of the new century—were to varying degrees communist in their sympathies, and a number were associated with the Melbourne branch of the left-wing Realist Writers' group and its publications.

A precursor to this group, in content if not entirely in style, was the radical nationalist Frank Wilmot, born in Collingwood in 1881, who began publishing poetry in the mid 1930s. Wilmot, firmly established in his loyalty to his native land, chose his pseudonym Furnley Maurice from the names of his two favourite Melbourne localities, Fern Tree Gully and Beaumaris. His frequently anthologised long poem 'The Victoria Markets Recollected in Tranquility' was published in *Melbourne Odes* (1934), and is a sympathetic attempt to combine the best principles of modern European and American poetry with firmly Australian subject matter.

Unique to Melbourne during this period also was a small but significant body of early migrant writing—largely Jewish—beginning with Judah Waten, born in Odessa in 1911, and David Martin, born in Hungary in 1915, and continued by Australian-born writers such as Morris Lurie, who was born in Melbourne of Polish Jewish parents in 1938. Lurie's comic novel *Rappaport* (1966) describes the life of a Polish Jewish antique dealer in the Melbourne suburb of Toorak.

Judah Waten arrived with his family in Western Australia in 1914 and lived in the Perth suburb of Midland Junction before moving to Carlton in Melbourne with his parents in 1925, where he lived for the next two decades. His series of autobiographical sketches *Alien Son* (1952) describe the early years in Western Australia, while his novel *Distant Land* (1964) describes the life of a Polish Jewish family that emigrates to Melbourne in the 1920s and finds accommodation in a second-floor apartment in Drummond Street, Carlton. Judah Waten moved to the Melbourne suburb of Box Hill in the 1940s.

David Martin, born in Hungary during World War I and brought up in Germany, settled in London before coming to Australia in 1949, where he joined the Australian Communist Party. Martin lived in East Coburg for many years before moving to the old gold mining town of Beechworth at the foot of the Australian Alps. His plays, novels, stories and poetry reflect a diversity of themes and locations, but two novels, *The Young Wife* (1962) and *The Hero of Too* (1965) (about a country town), are richly evocative of Australian life.

In *The Young Wife*, the Cypriot migrant Yannis Joannides, less successful than his shrewd and prosperous older brother, has a fruit shop in the Melbourne suburb of Brunswick. Early in the novel, he undergoes the experience, common to many early southern European migrants, of an arranged marriage. Anna, who arrives on a 'brideship' at a crowded wharf in Melbourne in the 1950s, is to marry a man she has never met.

> The grooms and lovers were getting out of hand. Ignoring the shouts of the ship's officers and evading the harassed policemen who guarded the gangways, they assaulted the steep flanks of the vessel like boarding parties. Some were trying to get a foothold in the open portholes of the crew's quarters, through which laughing men in singlets looked out. A few daring ones were climbing into the shore net that had already been dropped. One youth, handsome as Apollo, had taken off from a bollard and hand over hand was climbing up a cable at the bow, cheered on by the passengers and the waiting crowd.
>
> Standing a little apart, Anna was looking down from the for'ard end of the boat-deck where the press was not so severe. This upsurge of collective passion frightened her because she could feel the responding need of the women of the brideship who were calling out to the men below. She was too nervous to take her eyes off the pier for long, afraid that she and Yannis would not recognise each other. Never had she longed so much for her mother, and never had her mother been farther away than now.

Martin also documents the experience of Criton, a restless young Cypriot artist with a terrorist past, as he arrives on the same ship and attempts to make a life in the new land. He befriends Leo Pavoni, an Italian who is establishing himself with his common-law Australian wife in a house he is building on a block of land in a new suburb. Gradually the process of adaptation begins.

> He began to spend much time with the Pavonis. On Saturdays, when it did not rain, they worked on the house. This appealed to Criton. Sitting on the rafters and nailing down the roofing iron which Martha, laughing with the children and perspiring, handed up to him, he enjoyed his growing mastery of hammer and saw. His hands had hardened and he no longer felt stiff at the end of the day. He could look at Martha without the yearning that had tortured him for so long, and he slept through the nights without dreaming about the past.[2]

Frank Hardy, born in 1917, left school at thirteen to work at a variety of mainly manual jobs before becoming a part-time journalist in Melbourne while collecting material for his first novel, *Power Without Glory* (1950). Hardy joined the Australian Communist Party in 1939, and later became a member of the Realist Writers' group. He would subsequently immortalise his birthplace, Bacchus Marsh, as Benson's Valley in his humorous short stories.

In *Power Without Glory*, Frank Hardy's semi-fictional protagonist John West—loosely, and in the opinion of the prosecutor in the resulting subsequent criminal trial, libellously based on the Melbourne millionaire businessman John Wren—has his beginnings in the working-class milieu of Jackson Street, Carringbush—in reality Johnston Street, Collingwood. Hardy recreates the street in 1890.

> The shops, many of them shuttered and empty; the old houses; the rows of newer tenements and, sprinkled here and there, incomplete houses on which work had ceased; the TO LET signs; the group of ragged unemployed men standing outside the hotel near where the horse trams had changed steeds in the days before the cable was put down; the spindly children playing listlessly in the gutters ... the long queue of despair-haunted people waiting outside the Salvation Army Hall for their daily bowl of soup; the top-hatted, side-whiskered men standing outside the closed bank building, waiting and hoping against hope as they read the notice on the door, CLOSED FOR RECONSTRUCTION.[3]

This is the same suburb that Alan Marshall, born in Noorat in 1902, discovered when he moved to Melbourne in 1919 and worked as an accountant for a shoe company there. In the third part of his autobiographical trilogy *In Mine Own Heart* (1963) Marshall describes the factory in the 1920s: 'a two-storey, brick building sitting squarely on a Collingwood street corner ... an upward extension of the asphalt street ... [where] ... a hundred factories ... elbowed each other for room in cramped alleyways; steam and laden air welled upwards from their breathing windows and doors; their chimneys flung scarves of smoke across the narrow sky'.

> In the early morning the streets that during the night had been quiet and still became alive with people. The tapping of heels penetrated every alley. The people of the factories were hurrying to tend the machines that fed them—clickers, machinists, stuffcutters, lasters ...
>
> They came from trains, from trams—streams of people that divided at cross streets, dissolved into doorways. There was a great power in this movement of men and women from homes to work, a promise of momentous achievement. Yet it was their lives they were paying away in these grim buildings, a little bit each day, a constant depletion of all they had to offer—the great strength of them.
>
> They did not talk to each other. There was no time now, no wish to talk in those moments when one's sorrows were heaviest, when the future was seen as resting precariously on the result of today's labour, today's health, today's control.

Afterwards they went down, and for two hours wandered in the gardens, resting now and then when a seat with an attractive vista took their fancy. He told her much about Melbourne's notable public gardens, but most of the time they talked about themselves, with bits of conventional moralizing and philosophizing thrown in, after the fashion of elderly people.

JOHN MORRISON, 1982 (PAGE 162)

PARK, MELBOURNE, VICTORIA. PHOTOGRAPHER: PHILIP QUIRK

You know the streets between Gertrude and Victoria Parade? Well,
there are plenty of similar ones in our no-man's land. And, among the old
houses, in these straight and treeless streets and the back lanes and
alleys, the old houses of bluestone and red-brick, there's more life
there—real, squirming, dancing life—than to the square mile
of suburban Ringwood or Highett or Preston.

PETER MATHERS, 1966 (PAGE 179)

TERRACE HOUSE IN FITZROY, MELBOURNE, VICTORIA. PHOTOGRAPHER: PHILIP QUIRK

They would talk and laugh when a transformed street saw them surging homewards in the evening. Now they must hurry—girls with unbuttoned coats, the sides floating outspread, with hair still tossed from pillows, running girls with anxious faces.

It was almost 7.30. Hurry!

Rivers of men and youths flowed up streets and round corners—youths with belts around loose-hipped, grey trousers, with soiled grey trousers of flannel, with old coats ...

There were hatless youths with thick hair, youths wearing hats on the sides of their heads. There were old men with grease-polished suits, men carrying perished leather bags that rattled with tools or were fat with lunches, men on bicycles crowding the roadways.

The gaping doors swallowed them all. At 7.30 the tardy ones were jerked into a run by the sudden command of factory whistles. They shrieked from jets of steam on factory rooftops. There were wails and long-drawn blasts. They came faintly from distances. One followed another, answering, competing.

Beneath this shrill sounding came an answering murmur from the depths of buildings. It was more a tremor than a sound, the first movement of an unleashed power. It grew to a rumble, a growl.

Inside the factories pulleys sped into blurred circles, belts leaped upwards and fell thwarted. Machines began to clamour their answer to the whine of motors.[4]

Among the most accomplished writers to emerge from the social realist school in this period was John Morrison. Born in England in 1904, Morrison migrated to Australia in 1922, where he became a bush worker before marrying and settling in Melbourne in 1928. A member of the Realist Writers' group, he published his first stories in trade union magazines. In a later stage of his life, while working as a jobbing gardener, Morrison would ride his bicycle twenty kilometres from Mentone, on the outskirts of the city, to his work in respectable suburbs such as Kew, and then home at night. Morrison frequently recounts his stories in the first person from the point of view of the quiet observer: the anonymous man in the train or the tram, the nameless itinerant workman, the figure in the background unobtrusively tending someone else's flower beds. His tone is often quietly ironic.

In 'The Haunting of Hungry Jimmy', set in Mentone in the late 1950s and early 1960s, Morrison charts the progress of Jimmy Boon, a money-hungry working-class boy who finds himself in possession of a gracious old house with an old-fashioned garden.

I was glad that Neil was no longer around to witness what I anticipated as inevitable destruction. Every few days, in my journey to the city, I made a point of going down Plummer Road, following with a rising horror the domestic application of Jimmy's moronic ideas of home and garden design. The first shock came when I and several motorists were held up one morning by a wire line stretched taut across the road, one end of the line being attached to a tip-truck, the other to the first tree of Lawhill's encompassing cypress hedge. Trust Jimmy to find the shortest way home. When I passed again in the evening the hedge, pulled out tree by tree as cleanly as a dentist would pull out teeth, was gone, and Lawhill's secret garden lay exposed to the world.

Stopping the car, I sat for some minutes looking across at it, trying to fix an image of it before it vanished forever from the face of the earth. It had got a bit ragged since the death of the captain, but that only added to its snug, lush, old-world charm. In my racing imagination there was a shrinking fear in the two windows peeping through the screen of camellias, just as, in the uplifted limbs of the white tree, there was a suggestion of outraged innocence, like a lovely young girl suddenly caught naked.

I said then a silent goodbye to it, but week by week the respite—of the tree—was extended while Jimmy went to work on what he, no doubt, regarded as modernization.

Using cheap bricks, and his wife as a labourer, he built a low wall to replace the hedge. Diagonally across the big lawn he cut out and concreted a car drive, with a loop on the front of the house for turning, and access to a metal pre-fabricated garage on the side where the captain had had a picturesque fernery. In the centre of the loop he constructed a crude rockery topped by a white plaster stork that made a startling contrast to a pair of rainbow-painted gnomes pointing the way to the front steps. All the camellias were cut down, hedge-like, to the exact level of the verandah decking, and the house itself transformed from a quiet green to a brick red—which might have been nice enough but for a dazzling new iron roof. Along the sunny side of the garden every sizeable tree was removed, leaving only a few small shrubs as backing for what was obviously intended to be an annuals border in the best Melbourne suburban tradition. The general Jimmy Boon effect was finally topped off by white-painted, drip-catching clam shells placed under every garden tap.[5]

In 'Transit Passenger', which seems to be set in the late 1960s or early 1970s, and in which Morrison provides one of his most affectionate portraits of the ordinary byways of the city, his main character is an older man, a widower, lunching by himself in a city cafeteria.

By twelve-thirty he had changed his books at the Athenaeum Library, picked up his repaired watch at Dunkling's, shopped for a few trifles at Coles', and bought two singlets and a shirt at Myer's.

The rich smells of Myer's big ground-floor delicatessen tempted him to get something to take home, but there was little pleasure in eating a main meal alone, and when he came to the city he usually had something. So he made for Walton's, partly from habit—it had been his wife's favourite cafeteria—and partly with a vague idea of dropping in afterwards on one of the new cinemas at the end of Bourke Street.

Walton's was busy as usual, but he found a table. Leaving bag and hat as tokens of possession, he took his place in the queue and returned with a simple meal of vegetable soup, smoked fish with mash and green peas, prunes and custard. And coffee and biscuits to linger over with a cigarette.

He was commencing on the fish when he first saw her. A woman rather beyond middle age, wearing a smart little felt hat and well-cut brown overcoat, and with umbrella and handbag in contrasting browns looped over wrist. Not particularly noticeable; he wouldn't have observed her at all but for the fact that she happened to be looking in his direction as his eyes ran along the people in the race. She'd just picked up tray and cutlery, and between glances at the menu on the wall, seemed to be scouting for a place to sit. He'd completely forgotten her when a few minutes later he looked up to see her, tray in hands, confronting him from the other side of the table.

The man and his new acquaintance go for a walk, and Morrison offers a rare portrayal of central Melbourne.

They came out into the dusty yellow autumn sunshine and took a south bound tram in Swanston Street.

'I'm always glad to get out of the city centre,' he said when they were seated. 'Even Camberwell, five miles out, is better than this. Melbourne used to have a good smell. I remember when I first came down, a lump of a boy from the bush. I thought Melbourne smelled exciting, a mixture of all kinds of things: beer, paper, cooking, new clothes, hot stonework. Now it's beginning to stink like any other city. Gas, oil ...'

They got off at Domain Road and went first to the Shrine of Remembrance, strolling slowly around the base, then climbing to the top, from where he pointed to landmarks and places of interest: Port Melbourne where her ship was lying, the vague outline of the distant You Yangs, the nearer Dandenongs, the greens and whites and reds of the suburbs, the new multi-storeyed office blocks of the business centre.

Afterwards they went down, and for two hours wandered in the gardens, resting now and then when a seat with an attractive vista took their fancy. He told her much about Melbourne's notable public gardens, but most of the time they talked about themselves, with bits of conventional moralizing and philosophizing thrown in, after the fashion of elderly people.

'My daughter and her husband wanted me to go and live with them when Dad died,' she told him once, 'but I didn't even consider it. It hardly ever works out. Somebody always gets hurt.'[6]

John Morrison, who had worked on the Melbourne waterfront for a decade in the years before World War II, uses this milieu with authenticity in his almost documentary *Stories of the Waterfront*, collected in 1984. In 'Nightshift', published in the collection *North Wind* (1982), a group of wharf labourers travels by launch to the wharf where they will be working.

> As the little red light appears on the river the men crowd the edge of the landing-stage, each anxious to get a seat in the cabin on such a night. The water is very black and still, and the launch moves in with hardly a ripple. The night is full of sounds. Little sounds, like the rattle of winches at the distant timber berths; big sounds, like the crash of the coal-grabs opposite the gasworks. All have the quality of a peculiar hollowness, so that one still senses the overwhelming silence on which they impinge. In some strange way sound never quite destroys the portentous hush which goes with fog. Dick feels it as he follows old Joe over the gunwale and gropes his way through the cabin to the bows.
>
> 'It's quiet tonight, Joe. Can't be many ships working.'
>
> … The launch travels smoothly and swiftly. Quite safe. The mist is thickening, but there's a bit of light on the river here from the ships working on north side. Small ships, as ships go, but monstrous seen from the passing launch. Beautiful in a way of their own, too, with the clusters of lights hanging from masts and derricks. Little cities of industry resting on towering black cliffs. One can't tell where the black hulls join the black water.
>
> Nameless bows, but still familiar to the critical stevedores.
>
> 'That's the *Bundaleera*. Good job. She worked the weekend.'
>
> 'The *Era*. She'll finish tonight.'
>
> 'The *Montoro*. They say there's only one night in her.'
>
> Strange twentieth-century code of values. A collier which works Sundays is a good ship; a deep-water liner which works only one night is a bad ship.[7]

If Melbourne has no spectacular natural harbour to inspire its writers, then Port Phillip Bay evokes its own loyalties. George Johnston, whose early journalism and historical works were inspired by his love of ships and the sea, describes at length the wharves of Port Melbourne in his semi-autobiographical novel *My Brother Jack* (1964). His character David Meredith, as a young art student at the National Gallery School in the years following World War I, experiences a similar fascination.

> I began to leave home in the mornings an hour earlier. This carried with it the disadvantage of having to eat breakfast at the same time as Dad, but now instead of sitting on the steps in the lane waiting for the studio to be unlocked, I could go wandering around the waking wharves, and for the first time in my life I came to be aware of the existence of true beauty, of an opalescent world of infinite promise that had nothing whatever to do with the shabby suburbs that had engulfed me since my birth. The fine floating calligraphy of a tug's wake black on a mother-of-pearl stream in the first glow of a river dawn, the majesty of smoke in still air, the pale and tranquil breath of river mist and morning steam, the rising sun picking golden turrets out of derricks and samson-posts and cranes and davits, the coloured smoke-stacks and the slender gilt pencillings of masts declaring themselves little by little against the dark haze-banks that always in this waking time veiled the river flats, the faint images of ships far down the stream, coming in from Gellibrand, looming out of dew and light and sea mist, and then, at every bend and twist of the river, changing the shapes of beauty like a rare vase turned in the fingers of a connoisseur.
>
> It filled me with an excitement, almost an exaltation, that I could tell nobody about. I did not see it then as a way out of the wilderness, for the stuff of this material was too fragile to be considered as something which might be used, but I was quite sure that something important had happened to me. I moved through this newly-discovered world breathless and alone, like Adam in a new Eden, and I felt almost as if I had to walk on tiptoe wherever this shining place extended.[8]

'THE FINE EDGE OF SEEING':
Hal Porter's Melbourne

The detail!

The colour!

Except in dreams, neither detail nor colour has ever since been so detailed or coloured; the fine edge of seeing for the first time too early wears blunt. But the first seeing is so sure that nothing smudges it. Take Bellair Street.

Bellair Street, built about 1870, is a withdrawn street overhung by great plane-trees and is on the way to nowhere else. It is only several blocks long and, so far as houses are concerned, one-sided. This is because it is the last street, three-quarters of the way down, of several streets lying horizontally along the eastern slope of a ridge crowned with Norfolk Island pines, non-conformist churches of brick the colour of cannas or gravy-beef, and a state school of brick the colour of brick. The slope makes it necessary to ascend from the front gate of 36 along a path of encaustic tiles, next by eight wooden steps on to a front veranda which is therefore a long balcony balustraded with elaborately convoluted cast-iron railings. From this balconic veranda I look over the plane-trees towards a miles-off miles-long horizon composed of the trees of the Zoo, Prince's Park, Royal Park and the Melbourne University. There are the towers and domes of the University and the Exhibition Buildings, and countless nameless spires.

This prospect is less colonial Australian than eighteenth-century English in quality: billowy green trees, misty towers, even a shallow winding stream that starts and ends in obscurity like a painter's device. Southern Hemisphere clouds pile themselves up, up above, and take on Englishy oil-landscape tones, or steel-engraving shafts of biblical light strike down, or incandescent Mississippis of lightning. Between this romantic or dramatic background and the watcher at the cast-iron lace of the balustrade innumerable more sordid elements are disposed: paltry municipal parks like seedy displays of parsley; endless terraces of houses; endless perspectives of ignoble streets and, strange as palaces, many three-storeyed stucco hotels whose baroque façades topped with urns and krateres protrude here and there above an agitation of humbler roofs of slate or terracotta but largely of unpainted corrugated iron. Sometimes, brilliant and perfectly executed hailstorms load the gulches of the roofs with white. Sometimes, a sunset behind Kensington ridge is reflected in sunless distant windows like spots of golden oil. I seem to be often watching, now and again with Mother a shape behind my shoulder, but most often alone. This watching, this down-gazing, this faraway staring is an exercise in solitude and non-involvement. ... This landscape, transfixed and unpeopled, untouchable and mute and mine, is my first glimpse of a world I am to see far too much and yet not nearly enough of and into. Let my eyes, so sated and so deprived, turn from the scene and peer at closer matters.

Hal Porter, *The Watcher on the Cast-Iron Balcony*, 1963[1]

Perhaps of all portraits of cities to be recorded in Australian literature the most meticulously exact and detailed is Hal Porter's Melbourne, just as Porter's Bairnsdale and Gippsland are also meticulous and exact. And yet, like David Malouf's childhood Brisbane houses and Randolph Stow's Western Australian exteriors, one suspects that it is a Melbourne that could be experienced by no other writer.

Born at Albert Park in 1911, and living prior to World War I at 36 Bellair Street, Kensington, Porter recreates the house of his early years in the first volume of his autobiographical trilogy *The Watcher on the Cast-Iron Balcony*. In documenting his early life in Melbourne before the age of six with the eye of a child that encompasses everything and forgets nothing, Porter evokes his boyhood city with the vision and vocabulary of a grown man. 'At the moment of unpacking, I am astounded by the

164

size and complexity of this child's luggage,' he writes of his memories of the period. 'Even now, a middle-aged man, I cannot unpack it all.'

A writer who at times seems determined to test both his imagination and the possibilities of the language by reinventing in prose every landscape embedded in his seemingly indefatigable and photographic memory, Porter's descriptions are sometime reduced to a list of locations and attributes, but that list is never less than poetically sure.

> Going in steam-trains or cable-trams, sometimes in cabs, sometimes walking hand-in-hand with Mother or Father across streets where the crossing-sweeper pushes piles of horse-manure from before our feet, going to the places little Melbourne suburban boys of those years go with their mummies and daddies: the Museum, the Botanical Gardens, the Waxworks, Wirth's Circus, Punch and Judy shows, the Aquarium in the Exhibition Buildings, the Royal Park Zoo. The final stage in travelling to the Zoo is done in the last horse-drawn tram left over from an earlier age. Before I have seen a sheep or a cow, and many years before I see a kangaroo, I am familiar with elephants, camels, lions and leopards, with middle-class animals like the giraffe and the hippopotamus, old-fashioned nineteenth-century creatures. Indeed, all these entertainments are, in a sense, left-overs from Victoria's reign. Other animals like Charlie Chaplin are taking their places. The middle- and late- Victorian auras of the London originals on which they are modelled emanate from these pleasure places. Cole's Book Arcade, of which the lofty cast-iron galleries bisect two Melbourne blocks, has the common-sense yet engaging eccentricity that is Edward Lear's and the Englishman's. At the great entrance to the arcade with its architectural air of Waterloo Station, two small mechanical puppets, earnestly and rosily grinning like pot-boys from Dickens, jerk ceaselessly at crank handles which rotate into view successive boards advertising in rainbow colours The Largest Book Arcade In The World and its subsidiary attractions. These attractions have not always to do with books. I remember indoor cages of monkeys, tropical palms and tree-ferns, and an afternoon tea of cream horns eaten to the music of a small whining orchestra and, upstairs, in a first floor gallery supported by brass columns, tiers and seeming miles of gilt-poxed china figurines and curly vases with gilded handles.[2]

After completing his education in Bairnsdale, where he moved at the age of six with his family in 1917, and where he worked for a brief period as a journalist on the *Bairnsdale Advertiser*, Porter moved in early adulthood to Williamstown, the first port and one of the earliest settled areas of Melbourne. He was accepted as a student of drawing at the Melbourne National Gallery in 1927.

As he describes in *The Paper Chase* (1966), the second volume of the trilogy, the by-now self-consciously artistic young man sets out to discover the pubs of the city, where even his deliberate drunkenness represents a conscientious attempt to capture experience. Porter writes here as a painter paints, building up a sequence of minute brush strokes to create the impression of a whole.

> I have been drinking since the age of seventeen. More precisely, since the age of seventeen I have, now and then, drunk a little wine, a little beer, and experienced the lifting of a veil or two from the senses. The wine is drunk at Camillo Triaca's Latin Café in Exhibition Street, Melbourne, or on Saturday visits to a sow-fat, vivacious, French-speaking Madame Jorgensen, born in Vraa, Jutland, a Conrad character who runs a wine depôt on the Williamstown waterfront. At Madame's, with a sang-froid and fearlessness which, considering my youth and appearance (sometimes garbed in plus-fours, or a squire-of-the-village suit of sorrel-coloured Harris tweed), still amaze me, I brazenly mingle with, and openly hang on the words of, and unflinchingly stare at the brawls between Jack London seamen from every sixteenth of the globe: Creoles from Mauritius, Scotsmen in navy-blue Sunday-go-meeting suits, Scandinavians in cumbersome, high-necked sweaters, Japanese with teeth edged by what appears to be aluminium, all sorts of teak-textured but curiously innocent men tattooed with anchors, thistles, windjammers, roses, and S-curled scrolls bearing the word Mother. While I eke out two or three or, at devilish most, four port wines or clarets—Madame keeping a sly slit of maternal eye on me—the seamen

drink deep, offering me hair-on-the-chest potions Madame will not let me accept. It is handy to have her refuse for me, it saves the embarrassment of cissily refusing myself. At six o'clock Madame's jelly of a face switches off its professional animation and charm and sets like cement into a Hindenburg mask, and the symposium is over. Some of the men from the sea swagger zigzag across the plantain-ruptured asphalt to the railway station and the train that will take them and their bottles of muscadine to Melbourne, where are the brothels of Little Lonsdale Street, or the sly-grog-and-prostitute joints of Carlton. Others reel away, they and their reeling sunset shadows, through the fennel, along the old groaning wharves to orgies of dark singing and spiceless brawls in their own fo'c'sles.

In 1937 Porter resigned his position as a junior teacher at North Williamstown State School and established himself in a third-floor back room at 28 Collins Street in the city, where he found himself in the same milieu explored by Fergus Hume half a century before. He discovers the cafes and meeting places of Bohemian Melbourne of the 1930s.

Meantime, the glass-panelled door of the Café Petrushka opens and shuts on the entrances and exits of Melbourne's intelligentsia and decadentsia, leftists and gourmets, journalists and broadcasters, potters and actresses, lesbians and advocates of trial marriages, and high-school English teachers on the rantan ready and willing to buy tumblers of tea and scalenes of *halva* for any out-of-pocket intellectual who will bicker or agree with them on the Marx Brothers, Ernest Hemingway, Eisenstein, Grosz, Aldous Huxley, Walt Disney, Lytton Strachey, Liam O'Flaherty, Picasso, Delysia, Marie Dressler, any of the season's conversational counters.

There is no record of the famous, near-famous, flash-in-the-pan famous, or famous-to-be men and women who walked up the narrow incline of Little Collins Street, and opened the door of weeping glass, and dined at the small green tables, and yelled above Josephine Baker singing 'Le Petit Tonkinese', and the garrulous never-to-be-famous, and those most dangerous of all talking animals, the non-creative intellectuals. It seems reasonable, therefore, to record some of the diners or sippers of Russian tea I recall as being there during my few months' use of the place. There is Fay Compton nibbling a piece of *halva* with Michael Wilding as they are on their way downhill to the King's Theatre and *Victoria Regina*. There is Woizokowski, discreetly painted, making an entrance with his daughter Sonia and, trailing them, circumspectly a-twitter as a school crocodile, all eyes and legs, *the corps de ballet*. There is Loudon Sainthill eyeing the glass panel on which there will, late as ever, appear the white face of—is her name Diane?—a fey young musician dressed in black like a romantic midinette, with fog in her black hair, and a roll of music in her black-gloved hand. She will come. They all will, sooner or later. Helene Kirsova with her hair in pale braids, with pale eyes and flaming cheeks. Albert Tucker performing his squeaky baby-talk act for George Bell and a group of his students. Theodore Fink, driven by a chauffeur, and weighed down by wealth and a hundred-weight of rich overcoat. James Flett and a portfolio of water-colour pirates and self-portraits. Dr John Dale and Max Meldrum. Dargie and the beautiful Kath Howitt he is later to marry. Hayward Veal who is later to marry Minka and take her, and his paintings of Melbourne streets opalescently slimy with rain, to London. Tom Challen who has given up being a boy-prodigy violinist to grow into a plump and witty cartoonist for *Table Talk*, and is on his way to leaving his Collins Street attic and the ikon over his bed-head to be the cartoonist Tac on a Fleet Street newspaper. They are all on the way to somewhere else, to money or fame, to notoriety or nowhere, or to drowning in the whirlpool of talk-plans about their Great-Opus-to-Come, or in alcohol or the undercurrents of indiscipline or the sloughs of sex or the shallows of egoism.

And there is the literary world.

Bob Close has written a novel, not yet published in 1937, called *Love Me, Sailor*, and, his Borsalino rakishly set on the battleship-grey undulations of his hair, reads to us, while the portable gramophone sings 'Mean to Me' dimly from a record grey from years of rotations, the more censorable episodes. Alan Marshall has also written his novel. It is about life in a shoe factory and it, too, is not yet published. He reads the more gruelling episodes of this. I have written so little of mine that I keep my mouth zippered. It is all I can do to shut myself away from the many-doored, many-planed city long enough to sleep and, once in a while, to write a short story. These are set in the country because I have found out that it is unwise to write of a place while still in it, or of people and paraphernalia not separated from me by space or time. Hence I write of Japan in Western Australia, of Western Australia in Victoria, of Victoria in London and Rome, of cities in the country, of the country in the city. In fact, it becomes clear, as I listen to Close and Marshall and what they have done, that it will soon be time to move on again—Melbourne is not far away from Williamstown, the seagulls skimming the slate-tiled fairy-tale turrets and curlicued weather-vanes of the Melbourne fish market are the same ones that mince sideways along the wet and moiré sands of Williamstown Beach. I do not want to leave the bluestone pinnacles and Corinthian stucco. The writer has to, but I keep him a month or two more in the city by letting the artist take over for a while to prepare some work for an exhibition at the Athenaeum Gallery to be opened by Fay Compton.

In 1949, after a failed marriage and more years of schoolteaching in South Australia, as well as a variety of other jobs, Porter managed a hotel in St Kilda.

> St Kilda, once a fashionable and grandiose seaside suburb, a sort of Aussie Cannes with a better beach, has become tawdry, its one-guinea waves now cheap at a penny. Its mansions along the Esplanade—Belgian Gothic, Greek Revival, Florentine Renaissance, Moorish, even Spanish Mission—have become boarding-houses, or have been subdivided into flats, flatettes, and hives of bed-sitter cells smelling of gas-rings. In 1949, its drain-edges, and front steps, and the stairways descending to the beach zigzag through lawns and palms and gazania-covered rockeries, are still painted with blackout white, and the suburb has become a post-war working-class playground providing all the shoddy and instant pleasures the working-class go in for—a Luna Park, a Palais de Danse, a skating rink, fish-and-chip shops, hamburger shops, ice-cream kiosks, soft-drink stalls, pie-and-tea restaurants, milk bars, penny-in-the-slot machines. In the back streets, blocks away from the sea, are the houses of rag-trade Jews and fish-shop Greeks and fruit-shop Italians. Nearer the sea are several night-clubs haunted by car salesmen, petty criminals, Albanians and their factory-hand pick-ups, confidence men and their molls.

> When the wind flings the seagulls about like handbills, and the sea is frothing at the mouth and ejecting corks from fishermen's nets, and rotten oranges from P. And O. liners on to the beach, the spivs and gamblers and Esplanade larrikins and urinal-haunting perverts and burglars and pickpockets make for the bars. The bars include those of the George Hotel.

> The George's earlier reputation as one of Melbourne's half a dozen *cordon bleu* hotels still draws guests from the country; its old-fashioned lavishness and service still draw those who prefer these qualities in a hotel. Upstairs, the George is the *Titanic* that missed the iceberg, and has merely got shabby and patched. The patchers are at work all the time. I remember the housekeeper and her bowed-over staff, the sewing-machine ceaseless as creation, the sheets of fine pure linen being turned, the napkins and tablecloths darned, the chintz loose-covers mended, the bath mats re-hemmed.[3]

Porter subsequently went to Japan as a teacher with the Occupation forces, and finished his working life as a librarian at Bairnsdale, after which he became a fulltime writer in the early 1960s. He died in 1984.

THE CITY IN THE HOUSE

Beverley Grove, the house, the subdivision, the suburb … were immediate tokens and symbols of social progression. Of an advancement in caste, even.

So were our neighbours in the new houses around us. Mr Treadwell, our neighbour on the left, was a retired police court magistrate. Mr Phyland, our neighbour on the right, was a chartered accountant in the city. Wally Solomons, who had the house directly opposite us and an indolent, soignée, highly sophisticated and very beautiful wife called Sandra, who spent most of her time in black slacks and boleros, smoking pastel-tinted cigarettes through a twelve-inch holder, was the head salesman in the main showrooms of General Motors. On the corner, where Beverley Grove joined Park Crescent (these names were incised in Roman capitals into the cement paving blocks), lived Dr Felton Carradine, the dental surgeon.

The entire Beverley Park Gardens subdivision had been a big investment on the part of a building-contractor by the name of Bernie Rothenstein, who in the speculative days of Melbourne suburban estate development had a record for civic malefactions which was probably unmatched in the Southern Hemisphere, but which later won him a respected seat on the City Council and ultimately a knighthood: and although only three basic ground-plans were used for all the hundreds of houses in the subdivision, there were still no two houses in any one street, grove, crescent, drive, or avenue which could be said to really look alike. Each front elevation had its own distinct difference, in the design of the porch, the placement of the picture window, the run of the paths and whether plain or crazy … The interiors of all the houses were virtually the same, of course, to the very inch in things like roof-area and the dimensions of rooms and halls and passageways and cupboard-space, and the immutable sameness of their inventoried fittings. (Not until long, long afterwards did one pause to marvel at the deadening democracy of a system which could dictate, over nearly one square-mile of human habitation, that no man should have one more light-switch or power-point or water-faucet or sliding drawer than any other!)

George Johnston, *Clean Straw for Nothing*, 1964[1]

Melbourne in the first half of the new century, as described by its post-World War I writers, is quintessentially a suburban or industrial city, a city of interior lives, and often slightly claustrophobic. But if Hal Porter found in this environment a wealth of exotic detail in the texture and patina of city life, George Johnston, author of the trilogy of partly autobiographical novels beginning with *My Brother Jack*, creates a city that mirrors more closely his own internal dissatisfactions. Johnston presents a portrait of Melbourne which is less factually exact in relation to his own life than that of Hal Porter, and much less affectionate, but one which, judging by the immense popularity of the novel, nevertheless struck an emotional chord with an infinitely greater number of Australians.

Johnston was born in Malvern, Melbourne, in 1912. Shortly before the outbreak of World War I his father, a tramway sheds foreman, moved with his wife, a nursing sister, and four children to a house at 11 Buxton Street, Elsternwick, which the family christened Lochied after his mother's Scottish connections. This newly constructed five-roomed weatherboard house in a newly developed working-class district remained the family home for more than forty years—coming to contain, according to Johnston, an accretion of 'parents and an environment and tales told about ancestors and dusty vines growing over outhouses where remarkable insects might always drop out of hidden crevices'.

For Johnston's narrator, the young David Meredith, 'childhood, looking back on it, is like this—a mess of memories and impressions scattered and clotted and pasted together like a mulch of fallen leaves on a damp autumn pavement'.

With me it focuses most sharply around the small, rather fusty wallpapered hallway upon which the front door opened in that undistinguished house—weatherboard painted dark stone and a corrugated iron roof of sun-faded Indian red—which sat behind a wire fence, privet hedge, small square lawn of buffalo grass, and the name *Avalon* in gilt letters on a blackwood panel in a flat and dreary suburb far away in Melbourne, Australia. 'Far away' is meant in a temporal as well as spatial sense, for not only did all this begin to occur some ten thousand miles distant from where I am now, but this was more than forty years ago, not long after the First World War ended.

The hallway itself, in fact, was far from undistinguished, because a souvenired German gas-mask hung on the tall hallstand, looking like the head of a captured Martian, and the whole area of the hall was a clutter of walking-sticks with heavy grey rubber tips—the sort of tips on walking-sticks that relate to injury rather than to elegance—and sets of crutches—the French type as well as the conventional shapes of bent wood—and there was always at least one invalid wheel-chair there and some artificial limbs propped in the corners. Our sister Jean, who was the eldest of us four children, eventually married a returned soldier who had had his leg amputated, and this seemed to us, at the time, quite normal and expected. Jack and I must have spent a good part of our boyhood in the fixed belief that grown-up men who were complete were pretty rare beings—complete, that is, in that they had their sight or hearing or all their limbs. Well, we knew they existed, but they seldom came our way.

For a writer such as Queenslander David Malouf—another precise observer of interiors—his childhood home at 12 Edmondstone Street in South Brisbane can seem to encapsulate the whole of his family's personal history; but for George Johnston's David Meredith, growing up in the immediate post-war years and visiting wounded soldiers in hospital on Sundays while his mother installs more disabled men in the spare bedrooms, a working-class bungalow at 11 Buxton Street seems to contain all the history and horrors of World War I.

I don't know when it was that I began to suspect that these desperate Sunday feelings were really only an extension of a terror which I knew to be real, even though I did not understand it, but which was somehow related to the day the *Ceramic* brought my father back, and also to the leather-and-metal, stiff-jointed legs and the claw-like appendages to the artificial arms propped in the corners of our hallway, and to the faces of Aleck and Stubby posing for snapshots beneath their flat-brimmed New Zealand hats, and to Gabby Dixon weeping in the darkness of his bedroom, and to Jack and me in the sleep-out listening to the grown-up laughter around the cribbage-board in the sitting-room or to the more furtive slithering noise which was made by a big creeper vine we called the Dollicus as it moved in the night wind against the screen of flywire.

I never did talk to Jack about these mysterious, disturbing fears.

… In a sense, of course, I was too young for the war to have had any direct effect on me, since there was really nothing of it that I could remember. Yet what is significant to realise now is how every corner of that little suburban house must have been impregnated for years with the very essence of some gigantic and sombre experience that had taken place thousands of miles away, and quite outside the state of my own being, yet which ultimately had come to invade my mind and stay there, growing all the time, forming into a shape.

And it went on for years. There was no corner of the house from the time I was seven until I was twelve or thirteen that was not littered with the inanimate props of that vast, dark experience, no room that was not inhabited by the jetsam that the Somme and the Marne and the salient at Ypres and the Gallipoli beaches had thrown up. Stubby sitting by a window tearing with his teeth at the white threads of his doilies; Aleck in another room knitting his balaclavas or fumbling with quick-tapping insectine fingers for his tobacco pouch; the bumpy, squeaky sound of someone in a bedroom testing an artificial leg; the bathroom that everlastingly smelt of antiseptic and ointment and ether and Condy's crystals: and even outside, in the backyard sunshine, there would be Mother's white nursing veils and aprons blowing on the clothes-line in a smell of yellow soap, and underneath the fig-tree Bert sitting on an upturned packing-case, a long leather bib tied around his chest and his empty trouser-leg neatly folded up and fastened by a safety-pin, hammering away at half-soles and heels. There was no radio then, but we always had 'sing-songs' around the piano on Sunday nights, with Mother playing, and for years the songs were always the same—the 'old favourites' of the war years, 'Tipperary' and 'Over There' and 'Johnny Get Your Gun' and the rest of them.

Even beyond the house there is no escape, as Davy finds when he looks into the window of a deserted photographic studio.

The window was full of dust, dead flies and cockroaches, and a great many spotted photographs that had been there for years. The sun had faded them to a ghostly, deathly pallor, but they were all of young men in the uniform of soldiers, most of them wearing slouch hats turned up at one side, and all of them with Rising Sun badges on their tunics. They were mostly boyish-looking faces, none of them with the expressions that I had seen at home or in the hospital wards, so that I guessed the portraits had been taken years before, when they had enlisted or just before they had embarked for overseas.

I was staring in at these photographs one day—really I was only waiting for the signal of the train's whistle on the viaduct, which would be my cue to move on to the second-hand shop window—when a ragged, agate-eyed boy who was at my school but several years older and in the sixth grade, came down the street, kicking a tennis ball before him. He stopped alongside me but said nothing. He just stood there for a long time, right beside me, staring in with me at the pale photographs. Finally, in a flat voice and without even looking at me, he said, 'All them blokes in there is dead, you know.' He stared at the pictures a moment or two longer, then said, 'Well, hoo-roo,' and waved to me and moved off and kicked the tennis ball right down to the end of the street and trotted of after it, whistling.

I ran all the way home that day, trying not to cry, because I didn't know what it was I wanted to cry about, but I never after that looked in the window of the photographer's studio or the second-hand shop.

Beyond the house and its immediate surroundings, the suburbs of Melbourne stretch on all sides.

During this time, of course, one was becoming more and more aware of an overpowering exterior world that existed beyond the house and those who occupied it. This world, without boundaries or specific definition or safety, spread forever, flat and diffuse, monotonous yet inimical, pieced together in a dull geometry of dull houses behind sliver-painted fences of wire or splintery palings or picket fences and hedges of privet and cypress and lantana; and all these sad, tidy habitations had names like *Sans Souci* and *The Gables* and *Emoh Ruo* (which I always took to be a Maori name until I learnt that it was only 'Our Home' spelt backwards) and *The Rest* and *Nirvana* and, of course, other *Avalons* beside ours, for this was a very popular name which would occur once in very nearly every block. Most of the streets were named after long-dead councillors of the municipality or for battles in the Crimea or Boer wars. All the way through to the city proper there was nothing to break the drab flatness of this unadventurous repetition except the club flags flying over the grandstands of some football ground or other, or a particular factory smokestack that impressed by its height or shape or the amount of reek it gave off, or the grimy brick wall of the Rosella Jams and Pickles Factory with the cloth-capped girls working and chattering behind the railway-sooted windows.

A lifetime later I went back there and the horrible flatness of it all was just as real as ever, but far more depressing, since one no longer had the child's exaggeration of scale to help it out. In that earlier time it was always possible to invent what in reality did not exist. There was a public golf links about two miles away from where we lived, and in the middle of it an ugly grass-grown mound that could not have been more than thirty or forty feet in height. Yet we would walk there often, through the unmitigated melancholy of those suburban streets, simply for the adventure of scaling its sides and playing at Everest mountaineers and pretending we had toboggans.

Meredith is a teenager, apprenticed to a printing firm and submitting freelance articles to the Melbourne *Post,* when the 1930s depression hits.

They brought in the dole, and then the dole became 'the sustenance,' and around this time they un-locked the Defence Department warehouses and out of the mothballs they took the old surplus greatcoats and tunics and they dyed them a dull black—all that brave khaki of 1914-18—and against the contingency of a Melbourne winter issued them out as a charity to keep the workless warm. So that as the unemployed grew in number the black army coats became a kind of badge of adversity, a stigma of suffering.

One would see the shabby figures shambling along the suburban streets, carrying a loaf of bread and in a cloth bicycle-bag their meagre handout from the Sustenance Depot of tea and sugar and flour and potatoes, and a wisp of tobacco. Or there would be a queue of men the length of a block, most of them in the ill-fitting, shameful black, in apathetic competition for half a dozen casual jobs. As the situation grew worse desperate attempts were made towards alleviation, and the 'black coats' moved then in the more regimented bands of the 'sustenance-workers' and you would see them with their brooms and picks and shovels and council tip-drays working in slovenly unison on pointless municipal projects. ...

This was the time, too, of the first trickle in from Europe of that other human flotsam, Jews mostly and refugees from a new malignancy, and this, also, was misleading at first for the trickle had become a flood almost before one realised what was happening. Even the language of suffering, of course, had to be Australianised. The refugees became the 'Reffos,' just as the sustenance-workers had by this time be-come the 'Sussos.'

... In our suburb there was a constant, unnerving movement of these pathetic and yet somehow oddly sinister figures in their black tunics and greatcoats. Sometimes they would come to the door asking for an hour's work to cut the hedge or to mow the lawn or to stack firewood or even to run errands ... or some-times more bluntly just to ask for a handout of food or money. A few of the more resourceful among them had made themselves crude little hand trucks which they would push clatteringly around the streets, col-lecting old newspapers or scrap-metal or unwanted clothes, or with coal or kindling-wood to sell.

I remember the evening when Dad came into my room and said: 'I want you to print me up a sign. You do printing at Klebendorf's, don't you?'

'Lettering things, you mean? Yes.'

He handed me a bevelled oblong of hardwood, and said, 'This is for the front gate. The carpenter at the sheds fixed it for me. I want you to paint a sign on it. Can you print this up for me?' He passed over a crumpled slip of paper on which he had crudely printed out the words: BEGGARS, HAWKERS, AND CANVASSERS WILL BE PROSECUTED.

I lettered the sign for him in white on black, but then he made me take it out and screw it to the front gate. I wanted to protest but his face was so stern and implacable that I said nothing. While I was attaching the sign to the gate two middle-aged men in black greatcoats came across and stood there watching me. When in my nervousness and humiliation I dropped one of the screws one of the men stooped over and picked it up and handed it to me, but neither he nor his companion uttered a single word. When the sign was firmly fixed against the gate they just turned away and shuffled off along the street muttering to each other. It was not very long before these signs—or something like them—were on gates all over the suburbs.

When David Meredith marries his first wife Helen, he moves with her into what is commonly sup-posed, in the 1930s and 1940s, to approximate the Australian Dream: a 'double-fronted, ultra-mod-ern, red brick, three-bedroom villa' which '... stood on its sixty-foot frontage behind a low brick fence on its own small block of land beside a concrete drive leading to a separate fibro-plaster garage. It stank of cement mortar, raw floorboards, fresh paint, damp putty, and insulated electric wiring'.

Beverley Grove—or Glen Iris, the Melbourne suburb where Johnston himself, now a successful journalist, lived in Britten Street with his first wife in the 1940s—gradually becomes a symbol for Meredith of his own creative sterility. Meredith's wife Helen furnishes the house in Beverley Grove tastefully through hire purchase and her own efforts on the sewing machine, and buys pale modern furniture and French impressionist prints. However, Meredith is starting to incline more to the company and taste of his literary friends, notably the journalist Gavin Turley. He rather snobbishly prefers the ramshackle and bohemian Turley house in the more affluent suburb of Toorak—which, according to Johnston's biographer Gary Kinnane, was in actuality based on that of Johnston's friend the journalist Mungo McCallum, in Point Piper, Sydney.

We arrived there in the appropriateness of a late afterglow, and went through great creaky wooden gates hung from square stone pillars and into a dusky jungle of a garden with black thickets of azaleas and the biggest rhododendrons I ever saw in my life until years later when I looked across at the crimson forests buttressing Tibet from the Katmandu side, and there was a weird tangle of gigantic creepers and those huge leafy things that we always called elephants' ears and fat cacti standing on enormous thick hairy prickly stems like mammoths' legs. Curving through this dense wilderness of darkness and damp, decaying smells there was a crunchy gravelled carriage-drive scattered with fallen leaves, leaves that were long and stiff and curled-up and cardboardy, which had dropped from two tremendous Moreton Bay fig-trees that blocked out all the gloomy sky above us, and the leaves in the evening breeze were moving around with a dry, scaly, scurrying sound.

It was quite a walk up to the massive old entrance with the name of the house, *Bangalore*, chiselled in stone above a heavy, panelled door which had massive lion-knockers and big brass bell-pulls which, coming after the tangled garden, sharply reminded me of the old house where Helen had lived.

… They both came to meet us at the door, and Peggy took the coats away while Gavin showed us into the main room, and this was just about the most extraordinary mix and clutter and congested mess of a room I had ever seen. It had once been stately, and probably had been used for receptions, because the moulded ceiling was superb and there were marble pilasters on either side of the fireplace and above the ornate carved mantel a huge and magnificent French Renaissance looking-glass in bad repair which gave back a mysterious muddy reflection such as one might get from a stagnant pond, and from the ceiling-boss still hung the heavy gilded chains which must have once supported a chandelier. The immensely tall windows were hidden behind tarnished velvet curtains hanging from sagging pelmets, and there were damp mildewy stains down one wall, where the plaster moulding was broken away, and alarming cracks in all four walls, and the carpets on the floor were very thin and faded out, like old flowers found pressed between the leaves of a book, and tattered at the edges.

The room was furnished in the most astonishing jumble of good antique pieces and second-hand junk which had been rather amateurishly painted, and there was a great quantity of *chinoiserie*, lacquered screens and cloisonné vases as tall as an adolescent child, and jade pieces and a gong, and books and gramophone records scattered everywhere, and a mad proliferation of bric-à-brac, and a litter of papers and magazines, and more pictures on the walls than we had had in the front room at Avalon: great yellowed early Colonial landscapes of the Buvelot School, and dingy portraits staring out from their frames behind dark glassy varnish like night intruders peering furtively in through windows, and some works that were identifiably very good—a pair of lovely tiny Hilder water-colours, and a very good Elioth Gruner and a Heysen and a marvellous Tom Roberts—and in the very centre of all this incredible confusion was a gorgeous round table with a surface polished to the feel of soft old silk and on it were places set for four and white candles in Georgian candelabra.

'I'm afraid it's only the four of us,' Gavin said apologetically. 'To tell the truth, neither Peggy nor I really care much for big dinner parties. Or big parties of any sort, for that matter. Four's always fun, I think.'

'Oh, but that's splendid!' Helen said brightly …

One Sunday morning, after washing the car, Meredith gets on to the roof to fix the antenna of a new (bought on time payment) radio and record player unit, complete with cocktail cabinet. From his unaccustomed position, feeling temporarily unassailable, he surveys his tidy world.

… [My] elevation provided me with the first opportunity I had had to look out over all the Beverley Park Gardens Estate, and there was nothing all around me, as far as I could see, but a plain of dull red rooftops in their three forms of pitching and closer to hand the green squares and rectangles of lawns intersected by ribbons of asphalt and cement, and I counted nine cars out in Beverley Grove being washed and polished. In the slums, I reflected, they had a fetish about keeping front door-knobs polished, but here in the 'good' respectable suburbs the fetish was applied to cars and to gardens, and there were fixed rituals about this, so that hedges were clipped and lawns trimmed and beds weeded, and the lobelia and the mignonette were tidy in their borders, and the people would see that these things were so no matter what desolation or anxiety or fear was in their hearts, or what spiritless endeavours or connubial treacheries were practised behind the blind neat concealment of their thin red-brick walls.

... I stayed up on the roof because once I had worked this out a great many other things began to follow. Strange things. Terrifying things. Wondering things. (I could even stay up here for years, I thought, like some Stylite of the suburbs, on terracotta building tiles in place of a Syrian pillar, and ruminate on all the problems of the world. The ancient Stylites had liked desert places for their meditations.)

The realisation that I did not love Helen, and never had loved her, came to me quite dispassionately at first; so dispassionately that I was able to examine the revelation with a kind of clear careful logic, and find it sound, and put it aside for later. 'Later,' of course, would be another thing altogether, when I would want to blame *her* for the predicament we were in, and then passion and anger would need to be invoked. But not yet.

But, Meredith realises, as he stares out over 'the whole of the sterile desolation', this is the world he has chosen of his own free will—he has 'planned for it, approved of it, connived at it, worked for it, and paid for it'—at least on hire purchase. This, therefore, is to what he has mortgaged the next years of his life.

Comes another realisation.

There was not one tree on the whole estate.

Yet there must have been trees once, I thought, because when you closely examined the layout of the estate there were little folds to it and faint graceful rises and declivities, not anywhere near definite enough to be thought of as hills or gullies, but the place was not really *flat*, that was the point, and at one side, a little distance beyond Dr Felton Carradine's house, there was almost a real knoll. Once—I felt absolutely sure about this—there would have been trees growing here and there, and I pictured this knoll as having two or three good sturdy blue-gums or stringybarks on the crest, and slopes brown with bracken, and some sandy chewed-out patches where rabbits would have made little squats scattered with the liquorice-black pellets of their droppings and where they would have hopped about at dusk, flickering the pale cotton tufts of their tails. The place could have been really beautiful at one time in a tranquil sort of way, I thought—before Bernie Rothenstein came in with his bulldozers and graders and grubbed out all the trees and flattened everything out so that the subdivision pegs could be hammered in and his lorries could move about without hindrance—because there was a blur of higher ground much farther out, and beyond that the bluish bulk of the Dandenongs sat up there against a good bright sky in nice shapes and colours. And now there was nothing but a great red scab grown over the wounds the bulldozers had made, and not a single tree remaining, because by no stretch of the imagination could anybody count the spindly little sticks which had been stuck in at intervals along the footpaths, because they really were only sticks, and too hidden behind their ugly little tree-guards for anyone to know whether they were leafing or whether they were dead.

I climbed down the Solomons' ladder at once and went straight out to the car and just drove off.[2]

Meredith finally rebels against both the barren suburb and his unhappy wife by planting a well-established lemon gum tree in the front lawn; a fast-growing, hardy native intended to deliberately flout the neighbourhood social conventions that called for more demure and decorative exotic shrubs— and his wife's apparent social aspirations. But this is merely a delaying manoeuvre before he departs.

A NOTE OF REGRET:
Modern Melbourne

There is hardly anything in the world more desolate and deadening than Melbourne on a Sunday morning, unless it is Toronto, and a hot blustery north wind was driving leaves and scraps of old paper and dry chaff in eddies down the ugly street, and when we rang the hotel bell a fat surly thug in dirty shorts and a singlet came to the door and glowered at us. 'Who the hell d'you think you are, sport?' he snarled. 'It's Sunday mornin', ain't it? We don't serve Sunday mornin's.'

I indicated the sign by the door, and said, 'This place is called Batman's Hill, isn't it?' He scowled at me. 'And Batman founded this city, didn't he? And when he founded it he said, "This will be the place for a village." And how bloody right he was! Now let's have that breakfast.'

George Johnston, *Clean Straw for Nothing*, 1969[1]

In *Clean Straw for Nothing*, the second volume of George Johnston's semi-autobiographical trilogy, David Meredith leaves his wife and embarks on a tumultuous love affair with Cressida Morley, who represents for him a freedom of spirit which he himself does not possess. Johnston describes post-World War II Melbourne from the point of view of a restless and disaffected war correspondent. To Meredith, the city has lost its innocence in the years he has been away.

Melbourne is hateful, really. Butter and steaks are on the black market, and bottled beer, and cigarettes kept beneath counters, and taxi-drivers are churlish, remembering the free-spending time of the Yanks, and the city grins its rapacity and greed stalks through the town with a smirk on its face. Petrol rationing is worse than ever and the cars still lumber about with those grotesque charcoal burners built on behind, fouling the air. In the smoky swill of crowded pubs returned men force beery arguments and fights around their own roles in forgotten skirmishes, involved in loud lies and terrible uncertainties. This is a sour, cynical, nasty city. The night places are filled with costly and illicit shadows, where restless men and women promulgate anxious gaieties. A powerful tide of human dilemma runs beneath the skin of everything. I must stay within this tide, for it will move me along.[2]

Finally there is the incident of perceived provincialism at the Melbourne hotel—after an accumulation of small frustrations—which sends Meredith and Cressida Morley on their way north to Sydney, and eventually to an extended period of expatriatism in London and Greece. In much modern writing about Melbourne, beginning with its post-World War II evocations, there is a recurrence of this sense of transitoriness and impermanence, not only between Melbourne and other cities of the world, but frequently within Melbourne itself.

For writers from even further south such as Tasmanians Christopher Koch and Peter Conrad, born in 1932 and 1948 respectively, Melbourne merely represents the first stop on their eventual journeys elsewhere. Koch in *The Boys in the Island* (1958) brings Francis Cullen, his runaway adolescent protagonist to Melbourne in the late 1940s, which to his anticipating imagination represents the magical big city, the dream place that will be different from provincial Tasmania.

They stood in the bright-lit passenger lounge at Essendon, Melbourne's airport. It was seven o'clock on a cold July evening; they were on the Mainland. Staring about the sounding terminal, Francis had forgotten Shane. He had been trying to see what lay in the dark beyond the glass doors: he could not quite believe that he was out of the island.

Shane had left a note for his mother. He was running away: simply disappearing from the life that had been charted for him. Now, he picked up his bag; and on this brink of their separate departures, he and Francis could find no words. They were filled with separate and private dreams, flinging them hundreds of miles apart.

The stiff, blind smile flickered across Shane's face. He raised a hand in half-joking salute. 'Goodbye,' he said, and was gone, walking proudly across the gleaming, tiled floor to the doors leading onto the field. Francis was alone: a boy from an island.

In a few minutes, the airways bus would take him into Melbourne; meanwhile here he stood, still entranced by what he had just seen from the 'plane window: the great glow, as they had banked over the city, of countless lights storming soundlessly into the upper air. But soon he would no longer be one of the privileged race of travellers: the grey-suited men and beautifully dressed women with big hats and pigskin bags. Unknown Melbourne waited for him to fall into it, a shabby boy again.

But in the outskirts of Melbourne he does not see the city of his imagination.

What he saw were lines of shabby little weatherboard and brick houses in the mysterious weave of dark, illuminated by street lights with a sluggish intensity which made it appear that they gave off the light themselves. He saw no people. Skirted with grass-grown footpaths, the same as in many parts of Hobart, these were the mournful villas of fifty years ago, with mean, fussily decorative front verandahs; the same as he had seen all his life in the older parts of Hobart. Could they be the same?

They were the same. He did not quite know what he had expected, but it had not been this. They were same houses, and a chill of doubt went through the boy. Instantly, yet in the same instant rejecting it, he saw his foolishness. His dreams were nonsense. It would be bigger but the same. Districts were the same everywhere. …

These, he told himself, were the outskirts only; the real great difference would not be revealed yet. He searched the darkness fervently. But still, at this exact point of time he had dreamed of for years, there was a great doubt, final and unutterable as the darkness of the night itself.

And then the boy forgot it, because Melbourne was springing up darkly around the bus: the first big city of his life lay out the window. Buildings loomed, of a size he had seen only in pictures; between their bulwarks and towers and nineteenth century balustrades, trams bigger and more urgent than those in Hobart rushed and clanged, leaking a virulent yellow light from their windows, like the burning blood of the city itself; the city which was rushing in upon him, swamping doubt and delight together. He stared half wildly, half stupidly, wondering what he should think in this moment he had waited for so long: and he found he thought nothing. A fragment of flesh in a bus, carried between cliffs of electric-lit stone, there was nothing he could think. The city did not see him, and he knew that it never would; people didn't matter here, he knew this instantly. But when the bus drew into the kerb at the airways office, and he saw Lewie and Jake through the window, he forgot everything else.

Here they were to meet him, standing in a light shower of rain on the wide Melbourne footpath, hands in the pockets of their long overcoats, looking completely at home: pioneers before himself beyond the island. He waved at them joyfully.[3]

By the time Koch arrives at his fourth novel, *The Doubleman* (1985), in which his main character spends seven years in Melbourne working as an actor in the early 1950s, the city is dealt with in less than two pages.

I went there as a youth; I left as a somewhat disenchanted young man, harbouring that coldly passionate determination to change his life which so often comes at twenty-five. With the exception of six dull months working in an office, I actually managed to survive as an actor through every one of those years, getting by almost entirely through radio work. But survival was all it had been. I lived in single rooms: in bed-and-breakfast guest-houses smelling of floor polish and despair; in rooming houses in St Kilda, Caulfield and Fitzroy, where I shared dark bathrooms and ancient kitchens mysterious with grease. A flat was beyond my pocket; I never knew from month to month whether I could still go on covering basic rent and food. That depended always on the whims of radio producers—at whom, like all actors, I must smile and smile. They were the princes who totally controlled the livelihood we made from fantasy, and who could always take it away. An offhand tone from one of them produced a small pang of worry; a frown chilled the guts; no calls for a fortnight spelled doom. And yet we were always lighthearted; it was the great compensation of the job. We were in show-business, God help us, high above the grey channels of the workaday world, sharing our actors' jokes and show-business gossip and bitchery, each man keeping up appearances in the studios with his one set of elegant sports clothes, his one good suit; each of the women in her smartest suit, her sexiest dress, her most expensive make-up. Only the two or three top dogs went home to a good apartment, or a house in the suburbs; there wasn't the volume of work in Melbourne to keep a large community of actors prosperous. For that, you had to go to Sydney, up north in the sun; but Sydney, the nation's biggest city and the show business capital, whose sub-tropical hedonism made staid, cool old Melbourne envious and contemptuous, also meant fierce competition.[4]

To Peter Conrad, writing in *Down Home: Revisiting Tasmania* (1988), a memoir of his eventual return to his homeland from London, Melbourne during the period of his youth in the late 1960s also represents a mysterious other, but one even more mixed in its virtues.

Melbourne's banks fortify money, or build palaces for it to live in. One of them in Collins Street has the swank of a grand hotel, with a ceremonial staircase in its foyer; gryphons guard the porch of another and the hall inside is a castle of polished wood and gold paint, hung with heraldic shields. The columns sprout ears of corn, to honour the landed source of the wealth coffered here and to show how cash is a crop. The banks speak for a society rich in material contentment. The heir to the bourgeois grandees of Collins Street is the Fletcher Jones man, posing in relief on the side of a shop in Queen Street. He personifies the Melbourne I always imagined across the water. He's there to advertise a brand of trousers which were my initiation into adolescence: stiffly military, dowdy grey, pressed to a knife-edge and with turn-ups (adjusted every January to allow for your growth in the year ahead) which collected a museum of musty souvenirs—pebbles, dust, buzzies, paperclips, the occasional button and a blob of chewing gum. The Fletcher Jones man models them in nicotine-coloured stucco, with glossily varnished hair and jewelled cuff-links, one hand on a pleated hip, the other propped in the air before him in hieratic Egyptian attitude. This, I thought, was what being adult would consist of: rigid deportment; learning to be worthy of your Fletcher Joneses. Melbourne was where you got wisdom. The country which had its raw, unpropitious youth in Tasmania grew up to prosperity and smugness here.

When Conrad returns in adulthood in the late 1980s, Melbourne retains some—but not many—of the same qualities.

It's a Victorian city with optional palm trees. The railway station has a triumphal portico of clocks conserving the bourgeois regime of punctuality. Behind the main streets are narrower alleys, for stocking the emporia: the machine's workings are kept out of sight. The state library houses the brain which rules the system, in tiers of filed and stacked grey matter under a cranial cupola. Yet Melbourne is also the place where my cherished Anglo-Australia—the culture I pieced together from transplanted books—now goes under. In the years I've been away, the outpost of England I imagined we were has become a polyglot digest of Asia and the Mediterranean. Along Sydney Road in the suburb of Brunswick, nationalities collide: Gentiluomo Viaggi and Turismo shares the pavement with the legal firm of Numikoudis Ogilby; the Seattle butcher's shop evokes the meaty American west while Ali Baba's Variety Store displays a flurry of Greek, Lebanese, Turkish and Australian flags; tabouli salad and *baba gannouj* alternate with sugared Balkan snowballs, the Vietnamese sweets of Hông-Ngoc and indigenous 'Freshly-cut Sangers' (sandwiches in Ocker slang); the Bombay Club advertises its exclusive 'Crush Club (a Club within a Club)'. English literature with its gentle elegiac pastorals, against which Australia used to measure itself, retreats before other, more exotic mythologies. The Dedalos Garage on Sydney Road repairs hubristic high-flying autos gone astray in the urban labyrinth, and Dante's Espresso Bar boasts a fierily infernal Italian *rosticceria*. My main road back home in Hobart could manage no such supernatural diversions.[5]

This cosmopolitan influx of non-Anglo-Celtic migrants, largely post-World War II, which changes the character of the city so perceptibly for Conrad, accelerates the process of making Melbourne a less rigid society, and contributes to a transforming of demographic boundaries.

Just as George Johnston's David Meredith in *My Brother Jack* moves from the comfortable working-class suburb of Elsternwick to the new and sterile reaches of middle-class Glen Iris, so do migrants drift outwards from suburb to suburb as businesses build up and children are born, swapping inner city houses for more spacious bungalows in the outer suburbs. In Harry Marks' novel *The Heart is Where the Hurt Is* (1966), the Jewish migrant Sophie Liefman and her family, who came to Australia in the 1920s, survive the depression and move on.

> Fondly she thinks of the Carlton days when they first came to Melbourne. 'Little Jerusalem', Gentiles called it, before the great exodus to St Kilda. Hard, sad days, touched with many happinesses. Days already memories. But alive! So alive! People everywhere. Always someone to talk to. Streets vibrating with talk, whatever else had to be done. Over fences and cast-iron gates, in shops, out of shops, sitting at windows or on ribbon-like verandahs. And at her machine, before Max was born, in between coats …
>
> A dry cleaner's shop. A dream had become reality. Not overnight. Not without scrimping. Not, for Sophie, without misgivings. The choice was plain. Stay where you are, or move out and on. But was there really any choice? To move out and on. This was instinctive in them. They had to better themselves, for Max's sake. And that meant goodbye old friend to Carlton where, with kinder days, they shifted into larger, cosier rooms. So it was with most of the families. They felt the compulsion of the southern suburbs. As confidence grew with income, they were able to satisfy that urge. For the sake of the children. Yet, though the mind said go the heart said stay. Carlton was a family hearth. Balaclava, a social rung or two better, seemed to Sophie poor exchange.[6]

As Melbourne academic Laurie Clancy has noted,[7] the pattern repeats itself a generation later in modern Melbourne, if perhaps in different suburbs. Rosa Safransky, born in Poland and arriving as a child in Australia after World War II, describes in a story called 'Postcards', published in 1983, her narrator's father's arrival in Melbourne.

> When I was born, Europe became too crowded for my father, and he left. Two hotel rooms in Paris, with my uncle, mother and now me, was enough. Chips Rafferty decided him. Cattle crossing wild rivers, dancing, singing Aborigines. What could be less like Poland? My father had had enough of 'civilisation'. Enough of Auschwitz. A few postcards remain, written in Polish. My father's large hand addressed to my mother in Paris, 'Droga Elli'. My mother's neat microscopic scrawl to my father in Melbourne, 'Drogi Stan'.
>
> A tailor, he imagined himself coming to the wildest, most uncivilised part of the globe. To a nation of Chips Raffertys in slouching digger hats with crooked friendly grins. But 'civilisation' proved harder to leave than he imagined. He spent his first nights in Melbourne at the George Hotel, confronted by an intimidating array of silverware and a waitress who called him 'Frenchie'. Feeling ill at ease, he could only point to the dishes he could not name. A few of the diners turned and stared, wondering if there was something wrong with him.[8]

In a later story, 'Bonjour Brunswick' (1985), Safransky describes the new home in the same suburb that the Hungarian David Martin came to in the same period.

Au revoir Europe, bonjour Brunswick! 'Brunswick Creations', a two-storey shopfront, stands next to a surgical appliance store with a dismembered leg in its window. The doorbell rings and Emil's patterns jump from the wall but we go straight through to the back where my father stands ironing. My mother puts saucepans of food down on the sink, near the cracked Solvol soap in the wire rack. My father sits the iron down hard on the ironing board and holds the dress up to the light. He tests the zipper to see if it works. It does. Good, now he can eat! I go out into the backyard to look for the cats.

'Have you forgotten your father? What about a kiss?' He calls out and points to his cheek.

The backyard is a concrete path bounded by a high corrugated fence. A fuchsia tree stands in the corner and every few months, the butcher's cat gives birth to a new litter of kittens in its shade. The toilet stands just behind the fuchsia tree, so my father and Emil often cross paths out the back, my father clutching his newspaper, Emil a saucer of milk. My father glares at Emil caught red-handed feeding the wild cats. The exhausted butcher's cat is a savage Bengal tiger, waiting for me to open the door of the lavatory and run back into the factory. I peer through a crack in the wall and when everything seems safe, I throw open the door and run for my life, reaching the factory just in time to slam the door in its face.[9]

The same strong sense of place operates in Ania Walwicz's prose poem 'Australia'. Walwicz, born in Poland in 1951, arriving in Australia almost two hundred years after the first Australian literature of exile and loneliness was written, echoes the same immigrant lament, the same howl of outrage—at the emptiness of the soul, of the banality of the environment. 'Australia' could refer to any Australian city, until the last line places it, for Australians, irretrievably in Melbourne.

You big ugly. You too empty. You desert with your nothing nothing nothing. You scorched suntanned. Old too quickly. Acres of suburbs watching the telly. You bore me. Freckle silly children. You nothing much. With your big sea. Beach beach beach. I've seen enough already. You dumb dirty city with bar stools. You're ugly. You silly shoppingtown. You copy. You too far everywhere. You laugh at me. When I came this woman gave me a box of biscuits. You try to be friendly but you're not very friendly. You never ask me to your house. You insult me. You don't know how to be with me. Road road tree tree. I came from crowded and many. I came from rich. You have nothing to offer. You're poor and spread thin. You big. So what. I'm small. It's what's in. You silent on Sunday. Nobody on your streets. You dead at night. You go to sleep too early. You don't excite me. You scare me with your hopeless. Asleep when you walk. Too hot to think. You big awful. You don't match me. You burnt out. You too big sky. You make me a dot in the nowhere. You laugh with your big healthy. You want everyone to be the same. You're dumb. You do like anybody else. You engaged Doreen. You big cow. You average average. Cold day at school play- ing around at lunchtime. Running around for nothing. You never accept me. For your own. You always ask me where I'm from. You always ask me. You tell me I look strange. Different. You don't adopt me. You laugh at the way I speak. You think you're better than me. You don't like me. You don't have any interest in another country. Idiot centre of your own self. You think the rest of the world walks around without shoes or electric light. You don't go anywhere. You stay at home. You like one another. You go crazy on Saturday night. You get drunk. You don't like me and you don't like women. You put your arm around men in bars. You're rough. I can't speak to you. You burly burly. You're just silly to me. You big man. Poor with all your money. You ugly furniture. You ugly house. Relaxed in your summer stupor. All year. Never fully awake. Dull at school. Wait for other people to tell you what to do. Follow the leader. Can't imagine. Work horse. Thick legs. You go to work in the morning. You shiver on a tram.[10]

In the meantime, the displaced middle-class Australians return to the vacated inner suburbs. It is partly because of this, Laurie Clancy claims, that the suburb of Carlton, just north of the city centre, once predominantly Jewish and later Italian, has assumed a distinctive character in Australian literature.

Along with North Carlton, Parkville, Fitzroy and North Fitzroy, the area has developed the same sort of mythic ethos once ascribed to Balmain and its associated writers in Sydney.

For Peter Mathers, in his novel *Trap* (1966), which deals with the life of a part-Aboriginal character of the same name, the suburb of Fitzroy represents a fringe city within a city, one that does not officially exist, in which the underprivileged slip through the social services net and live between succeeding waves of urban migration.

> You know the streets between Gertrude and Victoria Parade? Well, there are plenty of similar ones in our no-man's land. And, among the old houses, in these straight and treeless streets and the back lanes and alleys, the old houses of bluestone and red-brick, there's more life there—real, squirming, dancing life—than to the square mile of suburban Ringwood or Highett or Preston. And one day soon, notwithstanding the current wishes of our illustrious mayor, this area will be discovered by the suburb-haters and wrested from the natives and hoisted level with Carlton and Parkville. And probably made twee and chichi—unless enough of the present locals can hang on. The suburb-haters'll find these one- and two- and three-level bluestone and brick places and drool over them. Our progressive mayor will be disgusted. But he'll probably not lose by it. For by then he and his pack of scabby suckhole mates'll own it all and make millions, despite their suburb ethic.
>
> Well, there's a big two-storied bluestone house, The Elms, in one of these streets—Cross Street. It's set well back from its corrugated iron front fence. The garden's hard dirt and prairie grass. Twenty-five people live in its fifteen rooms. The landlord is Circle Investments. The four Traps—or five or seven— have two rooms and a kitchen. Mrs Sally Trap is the listed tenant.[11]

In his story 'Landscape with Freckled Woman' from his collection *Landscape with Landscape* (1985), Gerald Murnane describes the beginning of just that process.

> At that time he had never heard of anyone wanting to settle in an inner suburb. In those far-off days of 1960, I said to the freckled woman in 1975, most inner suburbs were called slums. Young people were expected to buy blocks of land in new suburbs, in Chelsea Heights or Forest Hill or Banyule. Some couples when they were first married did live in South Yarra or Hawthorn, but only in rented flats in tree-lined streets while they saved for their newly built houses far away to the east. The young man looked for his inner suburb in the true slums north of the city.
>
> … He moved his suitcase of books and his grocery carton of notes and manuscripts to a room with use of bath (two pounds, five shillings per week) above the kitchen of a single-fronted house in Argyle Street, Fitzroy.
>
> The young man asked one of his few friends, a school teacher like himself, to drive him and his belongings from Malvern to Fitzroy on a Sunday morning. The friend helped to carry the young man's baggage across the tiny backyard, through a kitchen where a woman and three men were drinking beer in front of a television set, and up a narrow staircase that began near the kitchen stove and ended at the young man's door. When the carrying was done the friend looked at the bed with its bare, stained mattress. He looked at the cupboard and the table and chair and then he walked to the uncurtained window and looked across the backyards towards MacRobertson's chocolate factory. He asked the young man rather awkwardly did he know what he was doing. The young man knew at once that his friend suspected he was about to have what his friend would have called a nervous breakdown. The young man decided that his friend was no friend but only one more of the thousands who knew nothing of true landscapes because they had grown up in the belt of neat suburbs between Port Phillip Bay and Mount Dandenong.[12]

In this story, as in others of Munane's, a young man struggles to tell the truth about himself and his inner life of dreams, through which he tries not to lose contact with a landscape of the imagination which gives his life meaning. What he appears to be doing, in his attempts to place himself in a mundane inner suburban existence, is not to distract himself from his inner world by having to pay attention to the process of living. Inexorably, however, real life intrudes and he has to move on.

A similar, if more surreal, sensation of transition and dislocation is created by Peter Carey in his hypnotic short story 'Crabs', from his early collection *The Fat Man in History* (1974). Here the urban environment is anonymous, but a combination of 'floating' evocative detail—the leopard-skin car seats, the presence of Italian migrants, the sense of flatness and orderliness to the landscape despite the chaos of a broken-down society—gives a distinctively 'Melbourne' feel to the story. In some timeless future, after a car-dependent city finds itself in anarchy, life in the drive-in movie theatre where Carey's character Crabs and his girlfriend Carmen find themselves stranded becomes the only possible means of survival.

> He walks, very slowly, back the newly arrived Dodge. There are people in it. He ignores them. He opens the door and tugs the bonnet release catch. Someone pulls at his clothing. He knocks them off. He opens the bonnet and looks in, looking for the parts he will salvage. There is nothing there. No engine. A dirty piece of plywood has been placed inside to give the engine compartment a floor. Some small chickens, very young, are drinking water from a bowl in the middle.
>
> He lies back on the leopard skin and gazes at the sights outside. Carmen is beside him. She is snuggled up against him. She is saying a lot. Slowly Crabs begins to see what his eyes see.
>
> A large group of Indians, dressed in saris, are gathered around a battered blue Ford Falcon. One of them, an old man, squats on the roof. The Ford Falcon was delivered last night. A group of men, possibly Italians, lean against the front of Frank's Dodge. They are laughing. They seem to be playing a game, taking turns to throw a small stone so that it lands near the front wheel of a bright yellow Holden Monaro. Small children, black, with swollen bellies run past shouting, chased by a small English child with spectacles.[13]

Helen Garner, who in her earlier work is closely identified with the Melbourne suburbs of Carlton and Fitzroy, is one of the few modern Melbourne writers to treat the city with any degree of affection. However, unlike her predecessors Hal Porter and short story writer John Morrison, Garner achieves her ends while employing relatively little specific description of the exterior cityscape. In her novel *Monkey Grip* (1977), which deals with life on the fringes of the drug world in the early 1970s, Garner's evocations of individual houses and communal households in the inner city of Melbourne convey both the internal emotional states of her protagonist and a city firmly rooted in time and place. Here Garner describes a bicycle trip to Fitzroy baths.

> The bead fly curtain rattled and Clive stepped in from tending his pigeons. He stopped inside the door, grinning at us from under his absurd cloud of henna'd hair.
>
> 'Wanna take the kids to the baths?' he said, inadvertently puncturing the small balloon of awkwardness. Javo sat there smoking while we rounded up Gracie and the Roaster, took their bathers (smelling, like the children themselves, strongly of chlorine) off the line, and disentangled our bikes from the heap outside the kitchen door: my thirty dollar grid, and Clive's blue and silver Coppi racer, which he called his filly. The Roaster rode on Clive's bar, Gracie on my carrier. We bumped over the gutter and on to the softening bitumen.
>
> … 'Hang on, hang on!' I shout to Grace, and feel her fingers obediently tighten on my pants as we forge across a gap in the heedless double stream of traffic in Queen's Parade, and coast again (the Coppi ticking soothingly) the last few yards to the racks outside the Fitzroy baths.
>
> Broken glass glitters nastily all along the top of the cream brick walls. We chain our bikes to the rack. The Roaster grabs his towel and springs over the hot concrete to the turnstile. Gracie holds my hand with her hard brown one and we pick our way between the baking bodies to the shallow pool.
>
> The brightness of that expanse of concrete is atomic: eyes close up involuntarily, skin flinches. I lower myself gingerly on to the blazing ground and watch the kids approach the pool. The Roaster slips over the side and wades inexorably deeper; Gracie waves to me and squints, wraps her wiry arms around her belly, and sinks like a rich American lady beneath the chemicals.
>
> 'No-one will ever understand,' I say to Clive, 'but this is paradise.'

'Paradise enow,' he answers, neatly laying out a towel and applying his skin to its knobbly surface. No further need to speak. The sun batters us into a coma. I pull my hat over my eyes and settle down on my elbows to the day's vigilance.[14]

By her second publication, the paired novellas *Honour* and *Other People's Children* (1980), Garner's knack for evoking mood by the use of domestic detail is even more developed. In *Honour*:

> She stood still in the bare centre of the room, on boards, in dimness. The heat was breathless. A drop of water bulged and quivered under the tap.
>
> … She opened the door, stepped down into the dazzling yard, and walked along by the grey wooden fence and through the green, dried-out trellis door into the wash-house with its squat copper and pair of troughs under the window never meant to open. She placed her palms lightly on the edge of the troughs. They were grey, forever damp and cool, clotted of surface and rimmed lead-smooth in paler grey; she had been bathed when very small in troughs such as these, and her mother had let her play with the wooden stick that she used to stir the copper, a stock with a face on the knob. The wash-house smelled of wet cloth and blue bags, and she could not climb out of the high trough by herself, so she was obliged to sit there nipple-deep in cooling water waiting for her mother, gazing blankly out the blurred window panes to the corner next to the dunny where the tank stood on its wooden stand, up to its ankles in grass even in summer, and if you tapped its wavy sides it would not give out a note for it was full to a level higher than you could reach, and its water was clear and swirly with wrigglers, baby mosquitoes that would not hurt you if you guzzled fast enough, and she sang out, 'Mu—um! I've fi—nished!' but her mother did not hear, for she was outside in the yard at the clothes-line putting a shirt to her mouth to see if it was dry enough to be unpegged and taken in for ironing.
>
> A bike clattered against the front fence.

In *Other People's Children*:

> Whenever Ruth washed herself with Johnson's baby soap, she remembered when Laurel was a baby. They lived in a tall, dark terrace house with a yard full of useless sheds, behind which, when they moved in, she had found stuffed dozens of blood-soaked sanitary pads, dry and crackly and blackened. In the kitchen were two old troughs. She brought hot water in from the bathroom in a plastic bucket. The floor was of brick. Once the dog had surprised a rat among the paper bags in the cupboard under the sink: the dog bristled and roared, Ruth screamed, the rat thrashed about among the bags (they could only hear it) and shot suddenly into view through a crack, up the wall and out through a gap in the timber round an ill-fitted pipe.
>
> With the Johnson's baby soap she ran her slippery hands gently over Laurel's solid body; the water in the plastic tub lapped sweetly, her hands slid and met no resistance; the baby's head lolled in her palm, her hands moved effortlessly at the child's flesh.[15]

This evidence of a modern mellowing toward the city demonstrated in Garner's early work is also reflected faintly in the later writing of some of Melbourne's older literary figures. Alistair Kershaw, who was born in Melbourne in 1921 and left to live in France in 1947, writes in his introduction to *Hey Days* (1991), a memoir of life among the *avant garde* writers and artists of the city in the late 1930s and early 1940s, with a spirit of reconciliation.

> Poor unloved Melbourne! It never got a civil word from us. Its architecture was obnoxious, the climate was inclined to be peevish, you couldn't buy a drink after six p.m. (unless you were dining in a restaurant when you could have a high old time right up to eight p.m.), the tiniest deviation from conformity in the way you dressed earned you some very dirty looks, and we writers and painters and composers weren't treated with the respect we deserved. If the place didn't pull itself together, we told each other, one of these days we'd up-anchor and go somewhere where we'd be appreciated and Melbourne would just have to try to get on without us.

Well, some of our complaints were justified—no question of that; and some weren't. The drinking hours might have been a bit hard to take but later on some of us would find that, at any rate, the beer (when you could get to it) beat the stuff they served in London and Paris by miles. The shortcomings of the architecture could be matched in any other city. And compared to what went on in Europe, the climate was that of a tropic isle. Maybe artists weren't looked upon with holy dread (we never asked ourselves why the hell they should be) but at least they were left alone. By the public, that is. Now and again, some artist of the old school would reach the end of his tether and start mumbling about decadence and incompetence and fraud and we *avant-garde* kids occasionally roughed each other up a bit; but that was as far as it went. In Paris, the rival schools practically used knuckle-dusters on each other. French artists spent more time warding off attacks from the enemy camp than they did in actually painting or writing.[16]

While for George Johnston, writing not long before his death in 1970, there is also a note of belated regret for and appreciation of Melbourne not evident in his earlier work. In the third of his semi-autobiographical novels, *A Cartload Of Clay*, published posthumously in 1971, Johnston puts nostalgic thoughts of childhood into the mind of David Meredith, now living in the suburbs of Sydney and seriously ill.

It was funny that fifty years ago, when he must have actually suffered terribly from confinement and the smothering dullness imposed both by his home life and by his environment, he had never felt closed in or stifled or intimidated. Not the way he did now. Even the sad flat streets of Elsternwick could never smother the multitude of strange objects, lures, encounters, possibilities, riddles, discoveries and adventures that waited around every corner; the matchstick voyages along flooded gutters, the drowned earthworms on raspy grey ribs of sand after rain, the stacked bricks and wet foundations and boxed pits of quicklime on building sites (the prickly peril of walking bleached planks across these steaming terrors!), the thickets of brambles or rank weeds or thistles and the pine needles in the trampled sand and the Arthur Rackham roots of grubbed trees on vacant lots. Farther out, at the wasteland edges of the suburbs, were the ponds and council gravel-pits and the claypans of Cox's Hill, where an old derelict known as Snotty Dick lived in an abandoned railway cutting in a cave screened by corrugated iron and coarse hessian bags …[17]

If one thing is clear from this brief survey, it is that many of Melbourne's writers have not loved her enough. Perhaps it is best therefore to end on Bruce Dawe's—born in Geelong in 1930, later a resident of Queensland—note of regret, 'The Affair. For Melbourne', published in 1981.

On the train rolling north I thought of you
as of an older woman I could no longer afford
—it was July and the cold weather
said: 'It's all over, sport, and don't think
it hasn't been fun …' and I couldn't come up
with a single retort, I felt browned off and bored, looking bleakly
out at the West Footscray sidings, the night moving personally in
(although I could still remember when
you weren't half so expensive, but
we were both younger then, ah yes: the streets with straight seams
like stockings, the skirts of your suburbs
predictable and entrancing,
and cool, cool, your business premises those magnificent pillars
I would have embraced in broad daylight if it weren't
for the typists and stock-brokers …). Twelve years down the line,
what's left of our love? Very little. Only in dreams
do I wake up and say: 'I can afford you now!
I'm on my way! I'm on my way!'[18]

ADELAIDE

A Slightly Shady Utopia

In Adelaide everything had its place, you knew it would be the same for ever and ever. The winos beside the Torrens, the Aborigines on the grass in Victoria Square passing round their bottle. The matrons with petal hats, white gloves, raffia baskets waiting to meet each other on the Beehive Corner. The socialites dancing the débutante waltz in teardrop pearls and sweetheart necklines; getting married in magnolia brocade cut on empire lines; arranging the floral carpet, the floral butterfly, the floral map of South Australia before the War Memorial on Flower Day. Me, walking towards the Teachers' College, a building in the Spanish style, my skirt like a bell over stiffened petticoats.

And I had the right cummerbund, rope necklace, Paradise Pink lipstick; and we were told in Speech Education that the Australian accent was imperfect English, and Mr Lamb showed the film on sex education in Hygiene and the Methodists and Baptists were shocked. We did Educational Psychology and I worried I might be an introvert. Brown was the new basic colour because it was the colour the Queen wore most, and when you dressed up you always wore white cotton gloves—and then suddenly it was the sack look, and your shoes should be chisel-toed not pointy. Even your hairstyle had to keep changing: the wisp, the curly cap, the helmet, the angel-bob, the swirl, the choirboy.

At the teachers' college duffel coats and slacks were banned, but you could wear them to the art school where high fashion was early Beatnik and lips were white like worms. In life drawing the men wore underpants while the women posed naked, but it wasn't rude: the art school was run by the Education Department and they never employed exhibitionists or gigglers. We drew Venus and Apollo in the antique room and modelled giant Roman ears and eyes in clay. I wore a fur-fabric cape to the teachers' college dances and there were girls with pony-tails jiving in candy-striped dresses. I'd shaved my legs but the only time I danced was when it was the Ladies' Choice.

Barbara Hanrahan, *Earthworm Small*, 1989[1]

When the city of Adelaide was founded in 1836, Utopian societies were considered possible, and Adelaide was intended to be one. Edward Gibbon Wakefield, son of a London Quaker and briefly a diplomat, had propounded his theories of systematic colonisation as a method of extending civilisation, and libertarian social ideas current in America were permeating the rest of the English-speaking world. If some of Britain's other toeholds on Australia's attenuated coastline had originated as prison settlements or territorial outposts designed to secure possession, the establishment of the capital of South Australia was a conscious attempt to implement contemporary European ideas about the potential for moral enlightenment and natural evolution to the higher existential state considered inherent in humankind.

Wakefield in 1826 had been caught up in a sexual scandal involving the abduction from school and forced marriage of a young heiress, and it was during his subsequent three years in prison that he developed his ideas on colonisation and criminal justice. Some of these theories, expounded in his *Letter from Sydney* (1829), which many mistakenly believed to be the work of a colonist, played a part in the formation of an assisted migration scheme to New South Wales. His work *The New British Province of South Australia*, published in 1834, the same year that British parliament passed an act to permit settlement in South Australia, was influential in the establishment of the colony. Wakefield was aided by his editor Robert Gouger who, while himself in prison for debt, had learned something of the South Australian landscape from a fellow-prisoner who was a sea captain.

A number of societies were formed to follow up Wakefield's ideas, and the first English colonists sailed in 1836. Among them was Fidelia Hill, whose volume *Poems and Recollections of the Past* (1840) is probably the first book of poetry by a woman to be published in Australia. Adelaide was to be Australia's only truly planned city apart from Canberra, founded more than seventy years later, and the only one never to have received transported convicts. Wakefield himself never visited his creation although he eventually migrated to New Zealand, where he died in 1862.

Wakefield's brand of idealism mixed with a whiff of shady doings is symbolic of Adelaide's beginnings and of the place it would eventually become: a city sensitive simultaneously to its reputation for both provincialism and sophistication. A city offering the most intellectually invigorating of the current state cultural festivals, it is one where the taxi driver taking you to it will insist almost pugnaciously on his home town's superiority to Sydney. And at a precursor to that same festival, Adelaide will have the courage to be the first to recognise and promote a controversial new play by Patrick White, while at the same time supplying the bureaucrats that torpedo its production. Possessing restaurants which serve some of the best food in Australia, it is at the same time self-conscious to the point of preciousness about doing so.

'The Garden State' proclaims the tourist brochure.

'City of churches and axe murderers' decided one overseas visitor.

'A terribly sinister place' wrote novelist Barbara Hanrahan, no doubt referring to the city's reputation for displaying a decidedly respectable exterior while simultaneously producing a number of particularly vicious serial killers.

Adelaide itself is physically very much the result of the vision of British Army officer and surveyor-general William Light, who in 1837 devised the grid plan on which the city was built. The previously little-known coastline of this part of the continent had been explored by Matthew Flinders in 1802. His account of it, published in *A Voyage to Terra Australis* in 1814, along with the overland expedition of explorer Charles Sturt via the Murray River to the South Australian coast at Encounter Bay in 1829, brought the area to the attention of the rest of the colony. The city, named for Queen Adelaide, consort of William IV, was built a few kilometres up the River Torrens on a site first seen by Captain Collet Barker of the *Zebra* in 1832, shortly before he was speared to death by Aborigines at the mouth of the Murray River. This site caused some controversy between Colonel William Light, who selected it, and Captain Hindmarsh, the first governor, who believed the city should be closer to the sea. Light wrote of his reasons in his *Brief Journal* (1839) and left it for posterity to decide.

When twenty-two-year-old Edward John Eyre drove cattle overland from the east in 1838, he found a settlement where 'the buildings are few and for the most part of wood, reeds, and of wattle and daub'. He repeated his trip with sheep in 1839. After building a house in Adelaide he began an overland expedition to Albany in 1840 in an attempt to chart a stockroute to Western Australia, his account of which—*Journals of Expeditions of Discovery Into Central Australia and Overland From Adelaide to King George's Sound in the Years 1840-1*, published in 1845—would be seminal in the creation of Patrick White's novel *Voss*.

Eyre was followed in August 1844 by Charles Sturt, who had explored in the area a decade before. Sturt wrote encouragingly of the opportunities for settlement in his *Account of the Sea Coast and Interior of South Australia* (1849), in which he compared the climate favourably to that of Canada to compensate for the greater distance involved for prospective settlers from Europe.

Her lands, unencumbered by dense forests, are clear and open to the plough, or are so lightly wooded as to resemble a park, rather than a wild and untouched scene of nature ... As before observed, the aspect of South Australia, and indeed many parts of the neighbouring colonies, is essentially English. There, as in England, you see the white-washed cottage, and its little garden stocked with fruit trees of every kind, its outward show of cleanliness telling that peace and comfort are within.[2]

Such was the optimism of the period that the colonial poet William Howell was inspired to perceive in the fledgling colony 'spires and gilded domes', along with 'myriad happy homes' with which 'Approving Heaven' had apparently already crowned the settlers' efforts.

However, in *Clara Morison: A Tale of South Australia During The Gold Fever* (1854), Catherine Helen Spence describes the newly-arrived nineteen-year-old Clara's first impressions of Adelaide in the 1840s, as she arrives by cart on a hot day towards the end of summer, in somewhat less rosy terms.

> Everything looked as disconsolate as Clara's own thoughts. The grass was scanty, and so burnt up, that one wondered if it ever could have been green; there was not a flower to be seen; the sun was scorchingly hot; the wind, direct from the north, blew as if out of a furnace; the cart jolted, as if it would shake her to pieces, while the passengers abused the weather, and prayed for a railroad. Miss Waterstone's round face was streaming with perspiration; Clara's pocket-handkerchief became nearly black in her vain endeavours to keep hers clean; and the pale muslin dress and white chemisette and sleeves, which she had put on as suitable for the weather, were sadly crushed and soiled. The sight of green gardens in North Adelaide refreshed her eyes; and as the cart drove into town, her curiosity and interest in what she saw for a few minutes drove away her painful sensations. The streets, though straight, were most irregularly built upon; houses of brick, wood, earth, and stone, seemed to be thrown together without any plan whatever, and looked too incongruous even to be picturesque. The river was unworthy of the name; she had never seen a burn in Scotland so small. And when one of the gentlemen in the cart told her that on this river Torrens, the inhabitants of Adelaide were almost wholly dependent for water, she feared that there must be dreadful scarcity at times. She wondered at the complacent tone in which this gentleman talked of the colony, though he confessed that it was often as hot and as dusty as now; and that in winter the streets and roads were dreadfully bad—almost impassable.
>
> Miss Waterstone groaned audibly, from the effects of heat and exhaustion; and pitied Clara, who had been condemned to live in such a fiery furnace as Adelaide seemed to be.

Spence, herself born in Scotland in 1825, arrived in South Australia with her family in 1839 and lived at Gilles Street, and then at Halifax Street, in Adelaide. She became a governess at seventeen. Despite her later work as a novelist and journalist she saw herself primarily as an advocate of social reform, becoming, although she was unsuccessful, the first female political candidate at the Federal Convention of 1897. Spence, who rejected the romantic view of the bush presented in previous colonial fiction and tried to represent it realistically, was also interested in the problems of women who were obliged to earn a living.

After her arrival, Spence's character Clara, putting up initially at a guesthouse in Adelaide, applies for a job as a governess and walks for her first interview to a house four and a half miles out of town.

> The sun was overpoweringly hot, and when Clara got out of town, and had to walk between sections fenced with posts and rails, she longed for the green sheltering hedges of her own country. Here and there the corn was left on the field, though it had been reaped weeks ago, and she wondered to see how small and far apart the shocks were. Where the wheat had been reaped by the machine, and the heads merely had been taken off, the long stubble, which is reckoned of no value in Australia, had been either burned or was left standing till favourable weather came. She saw one large field which had accidentally taken fire, and watched the active exertions of all the people about to extinguish it by beating it out with boughs. It had been a very dry winter, and the crops in the plains near Adelaide had been very poor; so that she had no flattering view of the capabilities of South Australian soil. But with all this, there was an appearance of civilisation and comfort in the numerous cottages on the way, each having a small garden,

and generally a patch of vines, which were loaded with fruit; and what interested Clara still more, she saw many wells near the cottages, which encouraged her often to ask for a drink of water. She was unused to walking far for so many months, and the road was often so deep in sand, into which her feet sank at every step, that she was very thankful when a decent-looking woman asked her to sit down out of the sun and rest a bit. She wiped a chair for the lady to sit on, and went on with her washing. Several children were about her, eating bread and butter with their grapes. They had all dirty faces, but looked healthy enough; their clothes were neither fine nor altogether whole; the furniture was scanty, and altogether Clara did not see that over-powering contrast between the exterior of this dwelling and those of people in the same rank in Scotland which she had been led to expect. But the bread and butter, and the smell of meat baking in the camp oven, and the teapot, which the eldest girl was brightening a little for father's dinner cup of tea, were all very different, and looked as if, whatever crops might be, the labourer ran no risk of being starved.

Despite these mixed first impressions Clara, who after working unhappily as a servant finds a place with relatives where she is treated as one of the family, is soon more reconciled to the landscape of her new home.

They were both excellent walkers, and thought nothing of going four or five miles in search of a running stream or a romantic glen. They found their way up the range of steep hills which lie within walking distance of Adelaide; as long as flowers were to be found, they brought home nosegays; and when the advancing summer withered them all, they gathered green boughs instead. They would sit together under a gum-tree with a book, which they never read much of, but listened to the screaming of the paroquets and cockatoos, or the more musical chirping of the smaller birds. No one ever molested them in their rambles; there was scarcely a man to be seen, for at that time South Australia was in a very extraordinary position for a colony: there were far more women in it than men. Almost all the boys whom they met driving cows or sheep, were talking about the diggings, and how father got gold the very first spadeful he dug up, and how they were going themselves next month. The very young children were all busy with pannikins, making believe to wash for gold, and feeling convinced if they only had a cradle they should find plenty. Or sometimes Clara and Annie would wander through the quietest streets of Adelaide, and admire the beautiful irregularity of the buildings. Annie would point out a shortwaisted, broad-paling house, of which the bright red door and windows marked it as incontestably a German edifice. Next to that would be a two-story brick-house; then again, a low clay cottage, with dilapidated thatch; and close to that, a large iron-store, looking like a petrified tent.

'Is not Adelaide a delightful place?' said Annie, one day, when they had been across the river by the little wooden bridge, and round by the back of North Adelaide, and then having re-crossed the river by a fallen tree and stepping-stones, were returning home by the terraces. 'I do not remember Scotland distinctly, and my recollections of Edinburgh are merely of a wilderness of houses without any open spaces; so that I love Adelaide like a native.'[3]

In 1895, when Mark Twain visited the city on a world-wide lecture tour that took in Sydney, Melbourne, Adelaide, Ballarat, Bendigo and Hobart, he found 'a modern city, with wide streets and fine houses' everywhere covered with flowers and foliage, along with 'imposing masses of public buildings nobly grouped and architecturally beautiful'.

By 1898, even the visiting social reformers Beatrice and Sidney Webb, who cast a scathing eye over Sydney and Brisbane in the same year, deigned to find Adelaide 'the capital of a little principality, with its parks and gardens, its little courtly society ... and its general air of laying itself out to enjoy quietly a comfortable life ...'

The artist Stella Bowen, born in Adelaide in 1895, describes life in North Adelaide prior to World War I in her autobiography *Drawn From Life* (1941), seen from the viewpoint of an expatriate in England.

The land where I was born is a blue and yellow country, although when the sun pours out of a cloudless sky, there is very little colour to be seen. The blazing sky itself is almost empty of blue, and the yellow ochre of the dried grass is silvered by the glare. Even the shadows are less blue than painters like to pretend, and the air is so dry that the distance has the same quality as the foreground, and you can judge of space and perspective only by the diminishing scale in size. Local colour—flowers in the garden, or a little girl's dress—is sucked out and bleached by the sun, and the extreme visibility is tempered only by a shimmering heat-haze rising in the middle distance from the baked earth. The world is seen as a pattern of light and shade. ...

On the first day of an Australian heat wave, you shut and darken all the doors and windows of your home. Underneath the corrugated iron roof lies a layer of insulating seaweed. The hours pass in a dim and listless obscurity, until eight o'clock—or perhaps nine—when it has become cool enough outside to open up the windows. After dark, all the houses are empty, and from every garden comes the sound of quiet voices, relaxing into sociability.

Perhaps at midnight you carry your mattress on to the lawn, where a mass of pinky-white oleander flowers, sweeping to the ground beside you, reflect such brilliance from the moon that you must needs turn the other way. You wonder whether what your nurse said is true, and that the moon can change you into an idiot with a crooked face; but you are too tired to care much.

... I wish I knew the truth about that strangely dim and distant life in Adelaide before the war. I have reconstructed it in my memory as a queer little backwater of intellectual timidity—a kind of hangover of Victorian provincialism, isolated by three immense oceans and a great desert, and stricken by recurrent waves of paralysing heat. It lies shimmering on a plain encircled by soft blue hills, prettyish, banal, and filled to the brim with an anguish of boredom.

I must be wrong. There must have been more in it than ever met my eye. My poor small eye was placed very close to the ground, and my view was doubtless a worm's-eye view. But it was the only view I had ...

I was born in the sort of house that must inevitably end its days as a boarding house. It was sizable, rather gloomy, at a sufficiently good but not fashionable address, detached, with two little lawns and a summer-house in front and a back yard behind containing clothes lines, a stable and a coach house, a see-saw and a swing. Being in Australia, it had a front veranda with balcony above, and its roof was largely smothered in Banksia roses, bougainvillea and other greenery. There was a trellis of vines covering the path from the front to the back and figs, apricots, lemons and oranges grew around the back yard.

There was a pampas-grass in the garden, and an aspidistra in the drawing-room.

There were no modern conveniences. The nicest thing about it was the view. Being placed high on the edge of the town's oldest suburb, it looked down over low-lying park lands where cattle grazed, over the distant slums surrounding Port Adelaide, where factory chimneys trailed some wreaths, and delicately drawn shipmasts reared themselves against the most spectacular sunsets I have ever seen.

Trains going north ran across the middle of this view. They ran for three days as far as the desert, and then came back again. I hated the thought of that dead-end!

I never went more than a few hours north on that line myself, when visiting those of my girl friends whose parents had sheep stations in the fertile belt near the coast. The life on those stations was the most characteristically Australian life I ever saw. It had its own style and its own flavour and it could never have been the same anywhere else. In the towns, except for the modifications imposed by the climate, we were just pale imitations of something which was already moribund in England ...

Going to England was called 'going home', even by people who had never been there and whose father had never been. We all talked with varying degrees of Australian accent, of which we were ashamed when we became aware of it. We regarded a real English accent with positive reverence.

At Christmas, which came at midsummer with the temperature at—perhaps—100 degrees in the shade, we sat down at midday to turkey and flaming plum pudding, having sent each other cards depicting robins in the snow.

We were, in fact, a suburb of England.[4]

But Hal Porter, who arrived from Melbourne in 1940 to teach senior English at a boys' boarding school, and later at Prince Alfred College in Kent Town, found Adelaide vastly different to the rather 'English' state of Victoria.

It has not entered my head that, in voyaging asleep through the night, north-westward for about five hundred miles, I am to wake up in a part of Australia where everything is different, much or little, strikingly or subtly, from everything observed in the fraction of Australia I've known. The landscape seen through the sleeping compartment window an hour or so after sunrise galvanizes me. There are hills. Immeasurably ancient, abraded low and smooth, they seem young, boneless, pagan, sprawling like adolescent creoles in a languor of passion, but passion undeveloped and never to develop. Floating just above and across the sun-browned limbs, thighs, navels and half-ripened breasts of land, are shoals of blue-violet, colour disembodied, separated from matter, and unattached to the body of earth it feeds on, a weed in flower which, I am to learn, is called Salvation Jane. The hills are not stitched to the sky. There is no seam of horizon. The sky is seen to curve up from far far behind the hills as if the hills, the earth, are centrally contained in a globe of glass hanging plumb and steady from the rafters of infinity.

This first glimpse of a landscape less Anglicized and ordered than what I have been used to, more animal and flamboyant, arouses an excitement with this part of Australia which increases, and abates little, during my six years there. I am perpetually stimulated by dissimilarities and variations—climate, the quality of light, architecture, accent, vocabulary, vegetation, customs, sensibilities, and regional convictions. It is an Australia that is nothing but Australian but is tonally unlike the Australia I've known, Mediterranean and yet class-conscious almost à l'anglais, seductively half-somnolent and yet provincially wide-awake.

Minor cultural differences, he finds, are already considerably evolved.

In Adelaide I am fascinated by what is not in Melbourne—roast pigeons for sale; crimson radishes as big as quinces which one peels like a peach before eating; pies and green peas—floaters—to be bought at three in the morning from canvas street-stalls; tea-tree fences; sailcloth waterbags hanging on railway stations; the architectural oddity of upper-floor verandas extending only halfway forward over the fully forward lower veranda; the COWS ONLY notices in parklands, and the GENTS FOR SALE notices in fish-bait shops—I do not know maggots are called gentles. I pick up the local expressions parklands for park; Wyandotte for plonkie; handle, schooner, middy and butcher as names for beer-vessels; and notice that rich pastries, cakes, and buns seen nowhere else, many of German heritage, are displayed in pastrycooks', that a drink of German origin, hock-and-lemon, is a favourite in summer, and that the reception desk of a hotel is more often than not on the first rather than the ground floor.

In Adelaide, Porter finds himself meditating on the subject of class in a way that he has not previously done in his autobiographical writings. On finding that he prefers 'the privacy of the city' rather than 'publicity on a dusty shelf' in the bare and arid South Australian inland, he reflects:

Not the city for all that is there in the way of culture, philosophy, what is in vogue aesthetically—that is a barren and inorganic reality, a bucket of bones—but for the reality of late summer and early autumn when Adelaide, more than any place on earth, and as simply as pouring tea from a pot, pours from a lavish cornucopia into gardens and parks and markets and arcade stalls a cascade of carnations and grapes and melons, guavas and Michaelmas daisies and tomatoes, zinnias and belladonna lilies and tuberoses, lavender and quinces and cumquats and pomegranates, roses and roses and roses. This natural opulence is somehow middle class, like a Mrs Beeton dinner-party table, like the perfectly pitched vulgarity and comfort of the houses with their cellars, thick stone walls and deep porches, like the gloves and hats of the women, the self-sufficient statues in the squares, and the wedding-cake façades of the public buildings.

Looking back, I find that, as much as my life there is a mosaic one, it is a middle-class and upper-middle-class mosaic. This is purely a matter of circumstance, and the limitation gives me more chance to observe the fine distinctions of caste behaviour and class neuroses, not apparent to non-Australians, which exist in Adelaide, differ somewhat from those in Melbourne and, I am to discover, differ perceptibly from those in Sydney, Brisbane, and Hobart, less perceptibly from those in Perth. Such shades of social conduct are best seen in the middle or upper middle class. It is only when there is a solid and developed design of material safety that refinements in manners and morals appear, even in a democracy as free and easy as Australia's.[5]

However, when novelist and travel writer Cynthia Nolan visits in 1948, *en route* to Central Australia with her husband the painter Sidney Nolan, the city still seems something of a provincial curiosity, although once again the landscape is distinctive:

> We were flying from Sydney south-west towards Adelaide. Someone remarked that we had passed Young; sitting up, I looked out.
>
> Entrancing country lay below. Geometrically exact squares, anti-erosion ploughed, were fallow; neat homesteads showed pale blue roofs; single eucalyptus trees lining the roads threw long dark shadows; square dams and winding billabongs were coloured red, yellow, turquoise, grey and apricot. There was something here of the paintings of Klee.
>
> After lunch the land appeared rougher, uncultivated, dotted with a rash of scrub interspersed with strong darker trees. Gradually it became more thickly timbered, while occasional straggling spaces were covered by flood waters. Although far from resigned, I began to feel the small curious interest one experiences when on an unknown road, and was surprised to find the plane already descending to land at Parafield Aerodrome.
>
> By six o'clock we were in the city of Adelaide walking among quiet, respectable, dowdy people, until seeing a Chinese cafe we sat down to eat rice, chicken and pineapple. When we again stepped onto the pavement lined by emporiums and cinemas, and crowded by men and women hurrying home from work, the air was cool, the sky peacock blue. It was difficult to remember that had the explorer Edward John Eyre taken this same walk less than a hundred years ago he would have seen only mud huts and half-cleared scrub.[6]

'The Iceman Becomes Important'

The summer drifts on, and Nan outwits it with cucumber peel stuck to her forehead, home-made lemon-cordial, heart-shaped Fiji fan; with lettuce like mermaid's hair, condensed milk mayonnaise, Narcissus Blancmange, Dainty Chocolate Mould, Delicious Hawaiian Cake.

The iceman becomes important.

I wait for him to come round the corner with his cabbage-leaf hat, nose hung with an icicle, sad-eared horse. I cry 'Ice-o!' with the others, and dance after his cart and the wet trail of drips it leaves behind; watch him juggle the blocks from house to house; crow when he smashes one to shivers and gives us each a piece.

The sun is everywhere.

It is in the garden: peering huge-eyed over the berry bush, roosting behind the chimney, floating like a fried egg in puddles. It mocks me when I burn my bare feet on the earth and scorch my fingers on the iron fence. It peels my nose to jigsaw patches, gilds my skin with freckles, turns the hair on my arms to gold.

It is in the house: spangling the passage with leopard spots, turning the sheepskin rug tawny, casting zebra stripes through the shutters. It curdles milk, melts the butter, shows the dust, fades the curtains. It steals into vases and drinks their water; creeps up the cold tap and turns it into hot.

In the evening, after tea, we carried the cane chairs and the wireless with the long cord onto the lawn; sat under the fig-tree in the twilight. …

Nan shivered and fiddled with the wireless knob. We listened to the 'Australian Amateur Hour'. And complained about the mosquitoes; forgot the moon.

Barbara Hanrahan, *The Scent of Eucalyptus*, 1973[1]

As Hal Porter has meticulously preserved in aspic Melbourne in the pre-war years, and David Malouf has documented Brisbane in the 1930s and 1940s, so had Adelaide its chronicler in the South Australian writer Barbara Hanrahan. However, if Adelaide was presenting its 'middle class and upper middle class' face to Porter, by the later 1940s and 1950s, when Hanrahan was growing up south of the Torrens, the city was showing a different side to a shy girl going to Thebarton Technical College and studying at the South Australian School of Art. This was the suburb where Colonel Light had built his cottage in 1839, calling it Theberton after the house in Suffolk where he was raised. (The name was subsequently misspelled by the printer in Light's *Brief Journal*).

After the death of her father, Hanrahan spent a solitary childhood in Rose Street with her mother, who worked as a commercial artist, and her grandmother. She later went to London to continue her art studies, and while she was there, the death of her grandmother released childhood memories which led to the writing of her first autobiographical novel, *The Scent of Eucalyptus*. The novel, recognisably about Thebarton, was published in 1973. Hanrahan subsequently described Thebarton as 'a flat suburb set on a grid of often absurdly wide and generally treeless streets … a lower middle class place … of shopgirls and clerks, factory workers, tradesmen and the genteel poor'.

Like Stella Bowen, who also wrote from England but thirty years before, Hanrahan's narrator remembers Adelaide's summer heat. She also goes on rich voyages of exploration into the city centre with her grandmother, Iris, whom she called Nan.

We swept away in the tram, over the bridge, to the city; got off in Currie Street by the wedding-cake bank; turned into King William Street, and our exploration began.

(King William Street is slotted by Victoria Square. Here stands the dumpy monarch billowing on her pedestal, surveying her lawns, her trees, her flower-beds, her bronze heroes—Sturt and McDouall Stuart, become stay-at-homes for ever. Nearby are the dim halls of the Central Market where waxwork joints of meat jostle carnations and roses, hyacinths buried in shaggy moss, lilies and hydrangeas.) We walked past giant cheeses and slabs of marbled bacon, packets of biscuits and jars of pickles to the stall where Nan changed the *True Confessions*.

Beyond the market and the square and Queen Victoria is South Terrace, and beyond that the inevitable parklands, the dreary wastes of landscape-gardening and bowling greens and croquet lawns. At the other end, past the department stores of Rundle Street and the Little Italy warren of Hindley, is North Terrace. ...

We walked past the Public Library and a plot of grass edged with ragged palm-trees and an elongated shed that housed the skeleton that must surely have been a dinosaur's. We entered the Museum, and once again I felt lovely horror as the lions and tigers edged close with sharp white grins. I climbed the staircase to the very top, and hung over balustrades, and gazed to the bottom of the light well. I saw whales suspended in space, a wrinkled elephant, a rhinoceros, a bear. I found glass cases that lit up to reveal Willie wagtail, Robin redbreast, and speckled Jack. There were chocolate aborigines in modest loin-cloths, honey-ant men and witchetty-grub people. There was an Egyptian Room where I saw a mummy and a row of bandaged cats and a figure that looked like Reece.

We regressed to the world of 1890 when we entered the Art Gallery. Pink and silver ladies abounded, shrinking from harsh antipodean suns under parasols and trellised vine leaves. The coldness of satin gowns was matched by the chilling austerity of parquet floors. We were cowed by gold frames and too much varnish and the lizard glances of old men spangled with braid and seals and watch-chains.

(It is better if you probe a little further, for there are shearing sheds and jolly girls with ostrich plumes and Neapolitan violets; pictures with names like 'Sadder than a Single Star that Sets at Twilight in a Vale of Leaves'. There are nymphs and pipes of Pan and 'How We Lost Poor Flossie'.)

I came out with a pain in my back from all the vastness, dragging my feet. Further along the terrace were the spires of the University and the dome of the Art School, bristling with rococo cast-iron and Virginia creeper turning red. We saw art students come out with dandelion beards and skirts hemmed with safety pins. We flopped on the grass beside the pigeons and shop assistants with brown-paper lunches.[2]

But by the period covered by *Kewpie Doll* (1984), sequel to *The Scent of Eucalyptus*, the rich childhood satisfactions of Adelaide have turned to discontents: the world of the 1950s teenager—Saturday night dances in nylon dresses and rope petticoats—stifles the development of the artistic self. Training to be a teacher, engaged to be married at nineteen, for Hanrahan 'all the choices seem over'. Like Eve Langley before her, Hanrahan dreams of another life.

Silver planes flew in the sky to Melbourne and Sydney, ships went over the sea. I dreamed of being an artist, of going to Art School in London. I followed my grandmother about with my sketch book; drew her with her mouth open, fallen asleep in her chair. I sat on the verandah and read Van Gogh's letters, and planted sunflower seeds in memory of him, and poured milk over my charcoal drawings as a fixative because I read that it was what He did (but I must have done it wrong; because my drawings smelled).

Only the interior world of art and the imagination retains the power the transform the city.

I carried it with me under my arm in a folder, as I hurried twice a week to the Art School evening class in Printmaking. As I went down North Terrace under the trees' leafy network, all the pale statues seemed to be urging me on. There was a shrill piping of crickets and birds. Hot summer evenings, itchy with excitement. Girls in summer dresses coming out of the library, virtuous tap of heels. It was a different city, now, from the daytime one that belonged to the white-gloved matrons with their baskets. There was a smell of bruised grass. Velvety shadows crept into the fiery beds of salvia as I hurried to the evening class past the places that meant culture—miniature cathedrals and castles that lent a taste of old Europe

Her city was a long coastal strip squeezed between a range of low mountains and a flat line of sea. It was an elongated city, pushed beyond its self-imposed limits by fingers that groped, unsatisfied, along the endless plain. But although it spread horizontally, it had only half-heartedly broached the hills, because the hills were a different sort of place, out of it all, so intact from a distance that their cover looked like a grey-green fur …

NICHOLAS JOSE, 1984 (PAGE 194)

THE HILLS, ADELAIDE, SOUTH AUSTRALIA. PHOTOGRAPHER: CAROLYN JOHNS

If one has put one's stakes on, say, the flesh, and has gambled long and, inevitably, has finally lost, there is nothing in the landscape to free anyone, except the semi-crazed, from humanity and oneself. It is a country without consolations, and without a message.

HAL PORTER, 1966 (PAGE 198)

EUCLA, SOUTH AUSTRALIA. PHOTOGRAPHER: OLIVER STREWE

to the Terrace. Spires and turrets, lords and ladies in stained glass. Traffic lights blinked, the shop assistants streamed home, while other people spooned up peche Melba and banana split at the Black and White milkbar. The old Queen, erect in bronze, ruled over the Bay trams setting off from Victoria Square. Bronze explorers surveyed the distant bleached hills that were tattooed with indigo turning purple.[3]

At the end of *Kewpie Doll*, Hanrahan, aged twenty-three, takes a train to Melbourne to catch a ship to England. It is 1963.

Hanrahan's later novels—two of which are set in Adelaide in the nineteenth century and one in the 1920s—are concerned with what she terms 'contrasts, contradictions: beauty and horror, love and death, frivolity and menace; the precisely-detailed world of substance, the darker world of instinct, the queerness of mind split from body, the absurd fantasy of the "ordinary".'

In *Where the Queens All Strayed* (1978), Hanrahan invents an Edwardian Adelaide, inhabited by flawed and deformed characters, where all is not as it appears on the surface. This is a city where, to Hanrahan's character Thea Hodge, twelve years old in the year 1907, 'Mutilated body recovered from the Torrens' remains a headline never explained; where 'Teacher', beloved of his widowed landlady, secretly interferes with his young female pupils while an outdated portrait of Queen Victoria stares sternly down in the classroom; where the innocent advertisement of a fortune-teller denotes help for young unmarried ladies of an other than spiritual kind.

> Beside the bells were several fly-specked cards: W. Kelly, high-class American dentist; Madame Mora, the only Yorkshire palmist …
>
> But why did Meg want to have her fortune told, why here? It was a dirty looking building, most of the blinds were drawn. I wouldn't have gone to that dentist for anything in the world. And if she had to pick a palmist, why couldn't she have tried one well known? Every week in the Miscellaneous column there was a list of good ones: Professor Kennedy and Madame Phyllis, King and Queen of Astrologists; Zingara, mental scientist (magic bowl) … and the King and Queen were opposite the Trades Hall, Zingara next to St Paul's Parsonage—surely that guaranteed respectability?
>
> Then, for no reason at all, another name came into my head: Madame Lenard. And I couldn't, I wouldn't remember—but I did. Madame Lenard was wicked Mrs Amanda Gay who applied an instrument to Maisie Hood. I started running back along the Terrace. Towards the clean Kapunda marble of Parliament House and the Boer War statue.[4]

Murray Bail, born in Adelaide in 1941, in his 1988 novel *Holden's Performance,* presents by contrast a portrait of the city in the 1950s that defines its inhabitants by the psychology of its engineering.

> When the last of city's trams were removed with their poles and bells and the industrial paraphernalia of lines imposed in the mind's eye, it was as though a great net had been lifted clear of the city, letting in light. The trams had been inflicting all kinds of untold damage, running amok at will. Anyone living there beyond a certain length of time was in danger of becoming marked, ruled by inner grooves.
>
> Day after day, for years on end, the heavy oblong shapes had been lumbering backwards and forwards across the field of vision and conversations, or there'd be one travelling away in an absolute straight line, its pole waving, demonstrating the laws of perspective, which is how they entered the souls of generations.
>
> It was a small city and flat. There could be no escaping the trams.
>
> On summer nights it seemed as if the sky had been lowered to a false ceiling joined to the earth by a monopoly of constantly moving poles, emitting at set intervals their own brand of pale blue lightning.
>
> All this had been going on for as long as anyone could remember, and no one thought much about it.
>
> The city was laid out along the lines of a timetable. There were no hairpins or dog-legs, no French curves or crescents; diagonals were few and far between. Mondrian would have been pleased. It was a

city based on the original grid pattern laid down by the first surveyor, a tall colonel who'd come out all the way from England and knew where to place his knife and fork. Under the circumstances he was incongruously named Light—Colonel William Light. When the burghers muttered and sniffed in their balloons about the subversive elements or society's fabric, and other hot air, they should have looked at the directions of the streets and the presence of the trams. Instead, out of gratitude they had Light cast in bronze, and there he stands on a piece of high ground, a dunghill, the favourite of pigeons, one arm and forefinger pointing down to his regimented folly, Adelaide. ...

And in Adelaide, encouraged by the puritanical streets, the brown trams always went forward in straight lines, scattering traffic and pedestrians like minor objections or side-issues, and somehow this suggested the overwhelming logic of *plain thinking*. There always seemed to be a tram opening up a clear path to the distant goal of Truth. And so the people developed a certain ponderousness, a kind of nasal pedanticism; whole suburbs displayed maniacal obsessions with Methodism, with lawn manicure and precision hedge-cutting. These were a people who spoke slowly and distinctly, making frequent stops. There was a yes and a no, a right and a wrong. They liked to begin their sentences patiently with 'Look ...'. The real facts and direction of things, look, lay out in front: anyone could see that. They prided themselves on plain thinking. Personal anecdotes were trundled out from memory-sheds as evidence. Subtleties, complications and deviations were seen as unnecessary obstructions. And so talent here was brief: the spurt of a night tram between stops.[5]

Nicholas Jose, who was born in Britain in 1952 and grew up in Adelaide, describes the city according to the sociological status of its population. In his novel *Rowena's Field* (1984), the first part of which is set in Adelaide in 1970, Jose allows his characters to range between the opulent middle-class suburban homes in the foothills outside the city centre and the communal student households in the less prosperous suburbs beside the sea.

One of the better sort, Marcia lived in the foothills. Her city was a long coastal strip squeezed between a range of low mountains and a flat line of sea. It was an elongated city, pushed beyond its self-imposed limits by fingers that groped, unsatisfied, along the endless plain. But although it spread horizontally, it had only half-heartedly broached the hills, because the hills were a different sort of place, out of it all, so intact from a distance that their cover looked like a grey-green fur, except for the scar of quarry. Those who went to the hills were regarded as opting out, choosing a new sort of life apart from the city proper—just as those who clung to the first sandy yards of foreshore were nomads, surface dwellers liable suddenly to throw their belongings into a dug-out canoe and make off for elsewhere. The people who really belonged lived in the middle of the coastal plain. Without them it would have been dry and infertile, but they had made gardens grow.

The only problem down there was the lack of wind. Breezes blew in the hills, but on the plain the air was absolutely motionless, suffocating, a suspension of irritants. It was hell for anyone with breathing difficulties. As a compromise, therefore, some people had moved to the foothills where the air was clearer, yet where they were still near enough to preserve the old relationships on the plain. As a bonus, when they looked down at sunset they saw the dirty air turn to a rhapsody of colour, before it resolved at last into pure calming black. On the occasions when a brisk gully wind did sweep the air away, the city on the plain—Adelaide—the whole world—was brought into miraculous clarity. Those like Marcia Sonner looked down from their terraces and saw the murk give way to a city of dreams. Life was there in front.[6]

On the other hand, David Parker, in the prologue to *Building on Sand* (1988), a first novel of childhood and adolescence in 1950s and 1960s Adelaide and its environs, stresses once again the fragility of man-made structures in the face of the forces of nature.

The house was built on sand. When the winds blew straight in off the sea, as they did all winter, rocking and buffeting the house like a ship, and rain thrashed against the window-panes, and the sea thudded all night against the Esplanade wall, and the back paddocks flooded, my grandfather lay awake awaiting the fate of the foolish man in the Bible and dreamed of his bunker. He would dig deep into the lea-side of the hill, prop it up with railway sleepers, then bring in rocks for the foundation, setting the whole thing in concrete two foot thick. It would be solid, solid as a rock. And he would sleep all night.

However, the apparently solid fronts mask a frailty that applies equally to the facades of family life.

The houses along Seaview Road were set high on a long sand-dune overlooking the Esplanade. They got a clear view of St Vincent's Gulf.

When they were built of red brick in the 1920s they must have been imposing, most of them with their long lawns and flower-gardens in terraces that swept down to the street. But the wind and sea-spray made paint crack and peel, wood split, and nails rust, so the houses, even the best kept ones, always had the air of having seen better days. High narrow windows looked out over the waters that the builders or their fathers must have crossed in the last hundred years. Edward's own father was conceived on the voyage out from England and was born, in 1851, in a colony that was only fourteen years old. Paved terraces and half-enclosed verandahs all faced the sea where the older people sat and looked out emptily, dreamily, all year round, with endless need and satisfaction.

Occasionally, on hot days when the sea was flat pale-blue and welded indistinguishably to the sky, a mirage or a ship on its way to port would move imperceptibly across the invisible horizon—like a vast skeletal battleship floating through the lower air. The last time a more palpable warship appeared was in 1870, not long after the Crimean War when a Russian cruiser entered the Gulf. It steamed along the coast and—apparently seeing nothing but dunes—it turned around and disappeared again into the Southern Ocean. Soon after that came the concrete pill-boxes, walls two-foot thick and dug well into the sand like Edward's ideal bunker, with little slits where in two world wars binoculars combed the empty seas. The enemy warship had entered the imagination and remained.

… At the rear, the houses showed what they rested on more frankly: their yards ran down, for the most part naked of topsoil, to the still unsealed road built in the scare of the 1870s, Military Road. Here were lean-to chook-houses, wood and rubbish heaps, and a suggestion of sand-grass like thin fluff on a bald head. Dig anywhere deeper than six inches and you came to dry sand. Sand that fell off the spade like liquid leaving a fine white dust in the air. Sand that had no cohesion but remained separate, even under the magnifying glass, like the numberless grains in the Bible. Sand so dry and white and soft that it left no clear trace of the foot that trod it, but closed over almost as completely as the sea.

Parker, born in 1943, uses the beachside settlements of the South Australian coast outside Adelaide as a metaphor for the efforts of colonists, through the generations, to put down lasting roots.

What Edward did dig was a thriving vegetable garden. Nothing grew on the dunes but sand-grass, a few stunted shrubs and, mysteriously, native paddy-melons which grew wild, throwing out their skinny hairy runners for twenty yards or more in search of food and drink. Everything else could survive only in imported soil. So at the back of the house Edward dug out sand, tons of it, heaving it downhill like a man frantically bailing out a boat, until he had a level terrace right across the hill. Then he took new galvanised iron and made a retaining wall. It was two feet high and was held in place by stakes that had to be driven at least four feet into the sand. Edward sledge-hammered them like huge nails into the side of the hill until they were rock-solid. Then Edward and I set out with the wheelbarrow.

On the other side of Military Road were lush hedged-in paddocks of grass grazed by Jersey cows. They might have been the green fields of England. The grass grew on alluvial soil from the River Torrens which flowed into the sea half a mile to the south. Before the banks had been built up into a high concrete-walled outlet in 1937 the paddocks used to be a single sheet of water when the winter rains came. There were old photos at school of Jersey cows up to their udders in water. And even now there was a swamp fed by small creeks not half a mile inland to the east. The place was known as Jerusalem, and the very word filled me with a thrill of fear.

The soil in the paddocks was rich, jet black, full of huge fleshy earthworms and masses of grass-roots. Roots, some thick and succulent as parsnips and some slender as cotton thread, held the earth together. The spade cut straight in like an axe into wood and the soil was wet and heavy on the blade and smelt of decaying organic life, mushroom spores, juicy bulbs, seeds sweet as nuts. Edward loved the soil. As he held huge clods in his hands, he couldn't get over its richness, the living wealth of a whole river valley deposited at his back doorstep.

'Smell that,' he used to say as he pulled a fresh clod apart, carefully parting the network of hair-fine roots as if he couldn't bear to tear them. The smell seemed to answer to some obscure yearning.[7]

Building on Sand is a complex novel that traces the young Jude Watson's voyage of discovery, through the overheard and barely comprehended conversations of a family riven by secrets, to an understanding of human fallibility, including his own. As in Nicholas Jose's *Rowena's Field*, which also follows the progress of a young protagonist growing up in a family which presents a respectable facade to the world at a high cost, Parker's novel ends with the main character returning to Adelaide in a spirit of reconciliation after years of expatriatism.

Similar themes of respectability and the dark underside recur in two poems about the city by Friendly Street poets Andrew Taylor and Larry Buttrose. ('Friendly Street' was the name given to a series of poetry readings begun in Adelaide in 1975 using a disused factory as venue, the organisers of which collectively produce the annual *Friendly Street Poetry Reader*.) Taylor, born in Victoria in 1940, has lived in Adelaide since 1970 and lectures in English at the University of Adelaide. In his poem 'Adelaide', published in 1986 in *The Orange Tree,* an anthology devoted to South Australian poetry, he builds on and modernises some of the preoccupations of Barbara Hanrahan.

It falls apart like a well-dressed
party or a Sunday lunch with too many
interests into its carparks and bad
parts and parklands at night buzzing
with sex and by day especially
in spring a head hazard from magpies
and excessive exercise. Rigid
with tradition the older department
stores and a few empty churches
insist that Colonel Light's regimental
rectangles are right but the people
drive through red lights and rape
and murder though common enough
visit it from the suburbs where the bored
packs in Hindley Street have fled from.

It falls apart into those packs
and at weekends to the marauding throngs
of thong-shod families searching for
Athens or God help them Coventry
beside the dammed Torrens that leads
nowhere but it flows like a green spell
motionless through the desert. ... [8]

Larry Buttrose, born in Adelaide in 1952, in his poem 'The City on the Verge' from the same anthology, writes baldly of an urban population that avoids the questions posed by the landscape by avoiding its contemplation.

We are tough bastards to live on this land.
The ruler is death, the dominion sand and bone.
Our eyes cross a plain broken only by the bare spines
Of ridges, where the blue horizon thins,
Diffuses with distance, to imagination;
To be rediscovered so far off, in the coastal
Living rooms, in lines of blue flashes
On television screens—
Whose work is more a matter of battery and assault:
A static barrage to contain the threats of the land.

City at the edge of the desert. People who ignore
The sand and bone, or try to, though dust
Mats the floor, is the last coating on the paint
Of the car, dries the corners of the mouth in summer.
Here the soul seems to flee the mellow rind of the earth;
Any growth is discouraged: We pace a pen
Stripped to rock by sheep and wind—
The plains whose life we will not comprehend.

Here the cogs of industry slow, the market's iron laws
Have left us to rust, the chassis stripped, propped
On bricks at the back of the block, mossed with neglect.
House emplacements, dormitory deployments deemed fit
By developers, aspirant angles pressed to flat lines;
A laboratory of bread knife fights and scaling
Bedroom windows with your bare nails; the emptiness
Is a vessel for howling, and a vassal of the wind.
There is so much time. We feel the weight of freedom
And we will not apprehend it. The diversions are strange:
Tune guitars strings with the willow necks of women,
Massage the recti of boys with the broken heads
Of bottles, we are such artful lovers.

Buggers we are. Tough bastards we are.
We kick over the remains and slowly debate the transgression.
We inhabit the idea of a land, where the edges are soft
And the central essence slips by us nightly.
We do not see, nor hear. We are keyed to lines,
Blue lines on a screen; and the curse of the land is real.[9]

197

'A DREAM COUNTRY LESS TANGIBLE':
The South Australian Landscape

Inland and outback is the South Australia I know nothing about except from what is experienced during a never-again holiday on a sheep property where, restricted by the infinity of a white sky, and encircled by the limitlessness beyond a white horizon, I am glad for the walls and ceiling, the ante-rooms and galleries of night to be run up quickly after sunset, for this hideous form of wealth to be hidden.

Day has no ceiling, no walls. One inhabits an eternity populated by sheep and foxes where black geysers of crows explode out of the floors of dust, and anatomies of windmills shiver in the heat. What appals me most about much of inland Australia is its lack of consequence to all except its victims. Even victims, however, have tethers to come to the end of.

If one has put one's stakes on, say, the flesh, and has gambled long and, inevitably, has finally lost, there is nothing in the landscape to free anyone, except the semi-crazed, from humanity and oneself. It is a country without consolations, and without a message.

Hal Porter, *The Paper Chase*, 1966[1]

If there is a regional writing centred around Adelaide, it is a sparse one. South Australia, the driest state, with its limited coastal farming land and arid hinterland, is the most urbanised state of Australia, with over seventy per cent of its population living within its capital.

However, in *Where The Queens All Strayed*, when Barbara Hanrahan allows Thea Hodge to venture out of Adelaide in the early 1900s to visit her grandmother by horse and cart in the Adelaide hills to the east and south of the city, the outskirts of Adelaide seem reassuring enough.

We drove into the Hills. There were gum-trees and, at intervals, houses. Ladies in their gardens stopped hanging out washing and shaking the crumbs from tablecloths to wave. The driver jiggled his reins, and looked proud as he told us that the passing of the coach was the great event of the ladies' days. 'Good gracious,' they'd say, 'there's the coach, and the fire not alight' or 'Pussy not fed', or 'the kettle not on the boil'.

The road kept zigzagging. There were sharp rises and steep descents. I clung to Meg and worried that the driver might lose a rein, and the horses bolt. One dangerous declivity was known as Breakneck Cutting.

Soon the country changed. It was gentler, almost English. There were paddocks and orchards and farms. Mother's eyes shone. Though Father disregarded the view, and concentrated on his leg as if he felt a pain, she ignored him. She looked about her and smiled. She had grown up in the Hills; she loved them.

So often Mother told stories of the pioneers. How they'd come to a new country, and everything was strange. To feel safe they'd thought of home. The kangaroo grass might have been corn; the Adelaide plains, nicely wooded, resembled a gentleman's park. They missed the old things so much that they sent to India for slips of apple- and pear-tree. Granma had come to South Australia with a root of rhubarb wrapped in moss. She placed it in a box with her shoes, the leather of which kept it moist.

Mother had been a pioneer at Prospect. She was permanently homesick. Instead of transplanting rhubarb, she'd summoned up the Hills with words.

She said they were more like England than any other place in the state. As well as tea-tree and wattle there were ivy bushes and oak-trees, hedges of sweet-brier and monthly roses. The highest and most central point in the Hills was Mount Lofty. The Mount was Elysium. Up there were ribbon gardening and marble statues and the dwellings of fashionable society. Each mansion was surrounded by plants that would have died on the plains in summer. There were rhododendrons, camellias, even strawberries. In winter it was known for snow to fall.[2]

By the 1940s, when in *The Scent Of Eucalyptus* the young Barbara Hanrahan herself makes a similar trip by bus from Victoria Square to visit Aunty Poll and Uncle Will, her reactions resemble those of Catherine Helen Spence to the fertile farming country she had seen ninety years before.

> The bus roared off—the driver snatching lustful side-glances at girls on bicycles. Soon we left Adelaide behind, and the asphalt of the suburbs became grass. ... I saw olive-trees and a horse and a donkey. We passed little towns that clung to the foot-hills: just a church, a jumble of roofs and cast-iron porches, clumps of lilies and beds of smouldering pinks. The sky was hung with a mosquito-net of heat; the trees stood perfectly still. The road slipped on and on before us, threaded constantly with a line of wavering white.
>
> Then we were deep in the hills. The side roads dwindled to tracks under shadowy trees. I saw sheep dotted on a slope like the bumps in Reece's knitting. I saw stone houses—some like Swiss chalets, some peeping through ivy curtains—half hidden in thickets and purple valleys.
>
> Suddenly the bus wavered dangerously close to a sea of rippling leaves. The crisp waves whispered all about us; we faltered, then hurried on. And were pursued by a restless company of the stone houses. I saw fowls pecking under almond-trees, and far away the humps of other hills flung with paddocks: silver and sage, mustard and yellow-green.
>
> We passed iron gateways; a face at a window; children's fingers waving us on. Someone behind us put on perfume, and the bus was flooded with sweet voluptuous scent. Nan complained about the curve and said she felt sick.

For Hanrahan a walk through the mallee scrub country reveals the same magic that she finds in the cityscape.

> (The hills are gentle, with their pale trees, their stillnesses, their drifts of smoke. Soft-bristled bushes cling to my skirt, currents of strange insects wreathe my head. There are tufted ferns, black boys, and everywhere the wattle: intensely gold, on hair-like twigs; in plumes, amongst flannel leaves; fuming in a lemon fuzz; fallen to a shrivelled crust.)
>
> I crushed dried leaves in my hand. My feet walked over the earth, and I looked down at them and saw another world—miniature, microscopic: a rock was a hill, a crumpled leaf a plateau, a thorn a pagoda, groundsel and twining glycerine were forests; I was a giant.
>
> I watched ants pursue their scurry, frail stick-insects dance a gavotte, beetles stay earthbound for an instant. I saw grasshoppers, little flickering lizards, a delicate lacewing, a centipede ferociously nippered. There were secret hollows rank with the smell of mould, overgrown with brambles and toadstools. I walked over dead trails of blackberries sparked with fresh green, tiny seeds, hermit tatters of bark, twigs and mosses and fallen leaves.
>
> The hills are gentle, with their monotonous greys: green-grey and blue-grey, silver-grey and pink-grey. But the greys are sprigged with colour. There is the wattle, and there are chocolate lilies and honey-scented milkmaids; tinsel-lilies and spicy-leaved myrtle. There is the fringe-lily that blooms only for a day, the musky caladenia that hides under fallen trees, the spider orchid with its frilly lip. There are flowers that quilt the earth: bidgee-widgee and golden guinea-flower and lavender grevillea; clematis that climbs, apple-berry that creeps. There is the vam-daisy shaped like a dandelion; ivy-leaf violet that loves the dark, derwent speedwell and scented sundew.

But, as for Randolph Stow in Western Australia, this transitional region is not the inland the child has been led to expect.

> But where were the hills of the history book, stitched with the pathways of Burke and Sturt and Leichhardt?—the hills of the sun-burned earth and budgerigar grass, the azure skies and fiery mountains we sang about at school before the flag spangled with all the stars of the Southern Cross I was never sure of seeing? Where were the old dark people I did not link with the lost couples on suitcases at the railway

station? Where were the crocodiles and brolgas, the billabongs and snakes? Where were the flowers that wilted in blistered clay, the rusty waves of spinifex that looped the cliff?

The hills are gentle, their trees are pale: the scented paperbark with its peeling trunk, the snow-white ghostgum—warm to the touch. And prickly box that grows by rivers, silky-oak with orange flowers; blue-gum and red-gum, cider-gum and dwarf-gum; bottlebrush with tooth-brush spikes.

I looked about me for the sunburned land. In vain.[3]

When contemporary Adelaide or visiting writers venture further into the hinterland, however, their visits tend to be brief and uneasy. This is a landscape that tends to impinge on the collective unconsciousness of South Australian writers, whether they have witnessed it at firsthand or not. Twenty years after Hal Porter wrote in 1966 of the South Australian inland's lack of solace or meaning, in *Holden's Performance* (1988) Murray Bail also reflects on the ambivalent nature of the relationship between the city and the desert behind it.

There was something else about Adelaide, or rather the environs, which entered the mind; and it entered in the same manner it trespassed on the geometry of the city itself.

Beginning with the Hills in summer which rose up behind like a pair of agricultural trousers bent slightly at the knee, the country penetrated the city like no other city. A natural creepage of colourlessness breached the town plan, indenting and serrating the perimeter, at the same time vaulting deep into the most established suburbs of immaculate box-hedges, green lawns and culverts, and deposited vacant blocks of swaying chaff-coloured grass, one in every other street. The Dutch had better luck keeping out the sea. Whole tracts of land here had the country look. Colonel Light had surrounded the city centre with a band of open space mysteriously called 'parklands', and not even concrete benches and drinking fountains could soften it. Dust storms blew up there in the height of summer and small grass-fires started. Elsewhere, it appeared in its most contained form as a parched oval. Badly tended footpaths and lawn tennis courts reverted to 'the country' in a matter of weeks.

Everywhere a person looked the ragged edge of naturalness trespassed.

In the battle for people's minds it at first seemed to be an antidote to the streets … that habit-inducing pattern constantly underlined and repeated by the trams. But the stain of non-colouring spoke of the interior which, in southern Australia and the Northern Territory, was desolation. It was the struggle— and for what?—of the dry tangled bush and desiccated trees and the brave façade of the boulders that gave the country, unlike the deserts of dreams, its persistent melancholia. Within cooee of the town hall, blasted crows made their parched calls. What other city …? And the faces of the most optimistic smiling women in Adelaide eventually resembled the country itself: ravined, curiously wheat-coloured.[4]

Since the explorers wrote their journals in the last century, there have been few firsthand accounts of inland and coastal South Australia from the point of view of its inhabitants. One notable exception is Myrtle Rose White's *No Roads Go By* (1932), which describes the parched sandhills of the Lake Frome region where she spent seven years with her station manager husband after having established the only homestead in the area. Another nonconformity is Dal Stivens's imaginative novel *A Horse of Air* (1970), a description of an expedition to Central Australia ostensibly in search of the rare night parrot. Otherwise, visitors tend to leave with images of desolation. As the poet Douglas Stewart noted in a poem called 'The Fierce Country', this is terrain where: 'Man makes his mark across a fierce country/That has no flower but the whitening bone and skull …'[5]

Stewart, while editor of the literary page of the *Bulletin*, was commissioned in 1954 to write a film script for a film to be entitled *Back of Beyond*, for which he travelled along the Birdsville Track. The result of his reactions to the interior was a poem sequence, *The Birdsville Track* (1955) which begins with the town of Marree, once a halt for Afghan camel drivers, and traces a route strewn with meticulous observations of nature and eccentric identities to Birdsville, 480 kilometres north.

Marree, at the southern end of the Birdsville Track, eighty kilometres southeast of Lake Eyre, was described by Stewart as

> ... the corrugated-iron town
> In the corrugated-iron air
> Where the shimmering heat-waves glare
> To the red-got iron plain
> And the steel mirage beyond ...[6]

Colin Thiele, born in Eudunda in 1920, former schoolteacher and prolific writer and editor of over fifty books, has written of the opal fields at Coober Pedy, 937 kilometres north of Adelaide, midway between Port Augusta and Alice Springs, in his children's novel *The Fire in the Stone* (1973). He uses similar imagery to Stewart's to describe the underground homes of the miners:

> A flat, bare landscape it was for the most part, with undulations here and there and flat-topped hills and breakaways and windswept plains. An old land, eroded and wrinkled, worn down over endless ages, peneplain on peneplain, until even the hills were remnants of endless plains. And in the sides of the slopes, cut into every knoll and knob, were doorways and entrances and burrows as if the whole place was inhabited by five foot high rabbits ....[7]

In direct contrast, Thiele sets his other well-known children's novel *Storm Boy* (1963) in the Coorong, a series of lagoons and sandhills extending along the coast 180 kilometres south east of Adelaide, and dividing the Ninety Mile desert from the sea, a refuge of sea birds and pelicans. Thiele once again conveys an impression of emptiness, where humankind lives in close conjunction with nature, and the landscape is the dominating theme in their lives.

> Storm Boy lived between the Coorong and the sea. His home was the long, long snout of sandhill and scrub that curves away south-eastwards from the Murray Mouth. A wild strip it is, windswept and tussocky, with the flat shallow water of the South Australian Coorong on one side and the endless slam of the Southern Ocean on the other. They call it the Ninety Mile Beach. From thousands of miles around the cold, wet underbelly of the world the waves come sweeping in towards the shore and pitch down in a terrible ruin of white water and spray. All day and all night they tumble and thunder. And when the wind rises it whips the sand up the beach and the white spray darts and writhes in the air like snakes of salt. ...[8]

Nicholas Jose in his short story 'Roo Easter' (1986), also set in the Coorong, relies on similar natural imagery. But from the first sentence this story, which describes a young boy's passage to maturity during a family camping trip, also employs as a subtext an analogy with early explorations into the unknown. Outside the familiar urban milieu, white Australians are reduced to their elemental beginning on the new continent.

> Our convoy stopped and we waited while the men surveyed the terrain. Climbing a fence, my father signalled to the others to follow. We could see them conferring in the distance on top of a dune, and returning in formation three paces apart. They had been schoolboys together and fell back naturally into their old lore. My father chose a grassy dip for our campsite with a few crabbed ti-trees from which to sling the tent ropes....
> Over the rise, the trees and pasture stopped abruptly and flats of sand stretched wide in a moonstone vista, turning to stubbly saltbush southwards and to mud the other way. Water began as a squelching depression between the toes, surfacing here and there in patches of salty crystal until at last in the distance

it lay like a giant lizard's tongue. From Geography we knew all about the Coorong. The narrow ninety-mile lagoon was one of the weird places of the world. When I drew it on maps at school I imagined it was the river's shadow or a sheath of skin the river had thrown off. Although we acknowledged the early explorers, to Wal and me the Coorong was uncharted still. Beyond the flats were the sandhills, and beyond that massive breakwater of sand was surf which pulsed straight up from the South Pole.[9]

Perhaps the final comment on the landscape of South Australia is best left to Christopher Koch who, in an essay called 'The Novel as Narrative Poem' (1987), remarks on Charles Sturt's 4800-kilometre trek, carrying a boat, through the interior of South Australia to the edge of the Simpson Desert and back, during which he finally relinquished his dream of finding an inland sea. Instead, Koch notes, he found sandy undulations and deserts of gibber stones, and heat that caused birds to drop from the trees:

> There was no inland sea, only more desert. … There is a grand pathos in Sturt's expedition, and in that useless boat. The sea that wasn't there has taunted the Australian soul ever since; and despite the temperate beauties and fertility of our eastern and southern seaboards, we've been suspicious since then of paradise on earth, and of easy answers. And our novelists have lately tended to look for the dream country in less tangible zones.[10]

DARWIN

'The Port of Yellow Skeletons'

The capital of the Northern Territory is Palmerston on Port Darwin, a harbour little, if at all, inferior to Port Jackson. Palmerston is unique among Australian towns, inasmuch as it is filled with the boilings over of the great cauldron of Oriental humanity. Here comes the vagrant and shifting population of all the Eastern races. Here are gathered together Canton coolies, Japanese pearl divers, Malays, Manilamen, Portuguese from adjacent Timor, Cingalese, Zanzibar niggers looking for billets as stokers, frail (but not fair) damsels from Kobe; all sorts and conditions of men. Kipling tells what befell the man who 'tried to hustle the East', but the man who tried to hustle Palmerston would get a knife in him quick and lively.

A. B. (Banjo) Paterson, *The Cyclone, Paddy Cahill and the G.R.*, 1898[1]

Darwin, capital of the last region of Australia that the British settlers made their own, has always seemed the most isolated of Australian cities, and yet also the most exotic and cosmopolitan. It was to this far northern terrain that the ancestors of the Aborigines crossed from Southeast Asia to inhabit the continent more than 40,000 years ago; here that a Ming figurine of the God of Longevity was found in the roots of a banyan tree at Doctor's Gully in Darwin, suggesting that Chinese traders of the fifteenth century probably made landfall; here that the Portuguese, who were settled in Timor from 1516 onwards, marked their maps with the coastline of the south land they called *Jave-La-Grande*; here that the Makassan trepang fishermen came by *proa* to harvest sea slugs and trade with the Aborigines long before they set eyes on any other foreigners.

Matthew Flinders mapped the rivers and estuaries of the area during his circumnavigation of Australia in 1802, but for the next two decades the region was ignored by the British. From the early 1820s onwards several settlements, urged for strategic reasons by Sir Stamford Raffles, were begun and abandoned. Port Essington, on the Coburn Peninsula in the Arafura Sea—surveyed by Phillip Parker King in 1817 along with Raffles Bay, the Alligator River, and Bathurst and Melville islands—was made site of a settlement in 1824 by Captain J. G. Bremer and a party of convicts as a trading post to the north, but lack of fresh water soon caused him to retreat to Melville Island. The Melville Island settlement, Fort Dundas, survived from 1824 to 1829, while Fort Wellington at Raffles Bay, under Captain James Stirling and then Captain Collet Barker, lasted from 1827 to late 1829. A further attempt at settlement at Port Essington, called Victoria, was made in 1838 in response to the perceived threat of French territorial ambition.

When the explorer Ludwig Leichhardt and his party arrived overland from Queensland at Port Essington on 17 December 1845, suffering from malnutrition after being forced to live on flying foxes and the dried meat of their own horses, and with one member speared by Aborigines, the garrison, weakened by fever and the affects of the climate, was barely hanging on. Leichhardt and his men ate Christmas dinner with the commandant, John Macarthur, but left on the first available ship. Around the same time an observer aboard the H.M.S. *Fly*, in the area on a four-year surveying trip, described Victoria as 'barren, uninteresting and intensely hot'.

It seemed an utter banishment from civilisation to be stationed in so undesirable location ... [the buildings were] all wooden constructions imported from England, most of them in a state of decay from the ravages of white ants.[2]

A seaman on the same ship was quoted as describing the garrison as the 'port of yellow skeletons'.[3] This settlement was abandoned in 1849.

Port Darwin, the only navigable harbour in the area, had been named in 1839 by John Lort Stokes, an officer of the H.M.S. *Beagle*, after the young, unknown naturalist Charles Darwin who had sailed with the *Beagle* in 1836. A small settlement was established at Escape Cliffs near Hotham in 1864 and named Palmerston after the current Prime Minister of Britain. It also was abandoned three years later. A second Palmerston was established at Port Darwin in 1869, and placed under South Australian control.

A government house was built in 1870, and the Overland Telegraph terminus completed in 1872, probably rescuing the settlement from the fate of its predecessors until the Pine Creek gold rush of 1873 attracted more British settlers. The South Australian government urged the introduction of Chinese labour to further this consolidation, so that by the mid 1870s there were some 4500 Chinese in residence, and by the end of the same decade some 700 Europeans. Pearl fishing began in the mid 1880s, providing another boost to settlement, and the first Greeks began to arrive in the second decade of the twentieth century. Largely pioneered and built by Chinese and Greeks in the late nineteenth and early twentieth centuries, Darwin still has large populations of Greek and Chinese origin. Cattle stations were established in the hinterland in the Victoria River area from the 1870s.

Because of its tropical, malarial coastline surrounded by a hinterland of almost a million square kilometres, a third of which was semi-desert, most early accounts of the area were unfavourable. In the 1860s a British visitor to South Australia, Sir Charles Wentworth Dilke, who was making a survey of Britain's colonial cities, remarked that explorations in that direction had led to the discovery of nothing but 'natives, mangroves, alligators and sea-slugs'.[4] Hugh C. May, who worked for a mining company in the area in the 1870s, dismissed it as a land of 'heat, rain, mosquitoes and sandflies'.[5]

Somewhat contradictorily, Harriet Douglas, later Mrs Dominic Daley, and daughter of one of the earliest government residents in Palmerston in the early 1870s, in her memoir *Digging, Squatting and Pioneering Life in the Northern Territory of South Australia* (1877)—a title that must have sounded distinctly odd to foreigners—remembered days of endless horse-riding, picnics and dancing 'in a land of perpetual summer'.

It is a generally conceded opinion, and agreed to by all those who have visited the Northern Territory, that in point of beauty Port Darwin has few equals; only two other harbours were ever named, when a comparison with this one was sought for—those of Sydney and Rio [de] Janeiro. Having made the entrance of this magnificent haven, we found ourselves sailing into an immense space of perfectly smooth water, where, it has been said of this, as of other large harbours, the whole British fleet might lie at anchor.

The shores were clothed with masses of rich green vegetation down to the water's edge, and the cliffs overspread with thickly growing palms, in all the variety one would expect to see so far north. Ironbark trees, casuarinas and the bright green milkwood tree grew here in great luxuriance. It looked what it was,—a land of perpetual summer. We sailed along, passing smooth white beaches , on to which waterfalls from the overhanging cliffs shed glittering streams of crystal, dancing and shimmering in the sunlight. The air was warm and light, and a fair wind wafted us each moment nearer our future home. Beautiful it certainly was; but oh! so lonely and desolate, not a sign of human habitation could we yet discern; no living creature, not even a solitary blackfellow walked these lovely beaches. It was all just as nature had made it, just as it had remained from the beginning of time—untouched and untrodden by the foot of man; a region known only to the degraded tribes of savages, who had hitherto been the sole occupiers of this magnificent piece of country.

The scene of our exile—for such we deemed it then—though surpassingly beautiful in itself, was, from this very loneliness, hardly inviting to N. and myself, for we were at that time far too strongly attached to the pomps and vanities of this wicked world to appreciate being banished from all we had hitherto enjoyed so keenly.

At last we came in sight of the little settlement; it was situated in a gully on a broad tract of level ground between two steeply rising hills, having the sea on both sides. The 'camp,' to use the name so familiar to every one, and which to this day it has retained, consisted of a number of log and iron houses on either side of the gully. On Fort Hill to our right, a steep hill with a flat summit, one of the most prominent landmarks of the harbour, was a flagstaff, on which the Union Jack was flying. It was delightful to find the familiar flag in this far-away corner of the British Empire. Close to the flagstaff was a lonely grave—the last resting-place of a young surveyor who was treacherously murdered at Fred's Pass by the natives during the surveying expedition a year before. The opposite hill was covered with green shrubs, and at this moment it literally swarmed with black men and women. These unclothed spectators were the 'oldest inhabitants' of this part of the world—members of the Larrakiah tribe. The heads of the clan were amongst this eager and excited crowd. But as far as we could discern, there was nothing to distinguish them from the lesser lights of that barbarous horde of natives.[6]

Darwin subsequently inspired many writers to wax colourful and colloquial about the tropical outpost, particularly if they were visitors. In 1898, when A.B. Paterson visited the area briefly, having been commissioned by the Eastern and Australian Shipping Company to write a tourist guide, he wrote distastefully of the 'vagrant and shifting population of all the Eastern races' which filled the town:

The Chow and the Jap and the Malay consider themselves quite as good as any alleged white man. In Japtown (the Easterner's quarters) Chinese children by the dozen play about all day long in the dusty streets; gaily dressed cheerful little barbarians, revelling in the heat. The goldfields are all worked by Chinese labour; hundreds of Chinese fossick about the old alluvial claims; fifty pearling luggers go out every tide, carrying seven hands each, practically all coloured men—350 yellow, brown, and brindled vagrants moving backwards and forwards with the tide. And more boats building and more brindle-coloured Japanese arriving every month. To supply the needs of all these, there are stores of every kind in Japtown, and the storekeepers all deal with the East for their supplies. There is an Eastern flavour over everything; when the Palmerstonians want to gamble at the annual races they do it by Calcutta sweeps, an Eastern form of betting little known or practised elsewhere in Australia.[7]

In *Their Shining Eldorado* (1967), Elspeth Huxley followed Paterson's lead and imagined the town prior to World War II as the sort of place Conrad might have written about.

… a small, neglected tropical capital, with administrators cooled by punkahs and gin slings, hoisting and lowering flags over Government House; and of sweaty men with rolls of fat drinking themselves into stupors and keeping, or at least frequenting, native girls; of down-and-outs, no-hopers, traders in pearl and merchant seamen …[8]

And popular historian Ernestine Hill, in *The Territory* (1951), concluded that 'apart from a few old faithfuls' Darwin had only two classes of people: 'those paid to stay there and those with no money to go'.[9] However, Darwin would soon achieve its own chronicler, and one whose pen was as unrelenting as his subject matter.

In 1927 a young man called Xavier Herbert, aged twenty-six, arrived in the Northern Territory via Queensland and the Gulf country and worked as a drover on the Wave Hill–Camooweal Route. He was later employed as a fettler in the Rum Jungle, eighty kilometres south of Darwin (so named, according to legend, in the gold rush days when a group of government officials on important departmental business went missing in the area, to be discovered some time later by a search party, surrounded by empty bottles). In Herbert's sprawling novel *Capricornia*, the Rum Jungle is 'Black Adder Creek'.

Herbert, while en route to England in 1930, wrote what he described as 'a tough little book' called *Black Velvet*, dealing with sexual relations between black and white in Australia, for which he failed to find a publisher. He was persuaded to rewrite it by Sadie Norden, who later became his wife, and the manuscript became the first draft of *Capricornia*, with which he returned to Australia in 1933. From 1935 Herbert worked as a temporary superintendent of the Kahlin Aboriginal settlement, and later held a mining lease on the Darwin River. *Capricornia* was finally published in 1938.

In a first chapter savagely titled 'The Coming of the Dingoes', Xavier Herbert begins his fictional history of the Northern Territory.

Although that northern part of the Continent of Australia which is called Capricornia was pioneered long after the southern parts, its unofficial early history was even more bloody than that of the others. One probable reason for this is that the pioneers had already had experience in subduing Aborigines in the South and hence were impatient of wasting time with people who they knew were determined to take no immigrants. Another reason is that the Aborigines were there more numerous than in the South and more hostile because used to resisting casual invaders from the near East Indies. A third reason is that the pioneers had difficulty in establishing permanent settlements, having several times to abandon ground they had won with slaughter and go slaughtering again to secure more. This abandoning of ground was due not to the hostility of the natives, hostile enough though they were, but to the violence of the climate, which was not to be withstood even by men so well equipped with lethal weapons and belief in the decency of their purpose as Anglo-Saxon builders of the Empire.

The first white settlement in Capricornia was that of Treachery Bay—afterwards called New Westminster—which was set up on what was perhaps the most fertile and pleasant part of the coast and on the bones of half the Karrapillua Tribe. It was the resentment of the Karrapilluas to what probably seemed to them an inexcusable intrusion that was responsible for the choice of the name Treachery Bay. After having been driven off several times with firearms, the Tribe came up smiling, to all appearances unarmed and intending to surrender, but dragging their spears along the ground with their toes. The result of this strategy was havoc. The Karrapilluas were practically exterminated by uncomprehending neighbours into whose domains they were driven. The tribes lived in strict isolation that was rarely broken except in the cause of war. Primitive people that they were, they regarded their territorial rights as sacred.

When New Westminster was for the third time swept into the Silver Sea by the floods of the generous Wet Season, the pioneers abandoned the site to the crocodiles and jabiroos and devil-crabs, and went in search of a better. Next they founded the settlement of Princetown, on the mouth of what came to be called the Caroline River. In Wet Season the river drove them into barren hills in which it was impossible to live during the harsh Dry Season through lack of water. Later the settlements of Britannia and Port Leroy were founded. All were eventually swept into the Silver Sea. During Wet Season, which normally lasted for five months, beginning in November and slowly developing till the Summer Solstice, from when it raged till the Equinox, a good eighty inches of rain fell in such fertile places on the coast as had been chosen, and did so at the rate of from two to eight inches at a fall. As all these fertile places were low-lying, it was obviously impossible to settle on them permanently. In fact, as the first settlers saw it, the whole vast territory seemed never to be anything for long but either a swamp during Wet Season or a hard-baked desert during the Dry. During the seven months of a normal Dry Season never did a drop of rain fall and rarely did a cloud appear. Fierce suns and harsh hot winds soon dried up the lavished moisture.

It was beginning to look as through the land itself was hostile to anyone but the carefree nomads to whom the Lord gave it, when a man named Brittins Willnot found the site of what came to be the town of Port Zodiac, the only settlement of any size that ever stood permanently on all the long coastline, indeed the only one worthy of the name Town ever to be set up in the whole vast territory. Capricornia covered an area of about half a million square miles. This site of Willnot's was elevated, and situated in a pleasantly unfertile region where the annual rainfall was only about forty inches. Moreover, it had the advantage of standing as a promontory on a fair-sized navigable harbour and of being directly connected with what came to be called Willnot Plateau, a wide strip of highland that ran right back to the Interior. When gold was found on the Plateau, Port Zodiac became a town.

The site of Port Zodiac was a Corroboree Ground of the Larrapuna Tribe, who left the bones of most of their number to manure it. They called it Mailunga, or the Birth Place, believing it to be a sort of Garden of Eden and apparently revering it. The war they waged to retain possession of this barren spot was perhaps the most desperate that whitemen ever had to engage in with an Australian tribe. Although utterly routed in the first encounter, they continued to harrass the pioneers for months, exercising cunning that increased with their desperation. Then someone, discovering that they were hard-put for food since the warring had scared the game from their domains, conceived the idea of making friends with them and giving them several bags of flour spiced with arsenic. Nature is cruel. When dingoes come to a waterhole, the ancient kangaroos, not having teeth or ferocity sharp enough to defend their heritage, must relinquish it or die.

Thus Civilisation was at last planted permanently. However, it spread slowly, and did not take permanent root elsewhere than on the safe ground of the Plateau. Even the low-lying mangrove-cluttered further shores of Zodiac Harbour remained untrodden by the feet of whitemen for many a year. It was the same with the whole maritime region, most of which, although surveyed from the sea and in parts penetrated and occupied for a while by explorers, remained in much the same state as always. Some of the inhabitants were perhaps amazed and demoralized, but still went on living in the way of old, quite unaware of the presumably enormous fact that they had become subjects of the British Crown.

Herbert acerbically sums up the 'Port Zodiac' settlement, which he considers full of 'booze artists and pompous government officials', in 1904.

So slow was the settling of the Port Zodiac district that in the year 1904 the non-native population numbered no more than three thousand, a good half of which was Asiatic, and the settled area measured but three or four square miles. But the civilizing was so complete that the survivors of the original inhabitants numbered seven, of whom two were dying of consumption in the Native Compound, three confined in the Native Lazaret with leprosy, the rest, a man and a woman, living in a gunyah at the remote end of Devilfish Bay, subsisting on what they could buy with the pennies the man earned by doing odd jobs and the woman by prostitution. The lot of these last was not easy. Fish and game were scarce; and large numbers of natives of other tribes were available as odd-jobbers and prostitutes; and it was made still harder by the fact that they had to dodge the police to keep it, their one lawful place of abode in the land the Lord God gave them being now the Native Compound.

Herbert's character Mark Shillingsworth, a nomadic drunk who has come to Port Zodiac in that year with the intention of 'getting on and getting out', half-unwillingly succumbs to a relationship with a black woman, with the result that his son Norman (in Aboriginal English *Naw Nim*, or 'No Name') is born. The remainder of the novel follows the progress of Norman Shillingsworth until about 1930, along with that of a large cast of subsidiary characters. In the late 1920s, fifty or sixty years after the founding of the settlement, Herbert's vision of the social strata of the town is equally uncompromising. At the Darwin hospital, the Greek and Aboriginal staff work as the white nursing staff bring their guests back from a local cricket match.

During a momentary lull in the bustle, Con the cook heard a far-away sound, and cocked his ear, then looked to see if the ever-watchful leper girls were craning out. They were. So he concluded that the cars bringing the sisters and their guests were coming in, and said to his sweating comrades, 'Here dey comes—de great white pipples.'

A car raced through the yard, past the lazaret, round the mortuary and the Asiatic Ward, out of sight

Watching the vari-coloured grasses decorating the
cinnabar sand-hills I knew that few Australian painters yet
had found the light that floods this land; for me its impact
was overwhelmingly Eastern. … Seeing it one began to
feel the power of altjiringa, that dreamtime which
is the core of the aborigines' faith and legends. …

CYNTHIA NOLAN, 1948 (PAGE 216)

NORTHERN TERRITORY. PHOTOGRAPHER: OLIVER STREWE

*They were old, very old already, when life was new. They were worn
down and weathered before such as we were even heralded by creeping
things. To see them is to know that they could not possibly be less ancient
than that. A thousand million years at least it takes to make something so
rich and strange, so profound, so unbearably potent with dreams.*

CHARMIAN CLIFT, 1965 (PAGE 216)

STANLEY CHASM, NORTHERN TERRITORY. PHOTOGRAPHER: OLIVER STREWE

You get there already visually bruised and aching, tender in the sensibilities with the effort of belief in the awful innocence of your country so exposed to your inspection …

CHARMIAN CLIFT, 1965 (PAGE 216)

WATER LILIES, KAKADU, NORTHERN TERRITORY. PHOTOGRAPHER: CAROLYN JOHNS

*If you come to the Centre, as I did, by air ... the surprise of it is
infinitely more surprising than you are prepared for, even though
you have prepared yourself by much industrious homework on the
geology of the place, its flora and fauna, climate, characters,
myths, legends, yarns and tall tales. ...*

CHARMIAN CLIFT, 1965 (PAGE 216)

NORTHERN TERRITORY. PHOTOGRAPHER: OLIVER STREWE

towards the nurses' quarters. Then another car, another, then a fourth. The last was lost in dust. So was the Evening Star, as though the Great White Pipples could do what they liked with it.

Con said to Tocky, 'Better nick off and change yous clotheses, monkey. Here, Christy, gif dat jobs to Barney and start cuttin' dat breads.'

'I aint a monkey,' Tocky cried. 'But you'n ol' g'rilla.'

'Brrrr!' he snarled, acting the part and rushing at her. She squealed, and laughing bounced out of the kitchen and skipped away down the roofed-in gangway to an old white building where she and Christobel lived. This building had once served as the maternity section of the hospital. It could almost have served as an oven during the heat of a Dry Season day; for it was of corrugated iron built on a concrete floor; and by night and during the humid days of the Wet it could have served as an incubator. The few white women who were willing to breathe the same air as the wives of Greeks and other low-class people, were now confined in a modern building of asbestos. But the old place was not utterly disused; sometimes halfcaste women as needed more attention in travail than could be given them in the Compound hospital were confined there; such women were often so ill they died.

Before Tocky entered her little room at the rear she peeped round the latticed back veranda to see if Yeller Elbert was about, knowing from experience that he was not above creeping on her while she dressed. Elbert and his lubra, who was a kitchen-maid, lived, or camped, as they more aptly called their squalid state of residence, on the back veranda. Elbert was not there. So Tocky went into her room and switched on the light. The room was about fourteen feet square, terribly paint-worn, its furniture two rusty iron beds, a wooden cupboard, three kerosene-cans, a pickle-bottle holding a bunch of ponciana flowers, and a footworn fibre mat.

Tocky removed her smock, exposing her light brown body to the waist. Mosquitoes had been waiting all day for that. They rushed her. Stamping and slapping and swearing stockrider's oaths she rushed to the cupboard and took out a simple dress of flowered print. She dressed with haste, then surveyed herself as best she could without a mirror. Thus she was dressed in her best in a minute, though not so much because she lacked the feminine desire to make a long job of dressing as the means to exercise it. Dressed, she went to her bed and took from a hole in the mattress a mouth-organ. She sat on the bed, and, with shoulders dancing, played *Waltzin' Matilda*. So she was occupied till a burst of laughter was swept down to her from the nurses' quarters, a reminder of her duties.

At the nurses' quarters thirteen well-dressed people sat among palms and flowers in the bright well-furnished lounge, talking and drinking cocktails.[10]

In Xavier Herbert the Northern Territory, ostensibly one of the less fertile fields for great literature, nurtured a literary giant like a freak flower in waste ground, where the more manicured gardens of the south also grew prize blooms but in less overwhelming form. Herbert's other major work *Poor Fellow My Country* (1975), the longest Australian novel ever published, covers the years 1936 to 1942 and the stories of several dozen characters, and also takes as a general theme the devastation caused to the north and its inhabitants by white settlement. It is also notorious for its less than flattering account of the heroism of the inhabitants and administrators of Darwin during the bombing of the north by the Japanese in World War II.

Herbert died aged eighty-three in Alice Springs, having returned to the Northern Territory from his home in Redlynch, near Cairns in North Queensland, in 1984, hoping to complete a last book.

'THE STEAMY HOTHOUSE':
Modern Darwin

I clambered eagerly down among the slippery shrubs, slipping and sliding through the dense under-growth. I had never seen such greenness: an unnatural greenness, as if the leaves were a kind of plastic. Huge parrots yattered in the dripping fruit trees. Butterflies of brilliant colours—bright rainbow colours, chemistry set colours, coffee-table book colours—filled the air. Under any leaf I chose to lift small crea-tures seemed hidden: giant, clockwork insects, built from strange meccano, or grubs the size of small, juicy mammals.

Cartoon descriptions? How else to describe a cartoon world? The moths that thudded into the flyscreens that night were the size of bats—soft, powdery bats. And the bats that filled the mango trees in the darkening twilight were foxes. Even our garden lawn—most domesticated of foliage—needed mow-ing again almost as soon as it was done … like some lush, green five o'clock shadow.

Everything grew larger than life in the steamy hothouse of Darwin, and the people were no exception. Exotic, hothouse blooms.

Peter Goldsworthy, *Maestro*, 1989[1]

Modern Darwin has had few chroniclers since Xavier Herbert, but the settlement has survived against greater odds than those faced by the southern capitals. The Northern Territory, originally part of New South Wales, was annexed in 1863 by South Australia, but was placed under Federal Government control in 1911, when the name of its largest settlement reverted from Palmerston to Darwin. During World War II the town, along with Broome, was bombed by Japanese planes, and a large number of allied troops were stationed in the northern settlements in anticipation of a Japanese invasion. In 1974 Cyclone Tracy hit Darwin and destroyed some ninety per cent of its buildings (for the third time in its history), and much of its old-fashioned tropical architecture was lost in the reconstruction. In 1978 the Northern Territory achieved self-government.

South Australian-born novelist and short story writer Peter Goldsworthy finished his schooling in Darwin in the 1960s. Goldsworthy's experience of the town in 1967, pre-Cyclone Tracy, informs his novel *Maestro*, which describes the experiences of a boy who takes piano lessons from an alcoholic Viennese refugee, Keller. Once a well-known concert pianist, Keller leads a reclusive life in a balcony room at the local pub: 'a warren of crumbling weatherboard, overgrown with bougainvillea'.

Paul's father, a doctor, is not impressed with Darwin.

During the slow movement the rain began again. It fell abruptly, totally, a solid volume of water de-scending on the iron roof. The two of them kept playing—two parts, now, of a mismatched Trio—but after a few bars they abandoned the attempt, leaving the rain, deafening, solo.

My father loosened his tie. In those first weeks he still clung to the Southerner's uniform. Then he wiped the sweat from his brow.

'The arsehole of the earth,' he declared, loudly.

He dropped the piano lid with a thud.

'A city of booze, blow, and blasphemy,' he said, in the tone of voice he reserved for memorable quotes.

'Shakespeare?' my mother wondered.

He shook his head: 'Banjo Paterson.'

Nor is his mother.

I loved the town of booze and blow at first sight. And above all its *smell*: those hot, steamy perfumes that wrapped about me as we stepped off the plane, in the darkness, in the smallest hours of a January night. Moist, compost air. Sweet-and-sour air …

We spent the remaining few hours of that first night in a motel room, but I couldn't sleep. Sometime near dawn I jerked the mosquito netting aside, rose from the bed and peered out through the louvres. Always I'll remember that first morning: the brilliant furnace of the rising sun; the huge clouds that ruddered the sky. In every direction rain could be seen falling: vast, distant cubes of water dropping slowly, ponderously out of the sky.

From time to time a cube would descend from directly above: not so much rain as a solid mass of water, beginning and ceasing suddenly.

Mid-morning found us inspecting our new home: a bare shoebox of louvred walls and asbestos, perched above the wet shrubbery on high, thin stork-stilts.

'Is this it?' My mother tried to disbelieve.

'This is it.'

'Check the number again …'

'This is it,' my father repeated.

'Perhaps the key won't fit,' she hoped.

Later that morning I found her sitting on the edge of the bath, weeping silently: she had left a bluestone villa in the South for this.

As the hottest time of the year approaches, the people of Darwin begin to show signs of strain.

Outside in Darwin the rains came and went, but it was always Wet season in the front bar of the *Swan*: a monsoon of beer and sweat and smoke and noise. I pushed through the storm each Tuesday on my way to piano lessons, and out again afterwards. Overhead the big fans turned—slowly, slowly—stirring it all into one thick, exhausted atmosphere, seemingly unbreathable, uninhabitable …

And yet its inhabitants somehow survived. And only grew noisier, and thirstier, the longer they stayed. In the muggy heat of the evening the chinking of cold glasses drew them from every corner of the town, like tiny glass bells tolling for Mass, tolling for a communion of beer and tobacco.

A Mass taken in remembrance of nothing more than the previous night's beer and tobacco.

They sought forgetfulness, not remembrance. Who could blame them? Over dinner each night, my father recounted the day's horror stories from his work at the hospital. He spared us nothing: stories of wife-beaters, fugitives from justice, alcoholics and maintenance dodgers. Darwin was the terminus, he liked to say: the Top End of the road. A town populated by men who had run as far as they could flee. From here there was only one further escape. And each day on his rounds he saw any number of those hell bent on taking it.

As I pushed my way through the drinkers each Tuesday clutching my music satchel, I found it easy to place Keller among these fugitives.

Darwin in the wet season, Paul observes, has a definite psychological effect on its inhabitants, including his teacher.

As the world grew hotter, and the louvred walls of the houses were cranked open to maximum aperture, all privacy vanished. To walk the streets at night was to walk among rows of lined, illuminated screens, as if at some vast drive-in—a supermarket of drive-ins. A constant soundtrack of country and western music filled the air: plangent ballads of love, jealousy, murder and jail spilling out of the high, opened houses and pubs. Across each screen—raised on stilts above the shrubbery, louvres wide open—even the most fantastic stories my father related suddenly seemed possible, and visible.

In the entire town perhaps only the wooden slats of Eduard Keller's bedroom remained closed. Climbing the stairs to his shuttered room each Tuesday, I was able to tell myself I was finally beginning to gain some sort of understanding of the man. The *Swan* was a monastery, of a kind: a place of retreat, of renunciation of the world. A place of *partial* renunciation. His bottle of schnapps was never more than arm's length distance.

The November humidity seemed to draw out his worst, exaggerate his faults, and render him a caricature of himself. His face reddened further, his moist lids drooped—there was even the occasional crease in his starched white suit. I would to find him each steamy Tuesday hunched in his dark room, brooding, the schnapps bottle always at his elbow, his fat scrapbook of newspaper clippings once again before him.

As an adult, Paul revisits Darwin in 1977 in time to witness Keller's death. The climate remains a constant, but now a new generation of Asian migrants are starting restaurants in the newly rebuilt city's modern malls. However, some fundamental attitudes are unchanged.

In the afternoons I re-explored the town, trying to find some trace of the past, some ancient layer or deposit beneath this new city rebuilt of suburbs and supermarkets, shopping malls and overpasses. The shoebox houses on their stork-stilts had largely gone: stout, squat homes hugged the ground everywhere, built out of bricks that no amount of monsoonal huffing and puffing would bring down.

... From the street a car horn sounded, and then another, pitched slightly higher: a minor third. A soft, temperate rain was falling—slow, blunt, wet pinpricks—and the air seemed cooler than any Wet season I had known. I knew no-one in this rebuilt town, but wanted someone—anyone—to know that a Great Man had died, whatever the crimes he felt he had committed.

... A Thai restaurant in Smith Street provided some sort of comfort. A soup thick with ginger and hot chillis warmed my stomach, and also—the heat spreading to nearby tissues—seemed to cheer my heart. The smile of the waitress—shuffling with tiny stylised steps, incapacitated by some absurd traditional Asian garment—also warmed me, and for an instant I considered asking her what time she finished work ...[2]

Other contemporary accounts of the city are scarce. 'Louvres' (1984), by Les Murray—who hitch-hiked around Australia and visited Darwin in 1961—was inspired by the poet's subsequent visits in 1983 and 1984. The poem goes in part:

In the banana zone, in the poinciana tropics
reality is stacked on handsbreadth shelving,
open and shut, it is ruled across with lines
as in a gleaming gritty exercise book.

The world is seen through a cranked or levered
weatherboarding of explosive glass
angled floor-to-ceiling. Horizons which metre
the dazzling outdoors into green-edged couplets.

In the louvred latitudes
children fly to sleep in triplanes, and
cool nights are eerie with retracting flaps...

For drinkers under cyclonic pressure, such
a house can be a bridge of scythes—
groundlings scuffing by stop only for denouements.

But everyone comes out on platforms of command
to survey cloudy flame-trees, the plain of streets, the future:
only then descending to the level of affairs

and if these things are done in the green season
what to do in the crystalline dry? Well
below in the struts of laundry is the four-wheel drive

vehicle in which to make an expedition
to the bush, or as we now say, the Land,
the three quarters of our continent
set aside for mystic poetry.[3]

And in 'Jacques Tati at the Darwin Hotel' (1986), Western Australian-based poet Fay Zwicky also presents an impressionistic view of the city.

Palm-fringed patio,
a buzz of mauve, cerise,
a blaze of pink and gold and green,
cascades of luminous bougainvillea

and frangipanni fretworking
a weightless turquoise sky.

Mangoes drop their headlong smoothness
down the ropy vines, arched mangrove
roots, tangibles
unlimited.

'Minimum dress for this area will be
shirts, shorts, shoes, and long socks.'

The trustful waitress leaves me juggling
tiny plastic rectangles of butter
marmalade, honey. I ferret clumsily
around the toast.

And there's the coast![4]

One of the more notorious recent accounts of life in the Top End is supplied by Lindy Chamberlain, in her autobiography *Through My Eyes* (1990), and once more it is the account of an exile: this time in Darwin's gaol.

Sitting in the day room because it was too wet to go outside, we often saw the lightning hit the perimeter fence and the rolls of razor wire along the top of the cyclone fencing would illuminate like a Christmas tree as the lightning sparkled and spat right around from one section to another. Lightning often blew the bell on the gate …
  Some evenings I sat in my cell reading my letters and trying to study. If I turned my cell light off (so that only the security lights were on) I could look straight south and see in the moonlight a little bit of water and mile after mile of desert landscape. It was not true desert, but from a southern point of view, desert enough. I had flown over that land many times and I knew just how far away home was, south straight out of my window. I could look at the stars and know they were the same stars they saw at home and … say goodnight to my darlings at home.[5]

THE TOP END

We are travelling west of Alice Springs, and Sam is at the wheel;
Riding the diesel-grader I am watching its blade of steel
Roll back the dark-red sandy loam or grind the limestones grey,
And the wheels whirl in a red-dust swirl along the new highway.

We pass where Sturt-peas clothe the earth with a scarlet sweep of flowers,
And burst through green acacia-trees that send down golden showers;
The parakeelia's purple blooms are crushed in the dry, red sand
When the bright blade sweeps as the grader creeps over the stern, strange land.

The mulga, mallee, desert-oaks fall prostrate as we pass,
The lizards, pigeons, porcupines crouch low in stone and grass;
We brush the spinifex aside; tear down the bush-rat's shade,
And the desert mole in its sandhill hole digs faster from our blade.

The honey-ants are rooted out to roll upon the sand,
But ever the ramping, stamping fiend goes roaring through the land;
The tyres grind and the steel blade cuts the pads where camels trod
And claws at the ground of a stony mound where tribesmen praised their God.

We cross the desert rivers, formed when the world was new,
And churn to dust the fossil-bones of the giant kangaroo;
I wave to naked native kids upon Erldunda's plain,
And we fill our tank where the black men drank from rock-holes filled by rain.

We camp in Kulgera's weathered hills, scarred core of an ancient range,
Where the camp-fire flame throws out its light on a scene that is ever strange
As a dingo wails by the painted wall of a sacred cave near by
And the stars shine bright as we lie at night beneath a frosty sky.

We rise as mulga-parakeets go whirling through the dawn,
We see old star-man Manbuk rise from depths of midnight drawn;
We hear the grader's engine roar with Sam behind the wheel,
And I sing my song as we plunge along to the chatter of wheel and steel.

W. E. Harney, 'West of Alice', 1958[1]

If there is an authentic European regional writing associated with the 'Top End', then traditionally it is contained in the adventure tales and bush yarns of writers such as Ion Idriess, William Edward (Bill) Harney, Tom Ronan and Olaf Ruhen. Reaching its heyday in the first half of the twentieth century, the genre includes such titles as Idriess's *Lasseter's Last Ride* (1931), an account of an expedition to find a possibly chimerical reef of gold in central Australia, and other Idriess works such as *Flynn of the Inland* (1932) and *The Cattle King* (1936); Harney's *North of 23°* (1946), an autobiography; Ronan's ironically titled *Vision Splendid* (1954), an authentic fictional account of the cattlemen's life; and Ruhen's *Naked Under Capricorn* (1958), the story of the establishment of a cattle station in the north around the year 1900. Tom Cole's autobiography *Hell West and Crooked* (1988), which deals with life in the Northern

Territory in the 1920s and sold over 70,000 copies, demonstrates that the tradition is not dead.

Written in a plain, accessible, and often colloquial or humorous style, these books generally rely on the firsthand experience of their authors. Idriess travelled extensively in the Torres Strait region in the company of pearlers and missionaries and hunted buffalo in central Australia; Harney, born in 1895 in Charters Towers, Queensland, was a drover, bushman, soldier in the First World War, lugger captain, patrol officer in the Northern Territory and an amateur student of Aboriginal life and culture. According to popular legend, Harney educated himself while spending six months in jail in Borroloola on a cattle-duffing charge, in a tin shed which also held the local library, donated by the Carnegie Foundation at the request of a vice-regal visitor. Ronan, born in 1907 in Western Australia, worked from the age of fourteen with his stockman father in the northwest, and afterwards on the pearling boats of Broome. Ruhen, born in 1911 in New Zealand, was a deep sea fisherman as well as a journalist and novelist who travelled extensively in Australia after his arrival in 1947, and wrote about what he saw.

In addition, as with Western Australia and far north Queensland in their early pioneering periods, in the Northern Territory also it was quite often the well-educated wives, daughters and sisters of settlers, administrators and pastoralists, such as Jeannie Gunn and Harriet Douglas, who matter-of-factly recorded the social history of the region while the men recorded their more momentous adventures. Travel writers such as Ernestine Hill and Elspeth Huxley also described their experiences.

The Second World War brought other literary personages to the Northern Territory, including novelist George Johnston's brother, fictionalised as 'Jack' in *My Brother Jack*, who was stationed at Larrakeah military camp. In another altogether odd literary coupling, Frank Hardy and Sumner Locke Elliott were stationed together at Mataranka, 443 kilometres south, where Elliott wrote for an army magazine that Hardy had established. Elliott described the area of the army camp, 'seventy miles south of Katherine' as 'a lonely strip of barren and seemingly endless sandy waste of ant hills and stunted trees—thick, hot red sand in the winter time and a sea of mud during the dreaded "Wet"'. He based his play *Rusty Bugles* (1948)—subsequently banned by the police in Sydney for offensive language—on his six months there. Elliott also drew on these experiences in several fictional works, including his autobiographical novel *Fairyland* (1990), an account of growing up homosexual in Australia.

The poet Roland Robinson, born in Ireland in 1912, arrived in the Northern Territory as a labourer with the Civil Construction Corps during World War II, and later lived in Darwin. He subsequently worked as a fettler at Deep Well, 80 kilometres south of Alice Springs, which he celebrates in his 1962 collection of the same name. *The Drift of Things* (1973), the first part of his autobiographical trilogy, describes these experiences, and many of his later poems also originated in this period.

More recently the Centre and the Top End—despite their prosaic appellations—have begun to impinge on the Australian consciousness as more than a sparsely populated geological emptiness suitable only for cattle and mining. The imagery has changed from that of 'The Dead Heart' (a phrase coined by explorer J.W. Gregory in his book *The Dead Heart of Australia* (1906) to describe the area between Lake Eyre and the Simpson Desert) through being that of a distantly located tourist attraction to that of an area where the modern forces of politics and economics are observed to be in confrontation with the more fragile but perhaps ultimately enduring magic of the Dreamtime.

The subtle aesthetic power of the desert and semi-desert regions of Australia has always been apparent to a few observers, such as the explorer and poet Ernest Favenc, whose collection of poems *Voices of the Desert* appeared in 1905, but it is relatively recently that modern and urban Australian writers have begun to share the perception. Often it has been modern painters such as Sidney Nolan and Russell Drysdale who have shown the way.

Cynthia Nolan, travelling with her husband Sidney Nolan in 1948 on a pioneering trip across central Australia from Adelaide to Darwin in search of material for his paintings, wrote of the 'sweet elegance' of the country around the Finke River.

> Watching the vari-coloured grasses decorating the cinnabar sand-hills I knew that few Australian painters yet had found the light that floods this land; for me its impact was overwhelmingly Eastern. … Seeing it one began to feel the power of *altjiringa*, that dreamtime which is the core of the aborigines' faith and legends. One also felt the spell of unfamiliar trees; trees grey with slim black stems, trees closely branching from ground level to spread like fans … (yet) There was never any possibility of this being any other land; the trees and grasses we had never seen before were yet uncannily related to those we knew, while echoes of the tremendous sky, of the bare mountain ridges and scented arid plains had been familiar to us all our lives.[2]

When novelist George Johnston followed a similar route in 1965 with his wife Charmian Clift and Sidney Nolan, in connection with a film being made by the ABC on Nolan, he wrote about the journey in the third and uncompleted volume of his semi-autobiographical trilogy *Clean Straw For Nothing* (1969), with Nolan being transformed into the painter Tom Kiernan. Meanwhile Clift noted her own impressions of the landscape for her series of essays in the *Sydney Morning Herald* in the late 1960s, later collected in *The World of Charmian Clift* (1970), edited by her son Martin Johnston.

> If you come to the Centre, as I did, by air from Adelaide, the surprise of it is infinitely more surprising than you are prepared for, even though you have prepared yourself by much industrious homework on the geology of the place, its flora and fauna, climate, characters, myths, legends, yarns and tall tales.
>
> You get there already visually bruised and aching, tender in the sensibilities with the effort of belief in the awful innocence of your country so exposed to your inspection, drinking cold beer and eating more airline chicken and advancing your watch a half-hour. You are a little shamed and uneasy, as though you have taken an unfair advantage, and you think of the explorers crawling like maddened lice across that vast wrinkled anatomy, crazed by thirst and dreams and the radiantly tender pink blushed on beckoning hills. From twenty thousand feet the hills are like fat squishy tumours, or dried scabby ones. Benign compared with the incurable acid-wound of Lake Eyre, steaming corrosive white and vitriolic after placid Torrens, where, all unknowing, Swift set the longitude and latitude of Lilliput. Gulliver sprawls defenceless for your microscopic examination. Pitted pores. Dried out capillaries of watercourses. Culture slides of viridian clotting thick creamy yellow. Wind ridges raised like old scars, and beyond them the even arid serrations of the Simpson Desert, dead tissue, beyond regeneration.
>
> And yet, the tenderness of the pinks, the soft glow of the reds, the dulcet beige, and violet seeping in. The landscape, after all, is alluring beyond reason. Voluptuous, even. You could abandon yourself to it and die in a dream, like those savages of whom Kafka speaks, who have so great a longing for death that they do not even abandon themselves, but fall in the sand and never get up again. Such unearthly beauty, one knows—and still yearns—is fatal. It is a landscape for saints and mystics and madmen.
>
> And after that the vibrant shock of being terrestrial again, bucketing in toward the Alice in a Landrover with the air singing clear and thin and sweet and your lap filled with the strange flat podded pink flowers that are clumped beside the road and under the ghost gums and the wattle. You could not have foreseen that, nor, with all your homework on the history and geology of the Macdonnell Ranges, could you have foreseen the lilac beyond the red and the gold, floating in a boundless clarity where perspective is meaningless. You accept that the mountains are as old as the convulsions of the earth. They had suffered a sea-change before life crawled on the land. They were old, very old already, when life was new. They were worn down and weathered before such as we were even heralded by creeping things. To see them is to know that they could not possibly be less ancient than that. A thousand million years at least it takes to make something so rich and strange, so profound, so unbearably potent with dreams.[3]

In his novel *Maestro*, Peter Goldsworthy also notes his fictional impressions of travelling through the centre from Darwin to Adelaide, emerging from the desert into the more comfortable zones of the southern coast.

We fled south for several weeks that Christmas, arriving in Adelaide at the home of my grandparents, my mother's parents, after a five-day drive.

Of that interminable trip only odd dream images remain: a water sprinkler twinkling on a postage stamp of lawn somewhere in the desert; the sky for miles black with clouds of thirsty budgerigars; a taxi heading south into that same desert, pulling over at some arranged spot and disgorging its passenger—a bearded black tribesman who paid his fare and strode off into the hot dunes, barefoot, carrying nothing but spears.

And then we were through the desert and into the temperate wheat country, passing through the mid-north towns in which I had once lived—centuries ago, in an earlier life, it now seemed. The closer we approached to Adelaide, the slower our journey became, stopping more and more frequently in those small wheat-belt towns, visiting old friends, reliving past Gilbert and Sullivan triumphs …

Until finally, somehow, we reached the City itself.[4]

Goldsworthy would have passed through Alice Springs, the second principal town of the Northern Territory, which Bruce Chatwin described in his 1987 novel *The Songlines* as 'a grid of scorching streets where men in long white socks were forever getting in and out of Land Cruisers'. 1524 kilometres south of Darwin and roughly in the centre of Australia, Alice Springs was founded in 1888 and named for Alice Todd, the wife of the South Australian postmaster general who supervised the building of the Overland telegraph from Port Augusta to Darwin. The town was made famous through Nevil Shute's best-selling novel *A Town Like Alice* (1950), although the novel was not in fact about the Alice itself.

More recently, the town of Alice Springs has been perceived ambivalently by visitors. Charmian Clift, visiting in 1965, voices a note of regret.

It takes less time to make an Alice. Contemporary Alice, that is. They say it was different before the tourist boom. Old hands mourn, bitterly. It was nearly evening when we reached it, and the mountains were moving in close about hotels and motels and gem shops and rock shops, banks, garages, milk-bars, tourist agencies, boutiques, galleries and the Old Timers' Home. Is it Persian lilac that lines the main street? The scent in the evening air is enough to bowl you over. Fronded trees sparking delicately with little starry clusters, and between them a desecration of imbecile Op lighting, great lozenges of red and blue and green and yellow clownishly colouring the tourists stepping eager for bargains, souvenirs, and drinks before dinner. And through the tourists, the lilac-scented air, the hectic fun-fair illumination, the slow lurching drift and black shadow-weave of the disinherited, stripped of ancient dignity, degraded, subservient, aimlessly drunk on a Friday night.

A lady inheritor, sensible in drip-dry, shoulder-bag bulging—rocks? gems? berry necklaces? Mission grass-weaving? bad bark paintings? —postcards poised ready for the slot, pounces on a tall black trio teetering in the most curiously graceful progression. Cheap boomerangs, she wants. The real thing. Not the junk in the tourist shops. She is loud and articulate. Imperious. (I think of the last Afghan and the last camel train, also imperious, stepping out slow slow from the Alice and disappearing into the vast distance.) The black trio, thus accosted, are soft, slurred, incomprehensible and perhaps uncomprehending. They sway away and back again, surprisingly regrouped under the ghastly green of the street light, awaiting the lady's exasperated dismissal. Two barefooted women scuttle under the Persian lilac and across the street to their men on the opposite corner. They are high in the haunches, long in the heels. Their legs are like thin crumpled brown ribbons flying, their hair pale straw. Drysdale has drawn them often and compassionately, by tin huts and shanties, patient with the heavy burden of life. Now they are patient on the street corner, movement arrested, patiently waiting as if waiting was an end in itself.[5]

And Robyn Davidson, who lived there for two years while training her camels for her desert trek in 1977, described in her non-fiction account *Tracks* (1980), also found the town unfriendly and unpleasing.

My first impression as we strolled down the deserted street was of the architectural ugliness of the place, a discomforting contrast to the magnificence of the country which surrounded it. Dust covered everything from the large, dominant corner pub to the tacky, unimaginative shop fronts that lined the main street. Hordes of dead insects clustered in the arcing street lights, and four-wheel-drive vehicles spattered in red dirt, with only two spots swept clean by the windscreen wipers, rattled intermittently through the cement and bitumen town. This grey, cream and hospital-green shopping area gradually gave way to sprawling suburbia until it was stopped short by the great perpendicular red face of the Macdonnell Ranges which borders the southern side of town, and runs unbroken, but for a few spectacular gorges, east and west for several hundred miles. The Todd River, a dry white sandy bed lined with tall columns of silver eucalypts, winds through the town, then cuts into a narrow gap in the mountains. The range, looming menacingly like some petrified prehistoric monster, has, I was to discover, a profound psychological effect on the puny folk below. It sends them troppo. It reminds them of incomprehensible dimensions of time which they almost successfully block out with brick veneer houses and wilted English-style gardens.[6]

Once again, as so often occurs in Australian literature, there is a perceived lack of equilibrium between the built environment and the wilderness.

CANBERRA

'A GROUP OF SUBURBAN ENCLAVES
SEPARATED BY CIVIL SERVICE STATUS'

With only about twenty thousand people in the city and its district, it was just a group of suburban enclaves separated by Civil Service status and miles of open paddocks scattered with plantations, copses, wind-breaks and clumps of scores of thousands of imported deciduous trees. I'd read somewhere that in the earliest days of the fledgling Washington D.C., some very diplomatic European diplomat had allowed that it was 'a city of—uh—*magnificent spaces*'. The description fitted Canberra like a fingerstall. The mad-woman's-knitting of roads connecting and slicing through the enclaves was so circuitous, so confusing, that they'd become the basis of a local joke that the half of the population not employed in the Government offices consisted of touring motorists who'd despaired of ever finding their way out, and had settled there.

T.A.G. Hungerford, *A Knockabout with a Slouch Hat*, 1985[1]

In 1948, when Western Australian writer Tom Hungerford first encountered Canberra after returning from fighting in a commando unit in New Guinea and service with the post-war occupation forces in Japan, he discovered a certain innocence to the city perhaps most perceptible to a soldier just returned from World War II. Hungerford was working at the Australian War Memorial editing its annual publication *Stand Easy*, a series of soldiers' stories, an experience he later recalled in an autobiographical piece called 'Me and the National Capital'.

The Australian War Memorial stood in splendid isolation among the gullies of Mount Ainslie's ochre-coloured foothills. When I first saw it I thought it must be the most beautiful, most beautifully-sited, building in Australia, and I never tired of gazing at it. In that glass-clear upland air its vast, angled surfaces of plain grey stone changed from alabaster to honey to ice as the light of morning, noon and evening flowed over it and about it—and sometimes, it seemed, through it. It seemed to be growing out of its red and orange clay, one with the gum-trees on the slopes behind it and the lank, dry grasses of the paddocks which sprawled right up to its walls. I found a ground-lark's nest within fifty yards of the front entrance, and even closer to the back door a quail's with seven eggs in it. Within a couple of hundred yards of the south wall there was a slope where we used to fill our wastepaper baskets with autumn mushrooms fresh that morning and big as your hat.

In the mixture of mockery and affection that seems to characterise much recent and contemporary writing about Canberra, Hungerford noted that the prosperity brought to the surrounding district by the wool boom of the 1950s was still some years off. 'In 1948,' he wrote, 'the only local cash-crop was sowed and mowed exclusively in the Government offices scattered here and there like stranded whales on that sea of bleached and frozen paddocks.'

Around this slowly-beating heart of Federal pride there was, of course, a well-established bucolic fringe. For a century or more, the tinkle of its cowbells, the bleat of its jumbucks and the bubble of its rabbit and mutton stews had resounded among the secluded valleys and windy foothills receding through ever and ever deeper and more beautiful swathes of mauve, violet and purple from almost the edge of the city to the snowy peaks of the Australian Alps. Miles Franklin romanticised it, but the great times of the Monaro she wrote about were over long before my advent there. Most Australian settler families seemed to have passed quickly from their initial energy and affluence to either escape from the soil or to a sort of white-trashism on it, anchored by ignorance, isolation and poverty to worked-out acres and crumbling homesteads. The Monaro was no exception.[2]

T.A.G. Hungerford later worked as a kitchen hand and cleaner at the Eastlake migrant hostel to research a novel about the post-war migrant experience. The book, *Riverslake*, was published in 1953, and Hungerford also wrote about his time at Eastlake in a piece called 'My Turn in the Barrel' from the same memoir. Magda Bozic, who arrived in Canberra as a post-war migrant at about the same time that Hungerford was there, gave another side to the picture in her own memoir of the period, *Gather Your Dreams* (1984):

> To newcomers who were outside the supporting network of homes, clubs and pubs the capital city in those days was a smug closed shop. We, the outsiders, looked on the unknown and aloof citizens of Canberra like a protected species living in the undisturbed security of a national park.[3]

Canberra, once described by local farmers as 'a good sheep station spoilt', and by another nameless critic as 'seven suburbs in search of a city', has remained the city that Australian writers most like to joke about, perhaps not least because politicians and bureaucrats are traditional targets of writers. In the eyes of outsiders, Canberra is the Brasília of the Antipodes, without, perhaps, the dirt tracks between the skyscrapers. Canberra is more a city of beautifully laid out sweeps of road and highway that lead only to disappointing new suburbs, where the hopes of the inhabitants are signified by the young sapling (usually an Australian native) struggling to take root in front of a bleak town house. Canberra is seen as a city planned rather than allowed to grow organically, a city with a population generally born elsewhere, and a city many of whose inhabitants are often there for reasons other than personal preference.

A university and a diplomatic community as well as a seat of government, Canberra is also a young city, one that embodies the grass roots politics of a country that still relies largely on primary products for income and barely figures on the world's political stage. Like Washington, the Canberra of popular myth is a centre of alcoholism and suburban and political adultery. But the larger view is also of a graceful city of rectangles, ellipses and circles laid out on the flat expanse of the Limestone Plains by the Molonglo River, nestled under the range of the cool blue Brindabellas, laced with blossoming trees in spring, and scattered with the variously successful attempts of architects to keep their feet in a society with few cultural precedents to take hold of to maintain a balance. Like a formal European garden that must be viewed from an elevated point for proper appreciation, Canberra also seems to inspire writers, particularly visiting ones, to aerial views.

This is the city that career diplomat and poet John Russell Rowland celebrates in 'Canberra in April' (1965).

Vast mild melancholy splendid
Day succeeds day, in august chairmanship
Presiding over autumn. Poplars in valleys
Unwavering candleflames, balance over candid
Rough-linen fields, against a screen of hills

Sending invisible smokes from far below
To those majestic nostrils. A Tuscan landscape
On a larger scale; for olives eucalypts
In drifts and dots on hillheads, magpie and crow
For field-birds, light less intimate, long slopes stripped

Bare of vine or village, the human imprint
Scarcely apparent; distances immense
And glowing at the rim, as if the land
Were floating, like the round leaf of a water-plant
In a bright meniscus. Opposite, near at hand,

Outcrops of redbrick houses, northern trees
In costume, office-buildings
Like quartz-blocks flashing many-crystalled windows
Across the air. Oblivious, on their knees,
Of time and setting, admirals pick tomatoes

In their back gardens, hearty
Bankers exchange golf-scores, civil servants
Their after-office beers; the colony
Of diplomats prepares its cocktail parties
And politicians their escape to Melbourne.

Canberra's inhabitants lead exemplary lives.

This clean suburbia, house-proud but servantless
Is host to a multitude of children
Nightly conceived, born daily, riding bikes,
Requiring play-centres, schools and Progress
Associations: in cardigans and slacks

Their mothers polish kitchens, or in silk
White gloves and tight hats pour each other tea
In their best china, canvassing the merits
Of rival plumbers, grocers, Bega milk
And the cost of oil-fired heating or briquettes.

To every man his car, his wife's on Thursday
Plus one half-day she drops him at the office
(Air murmurous with typewriters) at eight-forty
To pick him up for lunch at home; one-thirty
Sees the streets gorged with his return to duty …

However, Rowland—even as in the manner of Australian poets of a bygone era he reverts to an older European reference (is Tuscany 'a Canberra landscape on a smaller scale'?) to achieve a simile—ends with a wish that the Australian landscape might yet redeem its comfortable inhabitants. The poem closes with an optimistic hope that, while 'beyond the decent lawn a visionary landscape wings the sight … And luminous distance feed imagination':

A sense of the pale curving continent
That, though a cliche, may still work unseen
And, with its script of white-limbed trees, impart
A cure for habit, some beneficent
Simplicity or steadiness of heart.[4]

Canberra evokes the same ironic, semi-humorous response even in the usually serious Judith Wright, who, in a seven-part poem entitled 'Brief Notes on Canberra', finds a whimsical note in the city's architecture:

> The tawny basin in the ring of hills
> held nothing but the sunlight's glaze,
> a blue-blank opaline mirage,
> sheep cropping, flies, the magpies' warble.
> Burley Griffin brimmed it with his gaze.
>
> Cloud-architecture in reflected image:
> arena, amphitheatre, gallery
> on gallery of quivering marble,
> rose from his mind—great circles, radials ...
> Over the clear-strung air his fingers played
> conjuring a rhetorical opera-city
> for that bald eagle, King O'Malley.
>
> Fantasies of power. The grey sheep nibble,
> dogs snap at flies. Shoddy officials
> argue his job away, confuse his plan.
> Mirages, changed to lakes, lap sewage.
> Cities are made of man.

The clouds:

> Canberra
> specializes in clouds—
> great haughty ones
> small frisky ones
> marble acropolises
> whiteheaded eagles
> tableaux, processions,
> galloping cavalry,
> cottonwool snowscenes
> with snowmen by Thurber.
>
> They act so extravagantly
> swirling their cloaks
> and striking great poses—
> Look at me. Look at me.
>
> Canberra residents
> don't seem to find them strange, but
> maybe the newspapers
> ought to review them.

The vegetation:

> It isn't that I don't like
> European trees.

Why, my great-grandfather came from …
Some of my best friends are …

But huddled together
in clumps and plantations
or lining the roads
like an official welcome
they look a bit lonely
slightly on guard, rather formal,
wishing the visit was over;

like the staff of an Embassy
at a party they don't really trust.

And even the weather:

I stepped out
into the day without thinking.

It rushed at me
took me by the throat
turned me back
and slammed the door after me.

—Blast you
can't you ever remember your coat and gloves?

Wright ends with a comment on the inefficiency of the city as a viable system:

Considered as an ecosystem
Canberra is impossible
No balance between input and output;
a monoculture community
whose energy goes entirely into organization.
Too little diversity
means instability
the scientists say.
No fooling.

Too many predators.
Too few producers.
Too little feedback
and very few refuges for prey species.

Somehow it continues to exist
as an ecological miracle.[5]

The city of Canberra came into existence in 1911 when the architect Walter Burley Griffin, a disciple of Frank Lloyd Wright, won an international competition to design the nation's capital. Griffin's plan for a dignified European-style city included three principal axes, terminated by grand vistas of the

Cloud-architecture in reflected image:
arena, amphitheatre, gallery
on gallery of quivering marble …

JUDITH WRIGHT, 1976 (PAGE 223)

NATIONAL LIBRARY, CANBERRA, AUSTRALIAN CAPITAL TERRITORY. PHOTOGRAPHER: PHILIP QUIRK

… A Tuscan landscape
On a larger scale; for olives eucalypts
In drifts and dots on hillheads …

JOHN RUSSELL ROWLAND, 1965   (PAGE 221)

RED HILL, CANBERRA, AUSTRALIAN CAPITAL TERRITORY. PHOTOGRAPHER: PHILIP QUIRK

surrounding mountins in their natural state, comprising a land axis; a water axis passing through five constructed ornamental basins; an architectural axis of monumental municipal buildings culminating in a 'capitol' on Capital Hill, all surrounded by a complex suburban network. Griffin was appointed director of the project in 1913, but departed after seven years of haggling with local authorities over its composition, and before a single government building had been built. The site—chosen as a compromise in 1908 after a long-running dispute between Sydney and Melbourne as to which should become the national capital, and over sixty nautical miles from the sea to prevent the possibility of naval bombardment—was first explored by Europeans in 1820 and settled by pastoralists from the 1930s onwards. It was called *Canberry* after its Aboriginal name; the last member of the local tribe is believed to have died in 1894. A foundation stone was laid in 1913, when the pronunciation of the name—also a matter of some dispute—was fixed. The first Federal Parliament building was ceremonially opened on 9 May 1927, of which occasion one anonymous poet wrote a little maliciously:

'... What inaugural tumult rent the skies today!
The martial music seemed to cleave its way
To empyreal heights, where human bird
With pinions wide, droned through the skies, to gird
Thy coming more triumphantly ...

How the 'Light Horse' came and went
Thundering straight in columned ranks
To greet thee—and incidentally to flank
The presence of the Princeling Son
Who came to represent his sceptred Sire
At thy auspicious birth ...'

In his novel *Plumbum* (1983) David Foster strikes a slightly more sincere but equally celebratory note.

People will tell you Canberra in the 1950s was like a big country town. Strictly speaking, not so: the cachet—that strangely remote and disorienting lustre—was there. No tan bark or pampas grass in those days, for living was modest; but I guess I should make clear at once I wish to praise Canberra, to celebrate her charms. Rubbishing the capital is all too easy. Besides, is there not, in each guilty heart of us, a longing for the bloody place? Mother Canberra?

Probably the most complete, ironic and affectionate portrait of Canberra yet written belongs to Foster's *Plumbum*, in which Foster, through his narrator, the drummer Felix Farquahar, traces the chaotic and almost paradoxical rise to international stardom of a local heavy metal rock band, born and nurtured in the Canberra suburbs where today, writes Foster, 'the clean, empty Mercedes buses roam spotless crescents in search of a fare'. (The novel's title is a play on *plumbum*, the Latin word for lead, a heavy metal.) Foster, born in Sydney in 1944, studied science at the Australian National University in Canberra and worked as a research scientist in the United States before becoming a writer, postman and subsistence farmer at Bundanoon in the New South Wales Highlands.

In *Plumbum*, Jason and Pete Blackman, the band's lead and bass guitarists, grow up with their father Arthur Blackman, lover of classical music and possible ex-Nazi, in the Canberra of the 1950s.

It's not sufficiently realised today that Canberra in the 1950s was a city of reborn people. This artificial metropolis attracted reffos of a certain kind the way the blue light in a butcher's shop does flies. And where was the enlightenment? Where the wisdom, where the insight?

One of Arthur's workmates began life in a Ukrainian hut. He never went to school, couldn't read or write, hadn't seen running water. When war broke out he fought for both sides and ended up in Auschwitz with Roman Polanski. Today you can find this man in the Commonwealth Club, hobnobbing with Oxford types. This is the man of whom I sing, self-made, a marvel of endurance. Such men have truth to tell us, and that truth is Canberra. Like it or not, when men have been through Hell, it is Canberra they desire, Canberra they create. If you find this hard to fathom, I guess you never went without. You probably don't remember the Great Depression. I'll bet you never even fought in The War.

Foster gives a thorough and ironic examination of Canberra's social and domestic mores.

A month after coming to Australia, Arthur moves his family from the immigration camp, a Nissen hut in Bathurst, straight to the old city. Hackett Gardens, Turner, between Haig Park and Boldrewood Street. Realising earlier than most Canberrans that nothing less than a brief stay in each suburb of the ACT can satisfy that lust for experience, Jason will make it his business to acquire an intimate knowledge of more houses than most. In the course of his protracted youth he will drain tinnies on the concrete patios of Belconnen Way in many a market gardener's dream of blond double brick with triple garage, shut aluminium windows in low-lying areas of Duffy, open embossed cedar doors on clinker brick colonials in Mawson, and creep across Grecian goat-skin rugs in the *en suites* of Garran by night, one ear open for the family Peugeot. He will lounge diplomatically in Mugga Way delighting in the envy of passers-by, who, unlike himself, will lack the key to admission there. He will plumb the heights and depths of Narrabundah, and so forth; but of all the streets in which he will leave his scent, none will be dearer to him than that first, unpretentious Hackett Gardens.

Turner, bordered by the flats of Northbourne Avenue, gritty O'Connor to the north, scholarly Acton to the south, has what is probably the highest incidence of *Reader's Digest* subscribers in the ACT. Almost every day in winter, the postman, his fingers benumbed by those frosty hedges, strives to thrust unsolicited gift offer envelopes into letter boxes too small to hold them. *Open Roads,* two and three to a household, rot on the permadamp lawns of Masson Street in the shadow of the Haig Park pines, while here and there, between arteries and veins in which live public servants and their widows, are small capillaries bulging to encompass a park, the better to daunt through traffic. Such a place is Hackett Gardens, though a glance at some of the homes there now shows, by subtle improvement (glassed-in verandahs, shingled car ports) that these are presently the property of people young and strong enough, if seemingly unwilling to reap the capital gains on Canberra's frontiers, where kerbing and guttering, oregon studs and builders' privies, are all that can be seen from the sealant-smeared windows. On the frontiers, town dogs worry sheep, and I once saw an Afghan hound, shot by an irate farmer and strung up on a barbed wire fence like a skinned fox. On the frontiers live pioneers, pushing back barriers like post-doctoral fellows, as soon exhausted and as readily replaced; for who reneges his stint out here is not a true Canberran.

Similarly, Pete Blackman, who has left home at sixteen, comes back to Canberra after nineteen years' absence and delivers his thoughts on the subject of street names.

'I was driving round Melba today when it suddenly hit me. There must be some pube in the NCDC on twenty-five grand a year just to sit at his desk and think up street names. And look at them. Amadeus Place, Kaleen. Amadio Place, Melba! Ashkanasy Crescent, Evatt! It all seems a bit arbitrary. Take Goossens Place off Copland Drive; shouldn't we have a Bach Place first? I would certainly have a Schubert Drive before I had a Copland Drive. Now Schubert Place, Kambah, that's Schubert the stock and station agent. It must be, it's off McKillop Circuit, near Snodgrass Grove.

'Snodgrass Grove! I award that third place behind Onkaparinga Circuit and Investigator Street. And then we got Grainger Circuit, Alfred Hill Drive; our guy's clearly a varsity man. Plain "Hill Drive" wouldn't do, you see; could be taken as topographical. These names however are quite acceptable, conforming to a general concept. But Sharp Place, Paling Place? Knight's move thinking there. Bit slack, I find that: bit uninspired. And Keats Place, Borovansky Place? The guy's a graduate engineer! And as for Zelman and Laver Place, the less said the better.'

Neither Mother or I bite, so Pete, without looking up, reads on.
… 'Lazarus Place, Wanniassa! Now there's a challenge to a public servant.'[6]

In *Holden's Performance* (1988), Murray Bail takes his Adelaide-born hero Holden Shadbolt (a proto-type Australian, like the car) to Canberra to work as a ministerial driver. Like Foster, Bail falls victim to a desire to *explain* Canberra, to analyse and define it. As with his description of the city of Adelaide in the same work, Bail's approach is more abstract, but no less ironic, than Foster's.

Few capital cities of consequence are located at the edges of a country. The instinct's to set them down towards the middle. Some nations have transferred their capitals holus-bolus after realising the mistake. A capital positioned at the edges of a country remains at the edges of the mind.

From the interior a capital can be seen by its citizens to be radiating power in all directions, a feeling reinforced by a psycho-geometric townplan of lawned circles and spokes, parliaments and palaces at the end of perfectly straight vistas. With so much symbolism invested a capital naturally is tenaciously defended. When the capital 'falls' a nation is weakened; almost too painful to contemplate. Here again, locating them away from the sea and adjacent countries is obeying a deep instinct, Moscow's experiences to the north merely being the most graphic among many examples.

Newly erected capitals—Delhi, Brasilia, Yamoussoukro!—have had the good fortune of drawing upon the combined experience of the others. Naturally each one has been set down in the interior, and following the examples of Paris, Washington, Pretoria, they've been lavish with space, for it denotes confidence, and devices such as artificial lakes and the fluted column suggest, to the innocent eye, tranquillity and permanence.

These last-to-arrive capitals have followed the eternal laws too literally, too eagerly, and with their concentration of two-hundred-foot wide avenues connected to circles like molecular structures, their deployment of obelisks, shadowless forecourts and memorials to the fallen, they display not so much obedience to the idea of a capital as an obsession for cleanliness and clarity, as though showing themselves and the world they have mastered the difficult local environment; cities then in the abstract, detergent capitals. Recalled in the mind's eye these new capitals have an aerial perspective. And because of the surrounding emptiness there is nothing to stop them spreading.

Like Foster's—and in common with most descriptions of Canberra—Bail's perspective is that of the motorist.

For the first two weeks Shadbolt drove with a map spread on his knees, and at the end of each day lay crucified in his curtained room.

The centrifugal forces of Canberra had entered his metabolism. For a good hour his body leaned this way and that, as if he was back on Frank McBee's motor bike. Closing his eyes he saw the slippery steering wheel bringing into view low white buildings and flagpoles, spinning away in vast semicircles of concrete. Even the word Canberra was circular in its loops and vowels. …

The pattern of orbs and semicircles was repeated on a smaller scale in suburbs. Wherever Shadbolt drove he discovered no relief from the crescents, dog-legs, returning boomerangs, cul-de-sacs uncompleted—never had Shadbolt come across so many dead ends. At regular intervals he passed a concrete velodrome where concave aerodynamics sucked in butterflies, bus tickets and inter-departmental papers, where civil servants and their masters could be seen going flat out in circles practising their craft. The centrifugal forces of Canberra …

Keeping his eyes on the damp road and simultaneously glancing down at the map on his knees, in case he was swept off course by a vicious circle, Shadbolt made his way over curving lines named after the first explorers who stumbled around in circles, and these formed a pattern with other lines named after the nation's artists and architects, the flying doctor and rural poets, characters out of fiction, opera singers and dead generals, the forgotten politicians, judges, backyard scientists, the longsighted graziers and businessmen—names assigned by a select committee with others thrown in to cover environmental factors—Waratah, London, Aboriginal myths—so that Shadbolt traversed the nation's culture rendered in material form, a ten-minute journey with plenty of dead ends, reaching by 7.30 am the Minister's surprisingly ordinary bungalow in Lamington Street.[7]

The novelist and journalist Blanche D'Alpuget, who lived in Canberra for some years, turns an equally acerbic eye on the interior decoration and denizens of Canberra's Lakeside Hotel.

> A banner across the front of the room said 'Welcome to the Atlanta Lions' Club'. Lions in sports jackets were gathered in a pride on the foyer's red armchairs while others wandered to and fro across the blue-green carpet to the gift shop, buying toy koalas and boomerangs. All Canberra's hotels were decorated in a way that suggested the furnishings had been ordered by catalogue, unseen, and that the wrong colours had turned up.[8]

Canberra writer Sara Dowse, on the other hand, relies more on images of unrelenting suburban domesticity to convey the dissatisfactions of a Canberra existence. Dowse, who moved from the United States to Australia in 1958 and to Canberra in 1969, in 1974 joined the Prime Minister's Department as head of its newly created Women's Affairs Unit, an experience she draws on for her novel *West Block* (1983), a series of connected stories examining the lives of a group of public servants. Here she presents public servant George Harland, a success in his career, but afflicted by an inner emptiness.

> Gold shantung curtains, fringed holland blinds, moulded coral carpet, rice-papered walls. For almost thirty years, since the time he had reached the top of the housing list, George had taken meals in this room, across the table from Trudie. Then came the satellites: Marion, Amy, Stanley. Patterns of wallpaper went up and came down, and new curtains; while the sun faded the carpet dusky peach in broad streaks on the floor. Sometimes he reflected on what this was doing to his digestion. As soon as he got used to one scheme he had to adjust to another. There was more to it as well. Tonight he was struggling with a modification of satay beef, from a recipe book he estimated Trudie had acquired about seven years before—when Australians began warming to the Indonesians. At least it went with the beer, a home brew he made from a pre-war Canberra formula.
> All things considered, he sensed they were due for a change.
> 'More *sambal*.'
> 'The coconut or the brown bean? There's both.'
> '*Please*, pass the *sambal*, Stanley. At this table we say please.'
> 'Mum's got it over there.'
> 'Oh, never mind.'
> Through the window came the night smells—jasmine, lemon and roses—stirred in the wake of the southern wind. His garden; and Trudie's. He thought he detected the sear of his sweat among those fragile odours. And from somewhere further, the whiff of newly sawn timber, pine shavings, paint and cement. It set Harland thinking about the curious effects of human effort, the belief that one might create a paradise, so long as one kept the definition narrow. He decided to pay no attention to Stanley.

For George Harland, as for David Foster's European emigrants, Canberra is initially a place of refuge. But for Harland and his wife the refuge proves ultimately illusory.

> He saw his wife and son conversing, heard the sounds, saw the gestures that signified their ease of communion. The breeze that had rattled the plants, released their aromas, had died, quite suddenly. Only the smell of Trudie smoking lingered. Harland's hair felt damp at the roots. *And there shall arise after them seven years of famine.* Genesis 41:30. He was pleased with himself for remembering. In the west of Victoria, where as a boy he'd worked on summer vacations, year after year the hot winds from the centre scourged the stunted mallee, blackening earth and tree, setting fire to the sky. Down to the very

sea, the sky looked like hell itself had risen. And when the majestic horror had gone, what was left was hell of a different kind. Simply more: grass, scrub, dust.

And here it had been the same.

With effort and planning and the help of some mountain streams they had beaten it. Everyone planted, in a peculiarly catholic way: silver birches and liquidambars, poplars, willows and rowan-trees, mixed with all manner of eucalypts and acacia; the claret ash and grevillea, banksia, oleander, jonquils and hyacinth, till the colours of spring splashed over the streets, and autumn cast such brilliant shadows they could never forget how they came to be. So when, after seven years or so, he had begun to be plagued by doubt and discovered, with pain, that Trudie had been too, he put the question to her. Where did she really want to live? 'Canberra,' was what the lady said, and what he wanted to hear, and that, he had hoped, had been the end of that.

On the occasions when Dowse does turn to externals she, like Foster and Bail, describes the city in terms of landmarks and streetscapes, marking the progress Harland makes across the windy, attenuated city by a litany of impressive-sounding names: Queen Victoria Terrace, Commonwealth Avenue, the National Library, the Captain Cook Jet, Lake Burley Griffin, the Black Mountain Tower. Again, however, little actual contact is made with the physical landscape: Harland and his driver remain cocooned.

On the lake's embankment sat Australia's National Library. A Parthenon, the brochures were wont to say. A flock of gulls lifted off the grass, wheeling upwards in a cloud, fluttered and dispersed.

The car hugged the bridge like a carapace. The sky was clear now; very black; the water too. Were it not for the breeze, the sky and the lake might have formed a seamless whole, one hollow sphere pricked with pinpoints of light. But below the blackness quivered; and the sour smell of algae, churned up by the Captain Cook Jet, rode on the air, recalling, by its contrast, the fields of lucerne that in earlier times had grown on the river flats.

They were approaching Civic. Harland cast his mind back to great cities he had visited, always on government business, always in stops too brief to permit any pleasure. Rome, Paris, Bombay, New York. He drew his lips over his teeth in half a smile. How smug he had been, travelling through them, how proud of the clean streets and tidy simplicity of Australian lives. The solace of wanderers, barnacle cities grew clinging to the hulk of a continent and here, in the middle, the quintessential statement—a city rising from the Limestone Plains like Atlantis from the sea. Dreamlike. From any angle, Canberra looked as though you could wrap it in a blanket and carry it on your back to the next watering hole. The triumph, he chuckled to himself, of the nomad.

They crossed the traffic lights at the intersection of London Circuit. To the left and the right stood remnants of an American dream, the shops and the offices arcaded in twin rectangular Californian missions, painted grey. Down the centre of the avenue lay a small park with cedars and elms and wooden benches dappled by the shadows of overhanging trees. The streets were utterly empty. But the effect on Harland was astringent. This lack of animation sharpened his senses, challenged imagination. It was stark, the mingling of fantasy and truth, with all the dissonant magic of forgotten desires.

Des turned the car into Alinga Street and stopped in front of the Civic. 'Will I hang round, Mr Harland?' On the other side of the street neon lights winked in the window of the Blue Moon Cafe.

'Thanks, Des,' Harland said. 'I'll find my own way home.'[9]

Seven writers who have overtly identified themselves with Canberra in a regional sense through their joint short story collection *Canberra Tales* (1988) are Margaret Barbalet, Sara Dowse, Suzanne Edgar, Marion Eldridge, Marion Halligan, Dorothy Horsfield and Dorothy Johnston. These writers generally place their stories of middle class domesticity, domestic violence and broken marriages against a common ground of the architectural patterns of the city. Once again, there is a tendency to 'map' the terrain. In Marion Eldridge's 'Capital Gains', for example, a story exploring the tensions within a family unit, a father drives his children through the outlying pine plantations and timber mills to the capital city for the first time.

Unexpectedly, they catch their first glimpses of the city. On their left, imposing order on a straggly hillside, are newly-constructed crescents and circuits and avenues, then rows of new brown brick houses, and a woman in a crimson headscarf pegging out a line of napkins. On their right on a distant mountain top a hypodermic plunges into the taut skin of the sky. …

To their surprise the city disappears and they are in the country again. Ray Skerritt indicates a patch of bright green trees pushing into a shallow valley between two hills.

'*Pine* trees!' snorts Trish.

Ray Skerritt laughs. His spirits lift. It's the sort of thing Alvie would say. They approach a factory with smoking chimneys. 'Timber mill,' he tells them. 'Wind down the windows.' The fresh smell of sawdust pours into the car. Piles of logs lie neatly stacked. 'Just keeping in touch,' he laughs as they grumble about people who can't leave their work at home.

The countryside is flatter now, open and dry. In the distance they glimpse the line of roofs of a country town, and low blue hills, and again the needle of the Black Mountain Tower, dancing low on their left this time.

The road turns, and the tower stands straight in front of them, a beacon. Not far now to the city. They all sit up, smooth their clothes, touch their hair. Their mouths get ready to say the things people say coming into Canberra: What wide streets! Isn't it clean! Isn't it pretty! Everything laid on! Nobody walking! Such clean air! So clean and neat!

From the back of a truck a roadside hawker is selling bunches of daffodils bright as egg yolks. Lurlene says she's starving. 'We'll be seeing embassies soon,' her father tells her. 'Try counting flags, why don't you?'

At last they are on a wide street where plane trees are coming into leaf. They peer eagerly, and see stolid red brick houses with red tiles and white window frames looking out on to yellowed, frost-bitten grass. Not a flag in sight. Trucks rush past them: a load of pine logs; a truckful of sheep; a concrete mixer painted harsh pink. They are suddenly quiet. Is *this* Canberra, houses no bigger than the doctor's in Hazelwood, sheep like sale day, logs like any mill town? Ray Skerritt, looking ahead, sees a red flag fluttering above a tall bluish-green hedge; he opens his mouth then closes it, falling like the others into a broody silence.

And then, as they come uphill to traffic lights, filling their windscreen are the rising walls of a huge building, the immense area it covers scaling down its height so that confronting them it appears comprehensible, graspable, biddable even, its dozens of great windows narrowing to the slits of Trish's medieval tower. Over it bend seven white and red cranes, their cables no more than threads of cotton binding the ice-blue sky.

Lurlene, counting, is the first to say something. 'Seven ladies with bustles!'

'Pollies pigging at a trough.'

'What are you on about, Billy?' his father exclaims, straining forward against his seat belt. 'That's it, the new Parliament House, that's why the whole shebang's here—' He gestures widely. 'All this—Canberra!'

'—where our money goes,' Billy adds, confident as an echo.

'Well you don't expect to get a national symbol for nothing, do you!' his father retorts.[10]

In 'Division of Love' by Margaret Barbalet, in what once again reveals itself as a common trait, the social status of different suburbs is precisely calibrated by material possessions and styles of architecture.

'I wonder why they chose Chapman,' Mary said.

'Exclusive Chapman,' Gerald said sleepily, frowning into the long shafts of sun. He steered carefully out of the high cul-de-sac, where as their party wound down, another, for teenagers further up the hill was gearing up; older cars passed them extravagantly as they drove down. The gardens around them were without exception, perfect. Soft lawns dripped moisture from hidden watering systems into the immaculate gutters. All the gardens had their edges carefully landscaped with firmly banked native shrubs so that the inhabitants would only have to step into the Volvo on the pavers and back carefully into the street, would never have to garden at all. Every house would mirror far-off views.

'They get plainer as you go down.'

'As always.' Gerald laughed with the wisdom of an ex-socialist.

'The blocks get smaller too, Some of those blocks at the top are really huge, fan-shaped. I guess that's why they moved here; to give the children space.'

'Backing onto Mount Neighbour.'

'And looking down on them. How very Canberra.'

They turned into Namatjira Drive and proceeded through perfectly ordinary Fisher.

'It's a long way to visit,' Mary said, conscious of her new tiredness as they turned at last onto the driveway of their weatherboard cottage on the other side of the lake.

'Do you think we'll visit them much?'[11]

In 'A Season Under Snow', also by Margaret Barbalet, the motif of road systems recurs. Canberra is not a pedestrians' city, but a terrain of transport networks that convey characters from the interior cell of home or domesticity to the interior cell of work or bureaucracy. This metropolis has none of the normal external functions of a city, except in providing an environment for cars.

Going under the concrete overpass to Yarralumla the view changed for a moment to something superb. He never thought about the colour of the Brindabellas, only their outline against the sky. And here it all was, a wall he drove towards but never through. Driving with him nearly every day, there would always be two other cars: the Commonwealth car with its prestigious low plates, and uniformed elbow on the window ledge, and the diplomat's car with darkened windows, speeding from lane to lane, leaving him behind.

He liked the way the best landscape emerged from under the monolithic concrete. And by now the road was clear and fast, especially when compared to the three lanes of cars creeping towards Civic on Adelaide Avenue. In winter he liked the lights as they floated overhead against a pearled and herring-bone sky: each one that slipped behind to another bar of music, brought him closer to the edge of that ache he never had time to feel. And on the ground they brought him closer to work, to neutrality, to the country of the ordinary, away from that dangerous embassy, emotion.

The music would never be finished when he pulled into the carpark. But he was used to this. When he saw the distant white buildings of Woden looming up, like towers of a walled city, at that same moment the line of the mountains would vanish.

Then he would be surrounded by cars and the music, merely static. The buildings kept the warmth from his car. He would turn off the radio, grab his keys and walk to the particular tower which housed his department.[12]

As in Dowse's *West Block*, much of this 'Canberra' writing is interior in both the emotional and the literal sense. Men are often barely visible. They and their occupations provide an unseen background to the more vivid domesticity of the women, and these occupations are rarely clearly spelt out, even where the protagonist of the story is male. One such has a nameless job in a nameless department; another photographs, for another nameless department, the process of building Parliament House. Canberra thus becomes a Kafkaesque domain of small emotional pockets and enclosed interiors concealed within the relatively lifeless urban network outside—the carefully regulated sprawl of the city and its outlying dormitory suburbs. The stories inform each other, perhaps due to the writers' self-professed habit of working together, much as did Michael Wilding and Frank Moorhouse's early tales of urban Sydney.

The newly built Parliament House, completed in 1988, emerges as a common motif in a number of stories. In 'New Parliament House' by Dorothy Johnston, life is measured out by its inching to completion, as once it was in Sydney in the 1930s by the construction of the Harbour Bridge. 'New Parliament House' refers back to another story, 'Most Mortal Enemy' by Marion Halligan, in which the motif of the taking of periodic photographic exposures of the work in progress, like time-lapse photography, is reflected in the structure of Johnston's work.

> Phoebe often drove past the construction site on her way to somewhere else. The moment you climbed up somewhere with a bit of a view you could see it: but it was mainly from the bridges, and Phoebe spent a lot of time crossing them. Around the lake stood the city's monuments, this one the largest, tumbling out of the hill that had been bulldozed to make room for it, and all its crowded haste of cranes and trucks and half-erected flag-mast.
>
> From these bridges, and at a certain speed, the view was best, always interrupted, never static, and different now that she had been up close as well. She had stood in the building's mouth and looked back over the bridges towards the small hub of Civic and the suburb where she lived. Without being con-scious of it, in her many trips backwards and forwards she had got to know many aspects of the new building, and she could call on these memories now, provided she did not make a conscious effort to do so. They were like the overlapping transparencies she had learnt to use during her days as a student teacher, and which she'd always got more pleasure out of making than using. Somehow, on the screen, their colours looked washed out by comparison with the bright, definite lines she had produced on clear plastic at home.
>
> She remembered seeing the first walls go up, the day when they became distinct, as walls, against the sky and the remains of the hill. She remembered the time her eye caught the sun reflecting off tinted glass and a finished granite surface. Her memory, when she allowed it to surface and re-surface in this way, often surprised her by its accuracy. She knew that if she wrote down the dates she remembered driving past the building in its different stages of construction, and then compared them with the dated photographs in the Exhibition Centre, then her memory would turn out to have been correct.
>
> It was chance and the uneventful passage of days which had taken her, on shopping trips, on outings for the children, backwards and forwards past the site so often she could not count the times.[13]

The underside of urban Canberra, the sanitised poverty of subsidised housing developments for the unemployed is only touched upon in this collection. In Marion Halligan's 'Belladonna Gardens' the author makes ironic use of street names: Amaryllis Crescent, Hellebore Close, Paeony Place and Camellia Walk belie the actuality they signify: an isolated suburb where a downwardly mobile single mother makes her escape into a *trompe l'oeil* fresco rather than television.

> So here he was in his taxi, ticking expensively through dun-coloured paddocks, taking him further and further from the briefly-glimpsed city. Finally it turned off among houses with struggling new gardens, went round several corners, and stopped near some brick walls.
> - Belladonna Gardens, mate. Which bit was you wanting?
> - Hellebore Close. I've got a map.
> But when the taxi had turned into the labyrinth he couldn't make any sense of it. The streets curved and turned, there were brick walls and letter boxes, and a jumble of dwelling boxes that seemed to have stuck together higgledy-piggledy where they had been strewn across the slope. Not a sign of Amaryllis, or Paeony, or Cineraria or Aubretia or Delphinium. Just dry grass brittle in the wind, and rusty cars, and tricycles, and so many rubbish bins it was difficult to see why there should be so much littering the ground.[14]

As these examples illustrate, Canberra, Australia's only inland capital city, seems to affect strongly its inhabitants by its nature and origins. Few of its writers seem able to escape—or wish to escape—using it as a geographic metaphor.

The city also represents a microcosm of literary Australia in the wider sense, in that many of its writers also seem afraid to take themselves too seriously, both as Canberrans and as writers. There is a note of apology or defensiveness in many of the works that deal with it; a tendency to poke fun, not only at the city but at its cultural aspirations. But perhaps this can also be regarded as a feature of an endearing larrikinism, as Laurie Duggan, born in Melbourne in 1949, demonstrates as he deftly turns on his fellow poets, writers and academics in a poem entitled 'After Strange Gods'.

Oh to be in Canberra
Now that it's July
Where poets write of Iceland,
Turn central heating high

Draft mid-term Aus-Lit lectures
For the Australian National U.
The leaves once bright have fallen
They do too …[15]

AFTERWORD

Sometimes, under the influence of drink or too many oysters, I feel a sense of amiable unreality to Sydney. It is after all a most unlikely place. Born of a concentration camp thousands of miles from the next town, maturing so improbably on the underside of the world, more than most cities it seems artificial of purpose. It really need not exist at all. No ancient cross roads or trading routes make sense of its location, it is not the centre of anything, its indigenes did not demand it and its annihilation tomorrow would not deprive the world of anything essential to the well-being or heritage of mankind. It is in short, say I to myself as I order another half dozen, a self-indulgent place, created solely to be itself.

Jan Morris, *Sydney*, 1992[1]

With the publication of British travel writer and novelist Jan Morris's book *Sydney* in the early 1990s, a minor milestone in Australian literary history was achieved. 'Having commemorated in a series of books the rise and decline of the Victorian Empire … I wanted to conclude my imperial commitment with a book about something grand, famous and preferably glittering left on the shores of history by Empire's receding tide', wrote Morris, optimistically recording her reasons for embarking on an essay about Australia's first and largest city as the last decade of the twentieth century was drawing to a close. Morris's light-hearted work meant that Australia's first European settlement could now take its place fully-fledged on the world literary stage as a major city.

Not since the days of Watkin Tench had an overseas writer devoted a full-length non-fiction volume to an Australian metropolis. While numerous Australian writers and journalists had written cultural and historical guides to Sydney, and well-known visiting writers such as Darwin, Trollope, Conrad and Twain routinely included descriptions of various Australian cities in their travel accounts, now Sydney had come out, so to speak, and the event was recorded in the world's social notes.

It may be, however, that it is only in cities such as Australia's that such an event would be noticed. As it is only in new and post-colonial cities that national identity becomes a preoccupation, so it is only in new and post-colonial cities that we find ourselves asking, as Morris does: what are we here for? Morris herself, not entirely seriously, provided an answer: Australia's first city—and thus by implication all Australian post-First Fleet settlement—existed to be itself, and nothing more. The rest of Morris's book is devoted to regional definition, or, more precisely, a description of the characteristics which differentiate Sydney and its inhabitants from other cities and citydwellers: as regionalism suggests variety, so also does it seek rules by which to categorise the differences it discovers. As Morris endeavours to define Sydneysiders (and thus by implication Australians?) against their parent cultures and the world, so many Australian writers have begun to concern themselves with even more minute and internal variations. But as the last decade of the twentieth century draws to a close, perhaps for these Australians it is also just as useful to differentiate, without making too many moral judgments, between what we are and what came before.

From the beginning, European settlers in Australia, largely urban dwellers by origin, embarked on the construction of cities, a relatively recent phenomenon in terms of human history. According to scholars, the move to urbanisation began about 2500 or 3000 BC in countries such as India, China and Mesopotamia, some 35,000 years at least after the predecessors of the Australian Aborigines crossed from the Asian landmass. The form and size of these first small, scattered settlements, combining market places and cultural and administrative centres, remained relatively static until the industrial revolution began in Europe some four and a half thousand years later, freeing a large proportion of the population from agricultural pursuits. Thus it has only been in the last, brief three and a half centuries—a mere 150 years before European settlement began on the Australian continent—that the modern city, and the way of life we take for granted, began. Today there are cities like these in every continent of the world, and today 42 per cent of the world's population lives in them.

Though the first, and relatively late, European settlement in Australia quite quickly superseded its original purpose—a gaol—for various reasons it spawned other factors of social attraction, and, like grains of sand trickling to the bottom of an hourglass, a flow of people from mainly the Northern Hemisphere continued to stream to the southern continent and continued the creation of a new urban, technological and industrial civilisation. The majority of these people settled in the cities, and lived in a way that was in total opposition to that of the country's earlier inhabitants. Australia is one of only a few such countries where two such contrasting societies—a modern urban technological society, and its ancient predecessor, a non-agricultural subsistence existence—have been so starkly and recently juxtaposed.

Thus, one people was nomadic; the other sedentary. One scattered itself sparsely all over the inhabitable parts of the continent in strict clan and territorial groups; the other clustered for the most part in continuously expanding urban settlements near the coast. One lived by hunting and gathering; the other by transforming raw resources in agriculture and manufacture. One lived almost completely within the parameters of the existing natural landscape; the other lived almost completely in a built or modified environment. One celebrated the landscape as the natural order, and saw itself as a part of its cycles; the other glorified the domination of humankind over nature, setting out to change and modify the landscape from the moment of arrival.

Today, seven major metropoles and a number of secondary ones litter the map. The more fertile and productive areas are covered by the net of agriculture, pastoralism and mining, while the previous fragile social order—where existence was physically precarious and depended on rigid adherence to a social system evolved to fit a difficult environment—is irrevocably disrupted. To what result?

One thing that is obvious from the literary evidence that has accumulated in the past 200 years since the second major wave of settlers arrived, is that the ancient social dichotomy, literary and otherwise, of 'the city or the bush' is still largely in force. The Australian cities have developed their separate characters according to the geographical and social factors that formed them, and the writers who have documented their growth have moved through a diverse process of alienation and nationalism to a close sense of identity with their own territory, and a celebration of their regional virtues. However, there has also been demonstrated a distinct lack of equilibrium between the built environment and the natural landscape in the areas where the two intersected, with the area of intersection often depicted as a zone of destruction. Now, as in other parts of the world, the Australian cities are also beginning to show signs of exhausting their own infrastructures and the natural resources on which they rely for their continuation. Perhaps this literary unease can and should be read as a metaphor or symptom of a deeper malaise in Australian society: a continuing lack of harmony, a lack of maturity in the relationship between humankind and the natural world on which it depends.

Perhaps in the coming century, as well as continuing the celebration of what we have become, and in a society still young and small enough to be able to observe clearly the lessons to be learnt from the juxtaposition of the old and the new, we will modify our own directions accordingly, so that settlement and wilderness, mutually dependent as they now irrevocably are, might reach a yet more harmonious existence.

ENDNOTES

INTRODUCTION

1 Patrick White, *Voss*, Eyre & Spottiswoode, London 1957.

2 Robin Boyd, *The Great Australian Ugliness*, Penguin 1960.

3 Hal Porter, quoted in *Australia's Writers*, Graeme Kinross Smith, Nelson 1980.

4 Bruce Bennett, *Place, Region and Community*, Foundation for Australian Literary Studies, 1985.

5 Hal Porter, *The Watcher on the Cast-Iron Balcony*, Faber, London 1963.

6 *Shorter Oxford English Dictionary*, vol. 2, Clarendon Press, Oxford, 1978.

7 George Johnston and Robert B. Goodman, *The Australians*, Rigby 1966.

8 Bennett, *Place, Region and Community* (4)

9 Rachel Henning, *The Letters of Rachel Henning*, Bulletin 1951–2, republished Angus & Robertson 1963.

10 Emily Manning (Emily Matilda Heron), *The Balance on Pain and Other Poems*, George Bell, London 1877.

11 Henry Lawson, 'Water Them Geraniums', *Joe Wilson and His Mates*, Blackwood, London and Edinburgh 1901, reprinted in *The Penguin Henry Lawson Short Stories*, ed. John Barnes, Penguin 1986.

12 Henry Handel Richardson, *The Fortunes of Richard Mahony: Ultima Thule*, J. G. Robertson, Sydney 1929 (combined edition Heinemann, London 1930).

13 Patrick White, *The Tree of Man*, Eyre & Spottiswoode, London 1956.

14 Patrick White, *The Twyborn Affair*, Jonathan Cape, London 1979.

15 Elizabeth Harrower, *The Long Prospect*, Cassell & Company, London 1958.

16 Peter Corris, *White Meat*, Pan Picador 1981.

17 Ernestine Hill, *The Great Australian Loneliness*, 1937, first Australian edition Robertson & Mullens Ltd, Melbourne 1940.

18 Dora Birtles, *The Overlanders*, The Shakespeare Head, London 1946.

19 Randolph Stow, *The Merry-Go-Round in the Sea*, Macdonald & Co., London 1965.

20 Kenneth Cook, *Wake in Fright*, Michael Joseph, London 1961.

21 Angelo Loukakis, 'For the Patriarch', *For the Patriarch*, University of Queensland Press 1984.

22 Kenneth Slessor, 'Country', *One Hundred Poems*, Angus & Robertson 1944.

23 Les A. Murray, 'Driving through Sawmill Towns', *The Vernacular Republic: Poems 1961–1981*, Angus & Robertson 1982.

24 David Malouf, 'A Traveller's Tale', *Antipodes*, Chatto & Windus, London 1985.

25 Rodney Hall, *Just Relations*, Allen Lane/Penguin 1982.

SYDNEY

THE VIEW FROM THE BOAT:
Nineteenth-Century Sydney

1 Murray Bail, *Holden's Performance*, Penguin 1988.

2 Arthur Phillip, *The Voyage of Governor Phillip to Botany Bay*, John Stockdale, London 1879.

3 Anthony Trollope, *Australia and New Zealand*, London 1873, reprinted as *Australia*, ed. P. Edwards and R. Joyce, University of Queensland Press 1967.

4 Rosa Praed, 'The Old Scenes', *By Creek and Gully: Stories and Sketches … by Australian Writers in England*, ed. Lala Fisher, Unwin, London 1899.

5 Phillip, *The Voyage of Governor Phillip to Botany Bay* (2)

6 William Charles Wentworth, *Statistical, Historical, and Political Description of the Colony of New South Wales*, G. B. Whittaker, London 1819.

7 Alexander Harris, *Settlers and Convicts*, C. Cox, London 1847.

8 Charles Darwin, *Voyage of the Beagle*, London 1839, reprinted J. M. Dent & Sons, London 1979.

9 Frank Fowler, *Southern Lights and Shadows*, London 1859, facsimile edition Sydney University Press 1975.

10 Richard Rowe, 'A Trip up the Hunter', *Peter Possum's Portfolio*, J. R. Clarke, Sydney 1858.

11 M. Barnard Eldershaw, *A House is Built*, George Harrap, London 1929.

12 Charles Wentworth Dilke, *Greater Britain: Charles Dilke Visits Her New Lands*, Macmillan, London 1885.

13 Francis Adams, *The Australians: A Social Sketch*, T. Fisher Unwin, London 1893.

14 Beatrice and Sidney Webb, *The Webbs' Australian Diary*, ed. A. G. Austin, Isaac Pittman & Sons, Melbourne 1965.

15 D. H. Lawrence, *Letters*, ed. Richard Aldington, Penguin Books, London 1978.

16 D. H. Lawrence, *Kangaroo*, Martin Secker, London 1923, this (corrected) edition Collins 1989.

17 Joseph Conrad, *The Mirror of the Sea: Memories and Impressions*, London 1906, reprinted J. M. Dent & Sons, London 1968.

THE VIEW FROM THE STREET:
1900 to World War I

1 Louis Stone, *Jonah*, Methuen, London 1911.

2 Charmian Clift, 'The Private Pleasures of a Public Market', *The World of Charmian Clift*, Collins 1970.

3 Stone, *Jonah* (1)

4 Lennie Lower, *Here's Luck*, Angus & Robertson 1930.

5 Jennifer Paynter, 'The Sad Heart of Ruth', 1988, reprinted in *Goodbye to Romance: Stories by Australian and New Zealand Women Writers*, ed. Elizabeth Webby and Lydia Wevers, Angus & Robertson 1989.

6 Christina Stead, *Seven Poor Men of Sydney*, Peter Davies, London 1934.

7 Kylie Tennant, *Ride on Stranger*, Gollancz, London 1943.

8 Marjorie Barnard, 'The Dry Spell', *The Persimmon Tree*, Clarendon Publishing, London 1943.

9 Ruth Park, *Poor Man's Orange*, Angus & Robertson 1949.

10 Elizabeth Harrower, *Down in the City*, Cassell & Company, London 1957.

11 Dymphna Cusack and Florence James, *Come In Spinner*, Angus & Robertson 1988 (abridged edition William Heinemann 1951).

12 George Johnston, *Clean Straw for Nothing*, Collins, 1969.

13 David Malouf, *The Great World*, Chatto & Windus, London 1990.

ESCAPE FROM SUBURBIA: The 1950s

1 Shirley Hazzard, *The Transit of Venus*, Viking Press, U. S. A. 1980.
2 Shirley Hazzard, *The Transit of Venus* (1)
3 Sumner Locke Elliot, *Water Under the Bridge*, Simon & Schuster, U. S. A. 1977./
4 Murray Bail, *Holden's Performance*, Penguin 1988.
5 George Johnston, *A Cartload of Clay*, Collins 1971.
6 Charmian Clift, 'Goodbye to a Skyline', *The World of Charmian Clift*, Collins 1970.
7 Hal Porter, *The Extra*, Thomas Nelson 1975.
8 Elliot, *Water Under the Bridge* (3)
9 C. J. Koch, *The Doubleman*, Chatto & Windus, London 1985.

PATRICK WHITE'S SYDNEY

1 Patrick White, *Flaws in the Glass*, Jonathan Cape, London 1981.
2 White, *Flaws in the Glass* (1)
3 Patrick White, *The Solid Mandala*, Eyre & Spottiswoode, London 1966.
4 Patrick White, *Riders in the Chariot*, Eyre & Spottiswoode, London 1961.
5 White, *Flaws in the Glass* (1)
6 Patrick White, *The Aunt's Story*, Eyre & Spottiswoode, London 1948.
7 Patrick White, *The Vivisector*, Jonathan Cape, London 1970.
8 White, *Flaws in the Glass* (1)
9 Patrick White, *The Eye of the Storm*, Jonathan Cape, London 1973.
10 Patrick White, 'The Night The Prowler', *The Cockatoos*, Jonathan Cape, London 1974.
11 David Marr, *Patrick White: A Life*, Jonathan Cape, London 1991.
12 White, *Flaws in the Glass* (1)

SUNLIGHT ON WATER: Modern Sydney

1 Peter Corris, *O'Fear*, Bantam Books 1990.
2 Kate Grenville, *Lilian's Story*, Allen & Unwin 1985.
3 Jessica Anderson, *The Impersonators*, Macmillan 1980.
4 Jill Neville, *Last Ferry to Manly*, Penguin 1984.
5 Robert Drewe, *The Savage Crows*, William Collins 1976.
6 Peter Corris, *White Meat*, Pan Books 1981.
7 Ruby Langford, *Don't Take Your Love to Town*, Penguin 1988.
8 Angelo Loukakis, 'For the Patriarch', *For the Patriarch*, University of Queensland Press 1981.
9 Anna Couani and Peter Lyssiotis, 'Parramatta Sestina', *The Harbour Breathes*, Sea Cruise Masterthief, 1989.
10 Judith Beveridge, 'Streets of Chippendale', *The Penguin Book of Australian Women Poets*, ed. Susan Hampton and Kate Llewellyn, Penguin 1986.
11 Obelia Modjeska, 'Observations from Rozelle', *Inner Cities: Australian Women's Memories of Place*, ed. Drusilla Modjeska, Penguin 1989.
12 George Papaellinas, 'Round the Crate', *Ikons*, Penguin 1986.
13 David Ireland, *The Unknown Industrial Prisoner*, Angus & Robertson 1971.
14 Gabrielle Lord, *Jumbo*, The Bodley Head, U. K. 1986.
15 Gabrielle Lord, *Salt*, Penguin McPhee Gribble 1990.

HOBART

A NATURAL PENITENTIARY: 'Bringing Hell and Heaven Together'

1 Allen Afterman, 'Van Diemen's Land', *Purple Adam*, Angus & Robertson 1980.
2 David Collins, *An Account of the English Colony in New South Wales*, London 1802, reprinted Reed 1976.
3 Charles Rowcroft, *Tales of the Colonies*, Saunders & Otley, London 1843.
4 Henry Savery, *The Hermit in Van Diemen's Land*, Andrew Bent, Hobart 1830, quoted in *Colonial Voices*, ed. Elizabeth Webby, University of Queensland Press 1989.
5 Marcus Clarke, *For the Term of His Natural Life*, published in book form 1874, revised edition 1882, this edition Lloyd O'Neil 1981.
6 John West, *History of Tasmania*, Henry Dowling, Launceston 1852, Angus & Robertson 1971.
7 Vivian Smith, 'Tasmania', *Vivian Smith: Selected Poems*, Angus & Robertson 1985.
8 Hal Porter, *The Paper Chase*, Angus & Robertson 1966.
9 Patrick White, *A Fringe of Leaves*, Jonathan Cape, London 1976.
10 Hal Porter, 'Hobart Town', *Elijah's Ravens: Poems of Hal Porter*, Angus & Robertson 1968.
11 Porter, *The Paper Chase* (8)
12 Gwen Harwood, 'Oyster Cove', *Selected Poems*, Angus & Robertson 1975.
13 C. J. Koch, *The Doubleman*, Chatto & Windus, London 1985.
14 Vivian Smith, 'Growing Up in Hobart', *First Rights: A Decade of Island Magazine*, ed. Michael Denholm and Andrew Sant, Greenhouse Publications 1989.
15 Robert Drewe, *The Savage Crows*, William Collins 1976.

THE LOST HEMISPHERE

1 Peter Conrad, *Down Home: Revisiting Tasmania*, Chatto & Windus, London 1988.
2 C. J. Koch, *The Doubleman*, Chatto & Windus, London 1985.
3 Conrad, *Down Home: Revisiting Tasmania* (1)
4 C. J. Koch, *The Boys in the Island*, Hamish Hamilton 1958, revised edition Angus & Robertson 1987.
5 Conrad, *Down Home: Revisiting Tasmania* (1)
6 C. J. Koch, 'The Lost Hemisphere', *Crossing the Gap*, Chatto & Windus 1987.

SUBURBIA VERSUS THE WILDERNESS

1 C. J. Koch, *The Boys in the Island*, Hamish Hamilton 1958, revised edition Angus & Robertson 1987.
2 C. J. Koch, *The Doubleman*, Chatto & Windus, London 1985.
3 Vivian Smith, 'Growing Up in Hobart', *First Rights: A Decade of Island Magazine*, ed. Michael Denholm and Andrew Sant, Greenhouse Publications 1989.
4 Peter Conrad, *Down Home: Revisiting Tasmania*, Chatto & Windus, London 1988.
5 Helen Hodgman, *Blue Skies*, Duckworth & Co., London 1976.
6 Helen Hodgman, conversation with the author, 1991.
7 Hodgman, *Blue Skies* (5)

8 James McQueen, *Hook's Mountain*, Macmillan 1982.

9 Carmel Bird, 'Getting My Grandmother's Sewing Machine Across Bass Strait', *Grand Street*, vol. 9, no. 1, autumn, New York 1980.

10 Carmel Bird, letter to author, 1991.

11 Carmel Bird, 'Kay Petman's Coloured Pencils', *The Woodpecker Toy Fact and Other Stories*, McPhee Gribble Penguin 1987.

12 Bird, 'The Woodpecker Toy Fact', *The Woodpecker Toy Fact and Other Stories* (10)

13 Bird, 'A Taste of Earth', *The Woodpecker Toy Fact and Other Stories* (10)

14 Carmel Bird, *The Bluebird Cafe*, McPhee Gribble 1990.

BRISBANE

'RAMSHACKLE HOUSES BUILT ON STILTS, AND AN AMIABLE, EXTROVERT RACE ...'

1 Tony Maniaty, *Smyrna*, Penguin 1990.

2 David Malouf, 'A First Place: The Mapping of a World', The Fourteenth Herbert Blaiklock Memorial Lecture, 26 September 1984.

3 Thea Astley, 'Being a Queenslander: A Form of Literary and Geographical Conceit', The Sixth Herbert BLaiklock Memorial Lecture, 1978, Wentworth Press 1978.

4 Malouf, 'A First Place ...' (2)

5 Jessica Anderson, *The Commandant*, Macmillan, London 1975.

6 David Marr, *Patrick White: A Life*, Jonathan Cape, London 1991.

7 Patrick White, *A Fringe of Leaves*, Jonathan Cape, London 1976.

8 Brian Penton, *Landtakers*, Angus & Robertson 1934.

9 Edmund Marin La Meslée, *L'Australie Nouvelle*, 1883, translated as *The New Australian* by Russel Ward, Heinemann, London 1973.

10 Rosa Praed, *Lady Bridget in the Never Never Land*, Hutchinson, London 1915, reprinted Pandora Press, London 1987.

11 David Malouf, *Johnno*, University of Queensland Press, 1975.

THE CITY IN THE HOUSE

1 David Malouf, 'A First Place: The Mapping of a World', The Fourteenth Herbert Blaiklock Memorial Lecture, 26 September 1984.

2 David Malouf, *Johnno*, University of Queensland Press, 1975.

3 David Malouf, *12 Edmondstone Street*, Chatoo & Windus, London 1985.

4 Jessica Anderson, conversation with the author, 1992.

5 Jessica Anderson, *Tirra Lirra by the River*, Macmillan 1978.

6 Jessica Anderson, *Stories from the Warm Zone*, Penguin 1987.

7 Malouf, *Johnno* (2)

8 David Malouf, *Harland's Half Acre*, Chatto & Windus 1984.

9 Tony Maniaty, *Smyrna*, Penguin 1990.

10 Malouf, *Johnno* (2)

11 Peter Corris, *Make Me Rich*, Allen & Unwin 1985.

'A PATCH OF COASTLINE': The Queensland Towns

1 Thea Astley, 'North: Some Compass Readings: Eden', *Hunting the Wild Pineapple*, Thomas Nelson Australia 1979.

2 Edmund Marin La Meslée, *L'Australie Nouvelle*, 1883, translated as *The New Australian* by Russel Ward, Heinemann, London 1973.

3 Thea Astley, *It's Raining in Mango*, G. P. Putnam's Sons, New York 1987.

4 Robert Drewe, *The Savage Crows*, William Collins 1976.

5 Thomas Wood, *Cobbers*, Oxford University Press, London 1934.

6 Thomas Wood, *True Thomas*, Jonathan Cape 1936.

7 Thea Astley, 'Being a Queenslander: A Form of Literary and Geographical Conceit', The Sixth Herbert Blaiklock Memorial Lecture, 1978, Wentworth Press 1978.

8 David Malouf, *Harland's Half Acre*, Chatto & Windus 1984.

9 Astley, *It's Raining in Mango* (3)

10 Astley, 'North: Some Compass Readings: Eden' (1)

11 Thea Astley, *An Item from the Late News*, University of Queensland Press 1982.

12 Vance Palmer, *The Passage*, Stanley Paul, London 1928, Robertson & Mullens, Melbourne 1944.

13 Charmian Clift, 'The Gulf', *The World of Charmian Clift*, Collins 1970.

PERTH

FROM EARLY SETTLEMENT TO THE ARCADIAN AGE

1 Jack Davis, *Kullark* and *The Dreamers*, Currency Press 1982.

2 E. L. Grant Watson, 'Caves Near Busselton', *Departures*, Pleiades Books, London 1948, republished as *Descent of Spirit: Writings of E. L. Grant Watson*, ed. Dorothy Green, Primavera Press 1990.

3 E. L. Grant Watson, *Descent of Spirit*, ed. Dorothy Green, Primavera Press 1990.

4 Quoted in *The Oxford Literary Guide to Australia*, ed. Peter Pierce, Oxford University Press 1987.

5 D. H. Lawrence, *Letters*, ed. Richard Aldington, Penguin Books, Harmondsworth 1978.

6 D. H. Lawrence, *Kangaroo*, Martin Secker, London 1923, this (corrected) edition Collins 1989.

7 T. A. G. Hungerford, *Stories from Suburban Road*, Fremantle Arts Centre Press 1983.

8 Kenneth Seaforth Mackenzie, *The Young Desire It*, U. K. 1937, angus & Robertson 1963.

9 Randolph Stow, *The Merry-Go-Round in the Sea*, Macdonald & Co., London 1965.

10 Randolph Stow, *A Counterfeit Silence: Selected Poems*, Angus & Robertson 1969.

11 Hungerford, *Stories from Suburban Road* (7)

12 Seaforth Mackenzie, *The Young Desire It* (8)

13 Stow, *The Merry-Go-Round in the Sea* (9)

A RISING TIDE OF HEAT: The Western Australian Landscape

1 Peter Cowan, 'Regionalism in Contemporary Australia', *Westerly*, vol. 23, no. 4, December 1978.

2 Peter Cowan, 'Collector', *Mobiles*, Fremantle Arts Centre Press 1979.

3 Randolph Stow, *To The Islands*, Macdonald & Co. London 1958.

4 Randolph Stow, *Tourmaline*, Macdonald & Co., London 1962.

5 E. L. Grant Watson, *Descent of Spirit*, ed. Dorothy Green, Primavera Press 1990.

6 Stow, *To the Islands* (3)

7 Randolph Stow, *The Merry-Go-Round in the Sea*, Macdonald & Co., London 1965.

8 Xavier Herbert, *Disturbing Element*, Cheshire, Melbourne 1962.

9 Nene Gare, *The Fringe Dwellers*, Heinemann, London 1961.

10 Stow, *The Merry-Go-Round in the Sea* (7)

11 Elizabeth Jolley, *Foxybaby*, University of Queensland Press 1985.

12 Tim Winton, *Shallows*, Allen & Unwin 1985.

13 Robert Drewe, *Fortune*, Pan Books 1986.

THE FRAGRANCE OF DUST: Modern Perth

1 Elizabeth Jolley, 'A Sort of Gift: Images of Perth', *Bulletin* 26 January 1988, republished in *Inner Cities: Australian Women's Memories of Place*, ed. Drusilla Modjeska, Penguin 1989.

2 Peter Cowan, 'Seminar', *Mobiles*, Fremantle Arts Centre Press 1979.

3 Walter Murdoch, quoted in John La Nauze, *Walter Murdoch: A Biographical Memoir*, Melbourne University Press 1977.

4 Bruce Bennett, *Place, Region and Community*, Foundation for Australian Literary Studies, 1985.

5 William Grono, 'Postcard from Perth', *On the Edge* (with Nicholas Hasluck), Freshwater Bay Press 1980.

6 Peter Cowan, 'The Island', *The Empty Street*, Angus & Robertson 1965.

7 Jolley, 'A Sort of Gift' (1)

8 Hal Colebatch, 'Sestina on Taking a Bus into Perth past the Narrows Bridge', *Spectators on the Shore*, Edgars & Shaw 1975.

9 Robert Drewe, *The Savage Crows*, William Collins 1976.

10 Jack Davis, 'Whither', *The First Born and Other Poems*, Angus & Robertson 1970.

11 Veronica Brady, 'A Postmodernist City', *Island Magazine* 1987, reprinted in *First Rights: A Decade of Island Magazine*, ed. Michael Denholm and Andrew Sant, Greenhouse Publications 1989.

MELBOURNE

'A FLAT PLACE, DIVIDED INTO A GRID': Nineteenth-Century Melbourne

1 Peter Carey, *Illywhacker*, University of Queensland Press 1985.

2 Kenneth Slessor, *Earth Visitors*, Fanfrolico Press, London 1926, quoted by Geoffrey Dutton, *Kenneth Slessor*, Viking Penguin 1991.

3 Peter Corris, *Deal Me Out*, Allen & Unwin 1986.

4 Chris Wallace-Crabbe, 'Melbourne in 1963', *Melbourne or the Bush*, Angus & Robertson 1974.

5 Chris Wallace-Crabbe, 'Melbourne', *In Light and Darkness*, Angus & Robertson 1963.

6 David Williamson, *Bulletin*, 11 November 1980.

7 Laurie Clancy, *Australian Book Review*, March 1981.

8 John Arnold, *Melbourne in the Minds of its Writers*, Allen & Unwin 1983.

9 Henry Kingsley, *The Recollections of Geoffry Hamlyn*, Macmillan, Cabridge 1859.

10 Rolf Boldrewood, *Old Melbourne Memories*, George Robretson & Co., Ltd 1884.

11 George Henry Haydon, *The Australian Emigrant*, Arthur Hale, Virtue & Co., 1854.

12 Catherine Helen Spence, *Clara Morison: A Tale of South Australia During the Gold Fever*, John W. Parker & Son, London 1854, this edition Wakefield Press, 1986.

13 Henry Handel Richardson, *The Fortunes of Richard Mahony: Australia Felix*, J. G. Robertson, sydney 1929, (combined edition Heinemann, London 1930.

14 Kingley, *The Recollections of Geoffry Hamlyn* (9)

15 Patrick Moloney, *Sonnets and Innuptum*, 1879, quoted in *The Oxford Literary Guide to Australia*, Oxford University PRess, Melbourne 1987.

16 Tasma (Jessie Couvreur), *Uncle Piper of Piper's Hill*, Trubner 1889, this edition Pandora Press, London 1987.

17 Martin Boyd, *The Cardboard Crown*, The Cresset Press, London 1952.

18 Ada Cambridge, *Thirty Years in Australia*, Angus & Robertson 1921, this edition University of New South Wales Press 1989.

19 Francis Adams, 'Melbourne and her Civilization, as they Strike an Englishman', *Australian Essays*, William Inglis & Co. 1886, quoted in *Colonial Voices*, ed. Elizabeth Webby, University of New South Wales Press 1989.

20 Martin Boyd, *The Montforts*, Constable, London 1928; reprinted Penguin 1986.

21 Martin Boyd, *Lucinda Brayford*, The Cresset Press 1946.

22 Fergus Hume, *The Mystery of a Hansom Cab*, Kemp & Boyce, Melbourne 1886.

THE NATIVE SONS

1 Furnley Maurice (Frank Wilmot), 'The Victoria Markets Recollected in Tranquillity', *Melbourne Odes*, Lothian Publishing Co., 1934.

2 David Martin, *The Young Wife*, Macmillan, London 1962.

3 Frank Hardy, *Power Without Glory*, privately published 1950, reprinted T. Werner Laurie, U. K. 1962.

4 Alan Marshall, *In Mine Own Heart*, Cheshire, London 1963.

5 John Morrison, 'The Haunting of Hungry Jimmy', *North Wind*, 1982, reprinted in *The Best Stories of John Morrison*, Penguin 1988.

6 John Morrison, 'Transit Passenger', *North Wind*, 1982 (5)

7 John Morrison, 'Nightshift', *North Wind*, 1982 (5)

8 George Johnston, *My Brother Jack*, Collins, London 1964.

'THE FINE EDGE OF SEEING': Hal Porter's Melbourne

1 Hal Porter, *The Watcher on the Cast-Iron Balcony*, Faber & Faber, London 1963.

2 Porter, *The Watcher on the Cast-Iron Balcony* (1)

3 Hal Porter, *The Paper Chase*, Angus & Robertson 1966.

THE CITY IN THE HOUSE

1 George Johnston, *My Brother Jack*, Collins, London 1964.

2 Johnston, *My Brother Jack* (1)

A NOTE OF REGRET: Modern Melbourne

1 George Johnston, *Clean Straw for Nothing*, Collins, London 1969.

2 Johnston, *Clean Straw for Nothing* (1)

3 C. J. Koch, *The Boys in the Island*, Hamish Hamilton 1958, revised edition Angus & Robertson 1987.

4 C. J. Koch, *The Doubleman*, Chatto & Windus, London 1985.

5 Peter Conrad, *Down Home: Revisiting Tasmania*, Chatto & Windus, 1988.

6 Harry Marks, *The Heart is Where the Hurt Is*, Victor Gollancz, London 1966.

7 Laurie Clancy, *Australian Book Review*, March 1981.

8 Rosa Safransky, 'Postcards', *Inprint* vol. 7 no. 4 1983, reprinted in *Coast to Coast*, ed. Kerryn Goldsworthy, Angus & Robertson 1986.

9 Rosa Safransky, 'Bonjour Brunswick', *Australian Short Stories II*, ed. Bruce Pascoe, Pascoe Publishing 1985.

10 Ania Walwicz, 'Australia', *Island in the Sun*, Sea Cruise Books 1981, reprinted in *The Penguin Book of Australian Women Poets*, ed. Susan Hampton and Kate Llewellyn, Penguin 1986.

11 Peter Mathers, *Trap*, Cassell, London 1966.

12 Gerald Murnane, 'Landscape with Freckled Woman', *Landscape with Landscape*, Norstrilia Press 1985, reprinted Penguin 1987.

13 Peter Carey, 'Crabs', *The Fat Man in History*, University of queensland Press 1974.

14 Helen Garner, *Monkey Grip*, McPhee Gribble 1977.

15 Helen Garner, *Honour and Other People's Children*, McPhee Gribble 1980.

16 Alister Kershaw, *Hey Days*, Collins Angus & Robertson 1991.

17 George Johnston, *A Cartload of Clay*, Collins 1971.

18 Bruce Dawe, 'The Affair, for Melbourne', *Sometimes Gladness: Collected Poems 1945–1982*, Longman Cheshire 1983.

ADELAIDE

A SLIGHTLY SHADY UTOPIA

1 Barbara Hanrahan, 'Earthworm Small', *Inner Cities: Australian Women's Memories of Place*, ed. Drusilla Modjeska, Penuin 1989.

2 Charles Sturt, *An Account of the Sea Coast and Interior of South Australia*, T. W. Boone, London 1849.

3 Catherine Helen Spence, *Clara Morison: A Tale of South Australia during the Gold Fever*, John W. Parker & Son, London 1854, this edition Wakefield Press 1986.

4 Stella Bowen, *Drawn From Life* 1941, this edition Virago, London 1984.

5 Hal Porter, *The Paper Chase*, Angus & Robertson 1966.

6 Cynthia Nolan, *Outback*, Methuen & Co. Ltd, London 1962.

'THE ICEMAN BECOMES IMPORTANT'

1 Barbara Hanrahan, *The Scent of Eucalyptus*, Chatto & Windus, London 1973.

2 Hanrahan, *The Scent of Eucalyptus* (1)

3 Barbara Hanrahan, *Kewpie Doll*, Chatto & Windus, London 1984.

4 Barbara Hanrahan, *Where the Queens All Strayed*, University of Queensland Press 1978.

5 Murray Bail, *Holden's Performance*, Penguin 1988.

6 Nicholas Jose, *Rowena's Field*, Rigby 1984.

7 David Parker, *Building on Sand*, Angus & Robertson 1988.

8 Andrew Taylor, 'Adelaide', *The Orange Tree: South Australian Poetry to the Present Day*, ed. Pearson and Churches, Wakefield Press 1986.

9 Larry Buttrose, 'The City on the Verge', *The Orange Tree: South Australian Poetry to the Present Day*, ed. Pearson and Churches, Wakefield Press 1986.

'A DREAM COUNTRY LESS TANGIBLE': The South Australian Landscape

1 Hal Porter, *The Paper Chase*, Angus & Robertson 1966.

2 Barbara Hanrahan, *Where the Queens All Strayed*, University of Queensland Press, 1978.

3 Barbara Hanrahan, *The Scent of Eucalyptus*, Chatto & Windus, London 1973.

4 Murray Bail, *Holden's Performance*, Penguin 1988.

5 Douglas Stewart, 'That Fierce Country', *The Birdsville Track*, *Bulletin* 22 October 1952, Angus & Robertson 1955.

6 Douglas Stewart, 'Marree', *The Birdsville Track* (5)

7 Colin Thiele, *The Fire in the Stone*, Rigby 1973.

8 Colin Thiele, *Storm Boy*, Rigby 1963.

9 Nicholas Jose, *Feathers or Lead*, Penguin 1986.

10 C. J. Koch, 'The Novel as Narrative Poem', *Crossing the Gap*, Chatto & Windus 1987.

DARWIN

'THE PORT OF YELLOW SKELETONS'

1 A. B. Paterson, 'The Cyclone, Paddy Cahill and the G. R.' 1898, reprinted in *The Collected Works of A. B. Paterson*, Angus & Robertson 1921.

2 Harden S. Melville, *The Adventure of a Griffin on a Voyage of Discovery*, Bell & Daldy, London 1867.

3 Charles Barnett, *Coast of Adventure*, 1941, quoted in *The Oxford Literary Guide to Australia*, Oxford University Press, Melbourne, 1987.

4 Charles Wentworth Dilke, *Greater Britain: Charles Dilke Visits Her New Lands*, Macmillan, London 1885.

5 Hugh C. May, *Diary of Hugh C. May, 1872–3*, quoted in *The Oxford Literary Guide to Australia*, Oxford University Press, Melbourne 1987.

6 Harriet Douglas, *Digging, Squatting and Pioneering Life in the Northern Territory of South Australia*, Sampson Low, Marston, Searle & Rivington, London 1877.

7 Paterson, 'The Cyclone, Paddy Cahill and the G. R.' (1)

8 Elspeth Huxley, *Their Shining Eldorado*, Chatto & Windus, London 1967.

9 Ernestine Hill, *The Territory*, Angus & Robertson 1951.

10 Xavier Herbert, *Capricornia*, Angus & Robertson 1938.

'THE STEAMY HOTHOUSE': Modern Darwin

1 Peter Goldsworthy, *Maestro*, Angus & Robertson 1989.

2 Goldsworthy, *Maestro* (1)

3 Les A. Murray, 'Louvres', *The Daylight Moon*, Angus & Robertson 1987.

4 Fay Zwicky, 'Jacques Tati at the Hotel Darwin', *Northern Perspective*, vol. 9, no. 1, 1986.

5 Lindy Chamberlain, *Through My Eyes*, Heinemann 1990.

THE TOP END

1 W. E. Harney, 'West of Alice', *Content to Lie in the Sun*, Hale, London 1958.

2 Cynthia Nolan, *Outback*, Methuen & Co. Ltd, London 1962.

3 Charmian Clift, 'The Centre', *The World of Charmian Clift*, Collins 1970.

4 Peter Goldsworthy, *Maestro*, Angus & Robertson 1989.

5 Clift, 'The Centre' (3)

6 Robyn Davidson, *Tracks*, Jonathan Cape 1980.

CANBERRA

'A GROUP OF SUBURBAN ENCLAVES SEPARATED BY CIVIL SERVICE STATUS'

1 T. A. G. Hungerford, *A Knockabout with a Slouch Hat*, Fremantle ARts Centre Press 1985.

2 Hungerford, *A Knockabout with a Slouch Hat* (1)

3 Magda Bozic, *Gather Your Dreams*, Hodja 1984.

4 John Russell Rowland, 'Canberra in April', *Canberra in April*, Angus & Robertson 1976.

5 Judith Wright, 'City and Mirage', 'Nobody Looks Up', 'Oaks, etc.', 'Going Outside' and 'Ecological Comment', from *Fourth Quarter and Other Poems*, Angus & Robertson 1976.

6 David Foster, *Plumbum*, Penguin 1983.

7 Murray Bail, *Holden's Performance*, Penguin 1988.

8 Blanche D'Alpuget, *Turtle Beach*, Penguin 1983.

9 Sara Dowse, *West Block*, Penguin 1983.

10 Marian Eldridge, 'Capital Gains', *Canberra Tales*, Penguin 1988.

11 Margaret Barbalet, 'Division of Love', *Canberra Tales* (10)

12 Barbalet, 'A Season Under Snow', *Canberra Tales* (10)

13 Dorothy Johnston, 'The New Parliament House', *Canberra Tales* (10)

14 Marion Halligan, 'Belladonna Gardens', *Canberra Tales* (10)

15 Laurie Duggan, 'After Strange Gods', *Adventures in Paradise*, Little Esther/Experimental Art Foundation, Adelaide 1982, second edition 1991.

AFTERWORD

1 Jan Morris, *Sydney*, Penguin Group 1992.

Acknowledgements

Permission to quote from the titles mentioned is gratefully acknowledged to the following authors, publishers, agents and copyright holders:

'Van Diemen's Land' from *Purple Adam* © Allen Afterman 1970, to Collins Angus & Robertson.

The Commandant by Jessica Anderson, to the author and to Penguin Books Australia.

The Impersonators, and 'Under the House' from *Stories from the Warm Zone* by Jessica Anderson, to Penguin Books Australia.

It's Raining in Mango, and 'North: Some Compass Readings: Eden' from *Hunting the Wild Pineapple* by Thea Astley, to Penguin Books Australia.

'Being a Queenslander: A Form of Literary and Geographical Conceit' by Thea Astley, to the author.

An Item from the Late News by Thea Astley, to University of Queensland Press.

Holden's Performance by Murray Bail, to Penguin Books Australia.

'A Division of Love' and 'A Season Under Snow' by Margaret Barbalet, from *Canberra Tales*, to Penguin Books Australia.

A House is Built by M. Barnard Eldershaw, to the copyright holder Alan Alford, c/o Curtis Brown (Aust.) Pty Ltd.

The Persimmon Tree by Marjorie Barnard, to Curtis Brown (Aust.) Pty Ltd.

'Streets of Chippendale' by Judith Beveridge, to the author and to Black Lightning Press.

'Getting My Mother's Sewing Machine Across Bass Strait', by Carmel Bird, to the author.

The Bluebird Cafe by Carmel Bird, to the author.

'Kay Petman's Coloured Pencils', 'A Taste of Earth', 'The Woodpecker Toy Fact from *The Woodpecker Toy Fact* by Carmel Bird, to Penguin Books Australia.

Drawn from Life by Stella Bowen, to David Higham and Associates, U. K.

The Cardboard Crown and *Lucinda Brayford*, by Martin Boyd, to Random Century Group, U. K.

The Montforts by Martin Boyd, to Phyllis Boyd, c/o Curtis Brown (Aust.) Pty Ltd.

'A Postmodernist City' by Veronica Brady, to the author and to Island Magazine Inc.

'The City on the Verge' by Larry Buttrose, to the author.

Illywhacker, and 'Crabs' from *The Fat Man in History* by Peter Carey, to University of Queensland Press.

Through My Eyes, by Lindy Chamberlain, to the author and to Harry M. Miller and Co. Management.

'The Private Pleasures of a Public Market', 'Goodbye to a Skyline' and 'The Centre' from *The World of Charmian Clift*, © The Estate of Charmian Clift, to Collins Angus & Robertson.

'Sestina on Taking a Bus into Perth past the Narrows Bridge' by Hal Colebatch, to the author.

Down Home: Revisiting Tasmania by Peter Conrad, to Random Century U. K.

Wake in Fright by Kenneth Cook, to Michael Joseph Ltd.

White Meat by Peter Corris, to the author, to Rose Creswell and to Pan Macmillan.

Deal Me Out and *Make Me Rich* by Peter Corris, to Allen & Unwin Australia.

O'Fear by Peter Corris, to Transworld Publishers.

'Parramatta Sestina' by Anna Couani from *The Harbour Breathes* by Couani and Lyssiotis, to the author.

'Regionalism in Contemporary Australia' by Peter Cowan, to the author.

'Collector' and 'Seminar' by Peter Cowan, to the author and to Fremantle Arts Centre Press.

'Melbourne' from *In Light and Darkness* by Christopher Wallace-Crabbe, to the author.

Come In Spinner by Dymphna Cusack and Florence James, © Florence James, James McGrath, to Collins Angus & Robertson.

Tracks by Robyn Davidson, to Random Century U. K.

'Whither' by Jack Davis, to the author.

The Dreamers by Jack Davis, to the author and to Currency Press Pty Ltd, Sydney.

'The Affair' by Bruce Dawe, to Longman Cheshire Pty Ltd.

West Block by Sara Dowse, to Penguin Books Australia.

The Savage Crows © Robert Drewe, to Collins Angus & Robertson.

Fortune by Robert Drewe, to the author and to Pan Macmillan.

'After Strange Gods' by Laurie Duggan, to the author.

'Capital Gains' by Marian Eldridge, from *Canberra Tales*, to Penguin Books Australia.

Water Under the Bridge by Sumner Locke Elliot, to Mr George Whitfield Cook.

Plumbum by David Foster, to Penguin Books Australia.

Honour and *Other People's Children*, and *Monkey Grip* by Helen Garner, to Penguin Books Australia.

Maestro © Peter Goldsworthy 1989, to Collins Angus & Robertson.

Lilian's Story by Kate Grenville, to Allen & Unwin Australia.

'Postcard from Perth' by William Grono, to the author.

Just Relations by Rodney Hall, to Penguin Books Australia.

'Belladonna Gardens' by Marian Halligan, from *Canberra Tales*, to Penguin Books Australia.

ACKNOWLEDGEMENTS

'Earthworm Small', *Kewpie Doll* and *The Scent of Eucalyptus* by Barbara Hanrahan, to Curtis Brown (Aust.) Pty Ltd.

Where All the Queens Strayed by Barbara Hanrahan, to University of Queensland Press.

'West of Alice' from *Content to Lie in the Sun* by W. E. Harney, to Mrs R. Lockwood.

'Oyster Cove' from *Selected Poems* © Gwen Harwood 1990, to Collins Angus & Robertson.

Transit of Venus by Shirley Hazzard, to Macmillan London Ltd.

Capricornia by Xavier Herbert, © Robin Pill, to Collins Angus & Robertson.

Blue Skies by Helen Hodgman, to Duckworth, U. K.

Stories from Suburban Road and 'My Turn in the Barrel' from *A Knockabout with a Slouch Hat* by T. A. G. Hungerford, to Fremantle Arts Centre Press.

The Unknown Industrial Prisoner © David Ireland, to Collins Angus & Robertson.

'New Parliament House' by Dorothy Johnston, from *Canberra Tales*, to Penguin Books Australia.

My Brother Jack and *Clean Straw for Nothing* and *A Cartload of Clay* by George Johnston, © The Johnston Estate, to Collins Angus & Robertson.

Foxybaby and *Palomino* by Elizabeth Jolley, to University of Queensland Press.

The Newspaper of Claremont Street by Elizabeth Jolley, to Fremantle Arts Centre Press.

'A Sort of Gift: Images of Perth' by Elizabeth Jolley, to Australian Literary Management.

Rowena's Field by Nicholas Jose, to the author.

'Roo Easter' from *Feathers or Lead* by Nicholas Jose, to Penguin Books Australia.

Hey Days © Alister Kershaw 1991, to Collins Angus & Robertson.

The Doubleman, and 'The Lost Hemisphere' from *Crossing the Gap* by C. J. Koch, to Random Century, U. K.

The Boys in the Island © Christopher Koch, to Collins Angus & Robertson.

Don't Take Your Love to Town by Ruby Langford, to Penguin Books Australia.

Letters of D. H. Lawrence, ed. Richard Aldington, to Penguin U. K.

Salt by Gabrielle Lord, to Penguin Books Australia.

Jumbo by Gabrielle Lord, to Random Century U. K.

'For the Patriarch' from *For the Patriarch* by Angelo Loukakis, to University of Queensland Press.

Here's Luck by Lennie Lower, © The Estate of L. W. Lower, to Collins Angus & Robertson.

The Young Desire It by Kenneth Seaforth Mackenzie, © E. A. Little, H. S. Mackenzie 1963, to Collins Angus & Robertson.

Hook's Mountain by James McQueen, to the author and to Pan Macmillan.

'A First Place: The Mapping of a World' by David Malouf, to the author.

Johnno by David Malouf, to University of Queensland Press.

The Great World, Harland's Half Acre, 12 Edmondstone Street, and 'A Traveller's Tale' from *Antipodes* by David Malouf, to Random Century, U. K., and to the author.

'All Over the Shop: A Writer's Beginnings' from *Smyrna* by Tony Maniaty, to the author and to Penguin Books Australia.

The Heart Is Where the Hurt Is by Harry Marks, to E. Marks.

In Mine Own Heart by Alan Marshall, to Longman Cheshire.

The Young Wife by David Martin, to Curtis Brown (Aust.) Pty Ltd.

Trap by Peter Mathers, to Cassell Pty Ltd, c/o Macmillan Publishing Company, U. S. A.

'Observations from Rozelle' by Obelia Modjeska, to the author.

Sydney by Jan Morris, to The Penguin Group, U. K.

'The Haunting of Jungry Jimmy', 'Night Shift' and 'Transit Passenger' from *The Best Stories of John Morrison* by John Morrison, to Penguin Books Australia.

Landscape with Landscape by Gerald Murnane, to Penguin Books Australia.

'Louvres' from *The Daylight Moon*; 'Driving through Sawmill Towns' from *The Vernacular Republic* © Les Murray 1991, to Collins Angus & Robertson.

Last Ferry to Manly by Jill Neville, to Penguin Books Australia.

Outback by Cynthia Nolan, to Sir Sidney Nolan.

The Passage by Vance Palmer, to Longman Cheshire.

Poor Man's Orange by Ruth Park, to Curtis Brown (Aust.) Pty Ltd.

Building on Sand © David Parker, to Collins Angus & Robertson.

'Around the Crate' from *Ikons* by George Papaellinas, to Penguin Books Australia.

'The Sad Heart of Ruth' by Jennifer Paynter, to the author.

Landtakers by Brian Penton, © The Estate of Brian Penton, to Collins Angus & Robertson.

The Watcher on the Cast-Iron Balcony by Hal Porter, to Faber & Faber Ltd, U. K.

The Extra by Hal Porter, to Jack Porter and to University of Queensland Press.

The Paper Chase, and 'Hobart Town' from *Elijah's Ravens* by Hal Porter, © J. Porter, to Collins Angus & Robertson.

'Canberra in April' from *The Feast of Ancestors* © J. R. Rowland 1965, to Collins Angus & Robertson.

'Postcards' by Rosa Safransky, to the author; 'Bonjour Brunswick' by Rosa Safransky, to the author and to Pascoe Publishing Pty Ltd.

'Country Towns' from *One Hundred Poems* by Kenneth Slessor, © Paul Slessor 1944, to Collins Angus & Robertson.

'Tasmania' from *Selected Poems* © Vivian Smith 1985, to the author and to Collins Angus & Robertson.

'Growing Up In Hobart' by Vivian Smith, to the author and to Island Magazine Inc.

INDEX